Laboratory Experiments for General Chemistry

Laboratory Experiments for General Chemistry

Third Edition

Harold R. Hunt

Associate Professor of Chemistry
Georgia Institute of Technology
Atlanta, Georgia

Toby F. Block

Laboratory Coordinator
School of Chemistry and Biochemistry
Georgia Institute of Technology
Atlanta, Georgia

George M. McKelvy

Demonstration Teacher
School of Chemistry and Biochemistry
Georgia Institute of Technology
Atlanta, Georgia

Saunders Golden Sunburst Series

SAUNDERS COLLEGE PUBLISHING
Harcourt Brace College Publishers

Fort Worth Philadelphia San Diego New York Orlando Austin
San Antonio Toronto Montreal London Sydney Tokyo

Text Typeface: Times Roman and Novarese
Compositor: Gloria E. Langer
Publisher: John Vondeling
Developmental Editor: Kent Porter Hamann
Manager of Art and Designs: Carol Bleistine
Cover Designer: Sue Kinney
Director of EDP: Joanne Cassetti
Production Manager: Arnold Lynch

Printed in the United States of America

Laboratory Experiments for General Chemistry, 3/e.

ISBN: 0-03-096746-5

Library of Congress Catalog Card Number: 93-084855

567 066 9876543

Preface

Anyone familiar with the second edition of *Laboratory Experiments for General Chemistry* will readily see that there is a big difference between the third edition and the second edition. Six new experiments appear for the first time in the third edition. They are: "On the Nature of Pennies," "Determination of the Densities of Soft Drinks," "Fun with Solutions," "Job's Method for Determining the Stoichiometry of a Reaction," "Chemical Equilibrium," and "Oxidation of Ethanol by Dichromate."

Further, many of the experiments that appeared in the second edition have been extensively revised. Portions of the experiment—"To Deliver or To Contain?"—have been introduced into "Getting Started in the Laboratory," and "To Deliver or to Contain?" has been deleted as a separate entity. One experiment, "Properties of Oxides, Hydroxides, and Oxo-Acids," has been completely rewritten; and three experiments—"The Burning of a Candle," "Absorption Spectroscopy and Beer's Law," and "Kinetic Study of the Reaction Between Ferric and Iodide Ions,"—have undergone major changes as well. A new system is studied in "Temperature Change and Equilibrium" and the work in "Thermodynamic Prediction of Precipitation Reactions" has been expanded to include an examination of the change in solubility with temperature for selected salts. New elements appear in "Constructing an Alien Periodic Table." The synthesis of aspirin has been deleted, although the properties of aspirin are still explored in "Properties of Aspirin."

In addition to "To Deliver or To Contain?", four experiments that appeared in the second edition do not appear in the third. They are: "Determination of the Atomic Mass of Zinc," "The Potentiometric Titration of Iron with Dichromate," "Half-life of a Radioactive Isotope," and "Kinetics of the Formation of the Chromium-EDTA Complex."

All of these additions, revisions, and deletions are in agreement with our unchanging philosophy about the place of the laboratory in the introductory chemistry curriculum. As we have noted in previous editions, we believe that chemistry is an experimental science and that laboratory work is an important part of any chemistry course. By doing laboratory work, interpreting results, and performing error analyses, students begin to appreciate the experimental skills and the intellectual processes necessary to obtain superior results.

While writing the first edition of this laboratory manual (and the second edition, as well), we were greatly challenged not only by our desire to expose students to the wonders of chemistry, but also by certain constraints that are placed on contemporary introductory chemistry laboratory courses. We assembled a manual that is effective for students who differ in scientific attitude, interest in chemistry, and scholastic background. We attempted to keep each experiment safe by limiting the use of toxic or suspected carcinogenic reagents.

While all of the above still hold true, we found, in writing this third edition, that we wanted to add our voices to those in the chemical education community who said, "Lab should be fun." "It's more important for students to learn to think than for them to get the right answer." "The concept is more important than the answer." As a result, we have eliminated or revised experiments that required students to spend long periods of time recording data from a spectrometer display or waiting for a blue color to appear in a reaction mixture. We have introduced new experiments (and modified old ones) so that students are challenged to devise and test hypotheses to explain the phenomena they observe in and out of the laboratory.

We have removed the restriction we placed on ourselves in the second edition that each experiment should be completed within a single laboratory period. Sometimes work must be extended because the system being studied requires it, and the future scientists of tomorrow must learn that research cannot always be accomplished on an arbitrary schedule. Where applicable, we have called for teamwork in accomplishing the goal of an experiment; the scientists and engineers of tomorrow are far more likely to be members of research groups than to be isolated individuals laboring in secrecy.

We have extended our practice, used extensively in the first and second editions, of having students work with household materials. Pennies, sour salt, and commercial aspirin tablets have now joined juice, bleach, rubbing alcohol, and cola drinks as objects of study. Soft drinks are examined in a new way and secret messages written on paper plates and solutions prepared in plastic sandwich bags are vehicles through which students are introduced to the laboratory manual as a research tool.

Making a new discovery in lab may well be the one thing that will excite a student about science. Yet, in large classes where students may not be able to receive individual assistance down the right path, "discovery" labs may leave many students totally in the dark, not knowing how to get started. We prefer to describe our new and revised experiments as "guided inquiry" exercises. Students are given information and guidance, but have the freedom to explore and, indeed, are required to make decisions on the quality of their results and the need for further experimentation. Thus, labwork should be a satisfying experience both for the student who has never before done any experimentation and for the veteran of Honors and Advanced Placement programs.

Like its predecessors, the third edition is a result of extensive testing by the authors, the teaching assistants, and the approximately 2,000 students who take introductory chemistry at Georgia Tech each year. We have heard good things from people who have used the second edition at two- and four-year colleges and technical institutes throughout the country. We hope that they will continue to be pleased with the third edition, and we welcome their comments as well as those of new adopters of the book.

Acknowledgements

We would like to express our appreciation to our spouses—Doris Hunt, Jerrold Greenberg, and Julie McKelvy—for their support during the preparation of this manuscript. We would also like to acknowledge the efforts of Kathy Wingate and Kathy Mims who scanned the second edition so we could begin to revise the manuscript electronically. Kathy Wingate also did a monumental job in managing our e-mailing and preparing the printed experiments for submission to the editor. Thanks also go to the men of the electronics shop at Georgia Tech's School of Chemistry and Biochemistry (Tony Bridges, Darryl Bailey, Richard Bedell, and Jack Hunt) for designing and constructing the conductivity testers used in Experiment 39.

Harold R. Hunt

Toby F. Block

George M. McKelvy

Atlanta, GA
April 1997

Contents

INTRODUCTION: LABORATORY CONDUCT AND PROCEDURES

EXPERIMENTS

APPENDICES

INTRODUCTION

Laboratory Conduct and Procedures

General Rules and Suggestions

You are in the laboratory primarily to learn by doing experiments. You will benefit most from the experience if you have prepared in advance by reading the assigned experiment and completing its pre-lab exercises. You are likely to waste precious laboratory time if you attempt to perform experiments without appropriate preparation. Always be sure to review (before lab) the sequence of operations to be performed in the assigned experiment and note those procedures that are time-consuming or hazardous. Then plan your work so that you may compete the experiment in a timely, safe manner.

Keep in mind that you share the laboratory and its equipment with other students. Therefore you should work safely and considerately. Loud noises, practical jokes, and similar offensive behavior have no place in the chemistry laboratory.

Do not attempt to repair or adjust malfunctioning instruments. Report problems to your instructor, who will see to it that instruments are repaired or adjusted.

Before you leave the laboratory, clean your work area. This includes your place at the bench, the reagent shelf, the sink, and other areas shared by the class. Dispose of waste material in an approved manner, following the instructions provided with each experiment.

Safety

In the course of your laboratory work, you will be exposed to new instruments, new procedures, new chemicals, and, possible, new hazards to health and safety. Fortunately, most laboratory accidents are avoidable. You can minimize the possibility of injury by observing the general safety rules given below. In addition to these rules, you should pay close attention to the discussion of possible hazards given in each experiment, and you should follow all procedures carefully. Study the First Aid section and be prepared to administer prompt and appropriate first aid should you or another student be injured.

1. Wear appropriate eye and face protection (safety goggles and/or face shield) whenever you are in the laboratory. This rule applies even when you are not conducting experiments; you could be the victim of another person's carelessness.
2. At the very beginning of the term, learn the location and

proper use of the fire blanket, fire extinguishers, safety shower, eye-wash fountains, and other emergency equipment that is found in your lab. This will permit you to use such equipment without delay in an emergency. You should also locate the exits from the lab and plan an escape route to be used in case of fire.

3. Do not perform unauthorized experiments. Do not work in the laboratory unless an instructor is present.
4. Regard all chemicals as hazardous; do not touch, taste, or intentionally inhale chemicals.
5. Do not eat, drink, or smoke in the laboratory. This will prevent the accidental ingestion of chemicals.
6. Never mouth pipet; always use a pipet bulb.
7. Do not use burners or open flames where acetone or other flammable solvents are being used.
8. Be sure beakers or flasks used for boiling liquids are securely supported by a tripod or iron ring. Use boiling chips to avoid bumping. When boiling small amounts of liquid in a test tube, always aim the open end of the tube away from yourself and other persons.
9. Do not use a glass tubing unless the cut ends have been fire polished to round their sharp edges. When inserting glass tubing or thermometers into rubber stoppers, lubricate the glass first with glycerin or water. Grasp the glass near the end to be inserted to minimize the chance of breakage. Use a towel between the glass and your hand to protect yourself from cuts—in the event that the glass breaks. Chipped or broken glassware of any kind should be replaced with new equipment.
10. Never add water to concentrated acids or bases because the heat of dilution may cause the corrosive solution to splatter. Dilute acids or bases by slowly adding small amounts of concentrated reagent to a large amount of water, with constant stirring.

First Aid

It is your responsibility to know the location of the emergency equipment. You should also know how to use such equipment. General first-aid recommendations are given in this section. Specific information concerning serious injury appears at the start of each experiment.

If you splash acids, bases, or other chemicals on yourself or in your eyes, wash the affected area immediately with large quantities of water, first giving attention to your eyes. Continue washing until all corrosive materials have been removed. The eyelid should be pulled up and away from the eye, by another person if necessary, while the eye is thoroughly irrigated with plain water. This procedure is most easily accomplished at an eyewash fountain. After chemicals have been rinsed away from the eye, seek medical attention immediately, to guard against permanent vision impairment.

Should your clothing catch on fire, remain as calm as possible. Walk (do not run) to the safety shower and pull the ring to douse yourself with water. Alternatively, you may drop to the floor and roll to extinguish the flames. Safety experts caution against wrapping yourself in a fire blanket. Doing so would hold the heat of the flame close to your body and might bring the heat of the flame up to your face in a chimney effect.

The blanket is present in the lab to provide a warm covering for victims of shock.

Minor burns are best treated by soaking the affected area in cool water. Minor cuts can be covered with adhesive bandages. Report such injuries to your instructor, who will assist you in treating them. (Because these injuries occur so frequently, both in and out of the laboratory, we will not discuss their treatment in the First aid sections of the experiments but will instead focus on more serious injuries that might occur.

Accidents are usually preventable. Be alert in the lab. Understand the experimental procedures and be aware of any hazards. Ask your instructor to explain operations you do not understand. Do not proceed until you know what you need to do.

WASTE DISPOSAL

As recently as 15 years ago, academic laboratories disposed of their chemical wastes in the most convenient manner: volatile solvents were allowed to evaporate; solids were dumped into the general trash; liquids and solutions were flushed down the sink. These practices are unacceptable today. There are now rules for disposing of hazardous waste. Unfortunately, these rules are often difficult and expense to follow. Thus, it is desirable to minimize the amount of hazardous wastes generated in academic laboratories. We have incorporated the following procedures into these experiments to accomplish this.

1. We have reduced the scale of experiments.
2. We have replaced hazardous chemicals with less hazardous (or nonhazardous) materials.
3. We have given instructions for converting wastes to nonhazardous form, by means of chemical reaction, prior to disposal.
4. We have given instructions for reducing the volumes to be disposed of, by concentrating solutions or by separating hazardous chemicals as precipitates.
5. We have given instructions for recovering chemicals for reuse, whenever practical.

Special instructions for the disposal of wastes are provided with the experiments. In most cases, the wastes generated in these experiments are nonhazardous and, unless local ordinances specify otherwise, can be rinsed down the sink according to the guidelines given below.

1. Solutions acceptable for sink disposal should be diluted one hundredfold with water before being slowly poured down the drain. This should be followed by flushing with fresh water. Alternatively, the solution may be poured slowly down the drain along with a stream of water sufficient to accomplish the desired dilution.
2. Unless otherwise specified, soluble salts may be disposed of as described in Step 1.
3. Small amounts of water-soluble organic solvents (such as methanol or acetone) may be diluted and flushed down the drain. Large amounts of

these solvents or any amount of volatile, highly flammable solvents (such as ether) should *not* be disposed of in this way.

4. Solutions of acids and bases should have their pH adjusted to the 2 – 11 range before being flushed down the sink. Small amounts (10 mL or so) of *dilute* acid or base may be simply diluted and flushed.
5. When in doubt, check with your instructor before pouring any chemical down the drain.

THE NOTEBOOK AND THE REPORT

Rules for Keeping the Notebook or Making Reports

Your instructor will determine if you are to maintain a laboratory notebook or if you will be submitting individual reports on your experiments. If a notebook is used, it will serve as an original, permanent record of your work. Thus, a permanently bound notebook will be required. Whether you maintain a notebook or not, you should follow the general guidelines given below for report preparation. You should also follow any additional directions by your instructor.

If you Keep a Notebook

1. The first two or three pages of the notebook should be left blank because they will be used to prepare a Table of Contents. The Table of Contents should be updated as each report is completed. If the notebook pages are not numbered, you should number them consecutively in ink.
2. All experimental observations should be recorded in ink directly in the notebook at the time of observation. Do not record figures on loose scraps of paper to be later copied neatly into the notebook. You may very well lose such scraps of paper and, even if you don't, copying destroys the value of the notebook as the original record of your laboratory work.
3. Do not obliterate or erase any data recorded in the notebook. If you wish to correct an error, draw a single line through the incorrect number or statement, and then write the correct entry beside it. If the reason for the change is not obvious, write a brief explanation in parentheses near the corrected data.
4. Do not remove pages from the notebook.
5. Try to be neat. Study the experiment and think it through in advance. Then decide on the clearest and most efficient way of recording your observations and make any necessary advance preparations, such as designing a table to be filled in at the laboratory. Some students strive for neatness by writing their formal reports, using only the right (or left) side of the notebook pages, leaving the other side for sketches and calculations.

If You Do Not Keep a Notebook

All experimental observations should be recorded directly on the Summary Report Sheets that accompany each experiment. Do not record figures on loose scraps of paper to be later copied neatly onto the Report Sheet. You may very well lose such scraps of paper and, even if you don't, copying destroys the value of the Summary Report Sheet as the original record of your laboratory work.

Do not obliterate or erase any data recorded on your Summary Report Sheet. If you wish to correct an error, draw a single line through the incorrect number or statement, and then write the correct entry beside it. If the reason for the change is not obvious, write a brief explanation in parentheses near the corrected data.

Whether or Not You Keep a Notebook

The report for each experiment should contain each of the sections listed below, arranged in the order presented.

1. *The TITLE of the experiment.*
2. *The PURPOSE of the experiment.* Here you should list the skills to be learned or the principles to be demonstrated by proper performance of the experiment.
3. *The PROCEDURE followed.* Briefly describe the equipment used and the procedure followed. It is not necessary to list the minor details of the procedure unless a new procedure has been substituted (with your instructor's approval) for the one given in the laboratory manual.
4. *The original DATA.* Data should be recorded directly in the notebook or on the Summary Report Sheets.
5. *Tabulated RESULTS.* In this section, you should give a table, or tables, showing all significant results obtained from your experiment.
6. *DISCUSSION.* In this section, you should show sample calculations relating your data to your results. Briefly discuss any errors that were made in the course of the experiment. You should also discuss the significance of the experimental results.
7. *QUESTIONS.* Many of the experiments in this laboratory manual contain questions for you to answer. Answer these questions as part of the report.

Because you have a limited amount of time in the laboratory in which to complete each experiment, as much of the report as is practical should be written outside the laboratory. The Title, Purpose, and Procedure sections may be written before the laboratory period. However, you must record all data (in the notebook or on the Summary Report Sheets) only at the time the original observations are made in the laboratory. You may complete the Results, Discussion, and Questions sections either in the laboratory or at home, depending on the time available.

LABORATORY EQUIPMENT

Common items of laboratory equipment are shown in Figures I.1 through I.3. Figure I.1 shows commonly used items of glassware. Figure I.2 shows other laboratory equipment. Study the figures until you are able to identify any unfamiliar items.

The Balance and Weighing

Despite the difference between mass and weight, we usually use the term "weighing" for the operation of determining the mass of an object using a

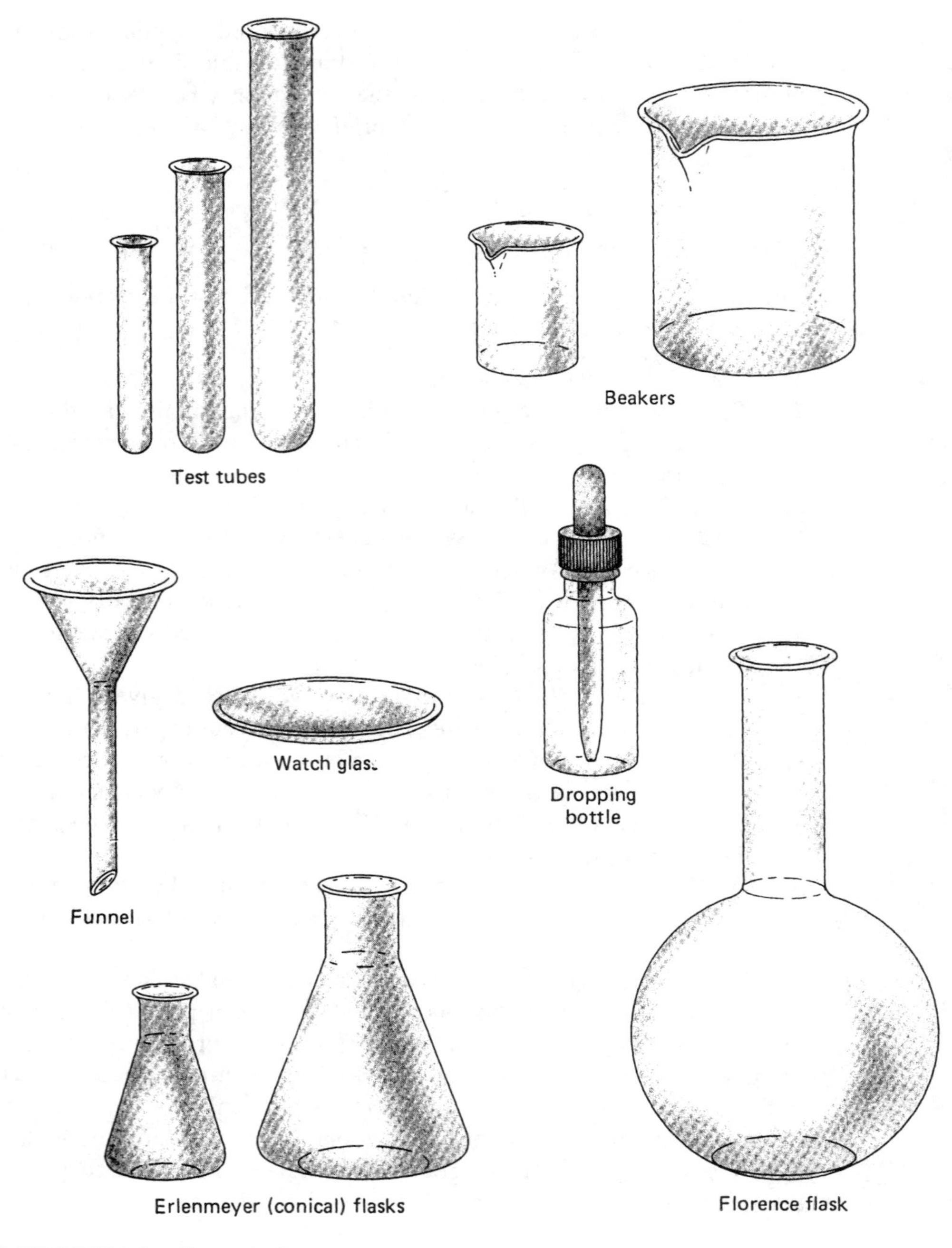

FIGURE I.1 General glassware.

laboratory balance. In weighing, we add or remove standard masses (often called "weights") until the gravitational attraction for the weights is equal to the gravitational attraction for the object. The mass of the object is then equal to the sum of the standard weights.

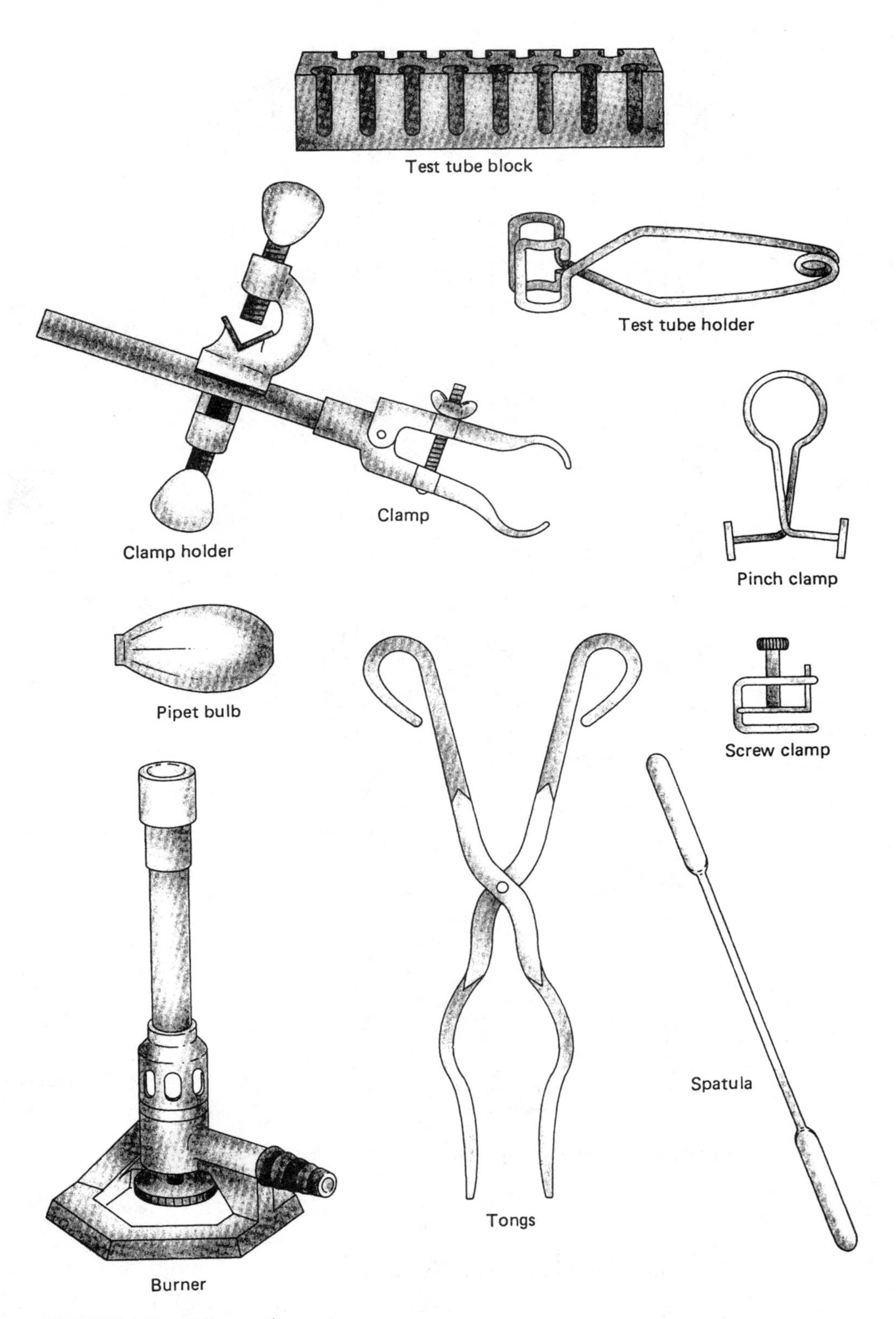

FIGURE I.2 Other equipment.

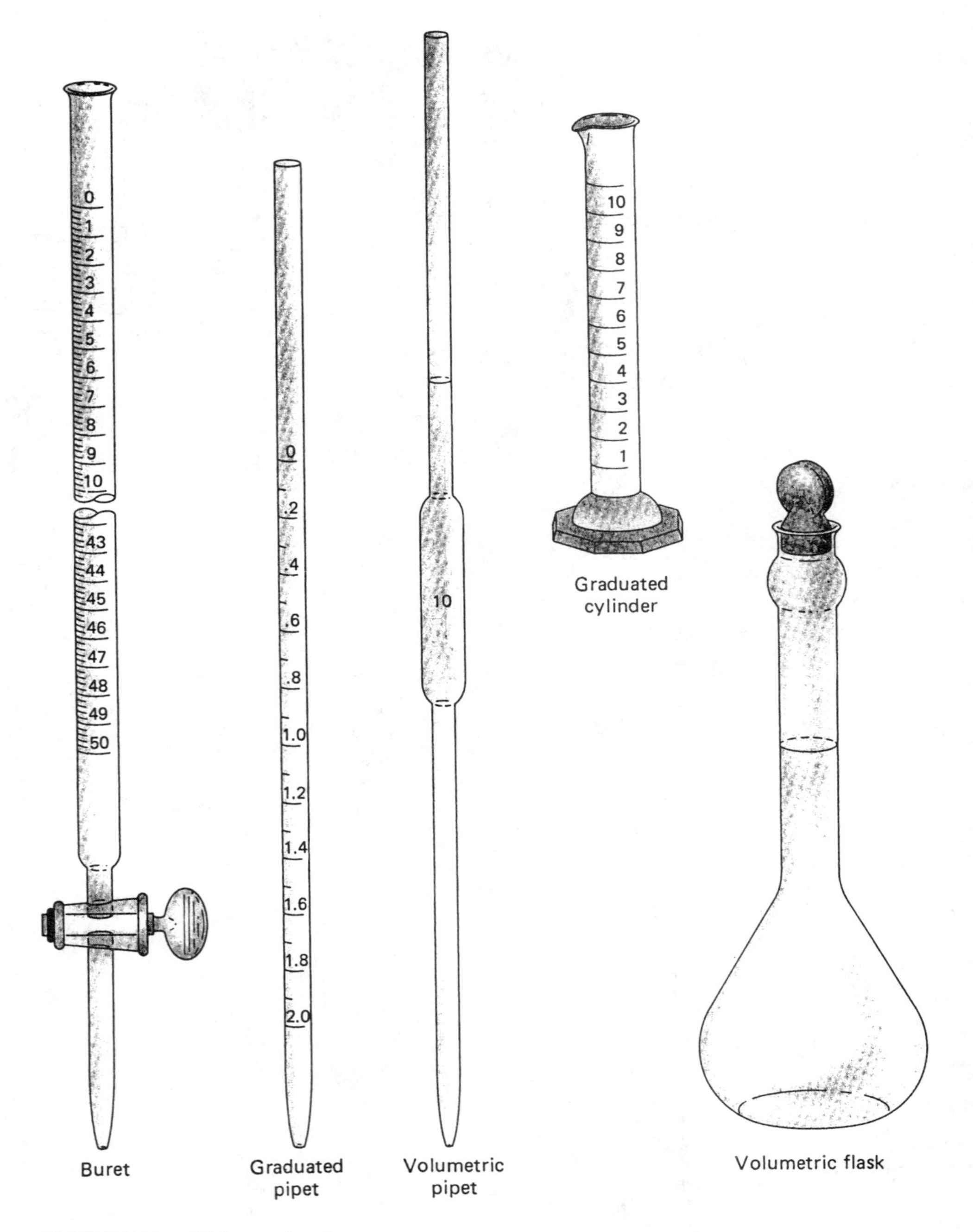

FIGURE I.3 Volumetric glassware.

Rules for Using Laboratory Balances

Several different types of laboratory balances are used in general chemistry laboratories. These range from triple-beam balances to mechanical analytical balances to electronic analytical balances with digital readout. Your instructor will provide you with directions for using the types of balances found in your lab. However, the following general rules are appropriate for all balances.

1. Do not attempt to repair or adjust a malfunctioning balance. Inform the instructor, who will see to it that the problem is corrected.
2. Never place chemicals on the balance pans. Solids should be weighed in a tared (pre-weighed) beaker, weighing boat, or watch glass. Liquids should be weighed only in a stoppered container, such as a flask or weighing bottle.
3. If you spill chemicals in or around the balance, use a balance brush to clean up the mess before you leave the balance. It is very important to remove all traces of spilled chemicals from inside the balance case because chemicals may corrode the balance and ruin it. Your instructor may be able to help you identify the spilled chemicals and can recommend an appropriate disposal procedure.
4. If the balance being used has knobs for adding or removing weights, turn the knobs slowly and return them to the zero setting before leaving the balance.
5. Do not weigh hot or cold objects.
6. When recording masses, do not lean on the balance table; learn to support your notebook or laboratory manual on your lap.

Volumetric Glassware

The most common items of equipment designed for measuring the volumes of liquids are illustrated in Figure I.3. Look at these items carefully and note the permanent markings provided to aid you in using the item correctly. In addition to their calibration marks, volumetric glassware is often marked to show the calibration temperature and intended method of use. The symbols **"TC"** and **"TD"** marked on a piece of volumetric glassware indicate, respectively, that the apparatus has been calibrated to contain or to deliver the specified volume of liquid. For example, a volumetric flask will be designated **TC**, a buret or pipet will be marked **TD**, and a graduated cylinder may be marked **TC** or **TD**, depending on its intended application. The volumetric pipet and volumetric flask have a single calibration mark and can be used for measuring only the specified volume of liquid; the graduated cylinder, measuring pipet, and buret are marked with a scale that permits volumes from 0 mL up to the capacity of the apparatus to be measured.

Cleaning Glassware

Before using a pipet, buret, or other volumetric item, you should first fill it with distilled water and then allow the water to drain out completely under the influence of gravity. If the glass is clean, only a thin, even film of water will remain on the walls. Where there is grease or dirt on the glass, large irregular drops of water will be observed. Because the calibration cannot take adhering drops into account, it is necessary to clean the glass before accurate measurements can be made.

Glassware will generally remain clean if it is washed and rinsed immediately after use, before chemicals have had an opportunity to dry on the glass. When cleaning is required, soak the item in aqueous detergent solution, then rinse it in ordinary water and then in distilled water. You may use a soft brush with graduated cylinders or volumetric flasks, but do not use a brush with expensive burets because the wire shaft of a long, stiff brush may scratch the inner walls of the buret.

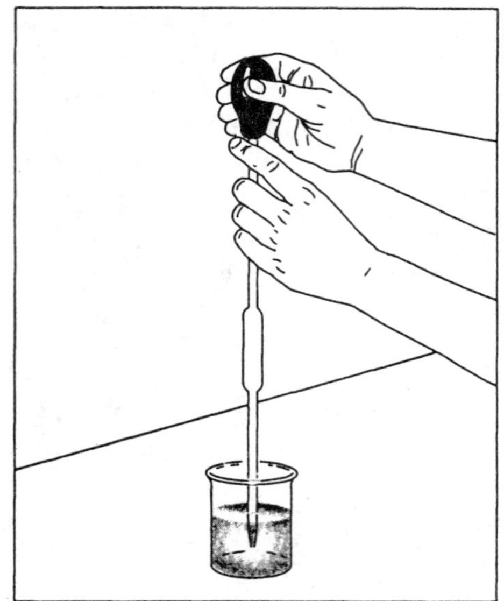

A. Use suction from a rubber bulb to draw liquid into the pipet.

B. Dry the outside of the pipet using a tissue.

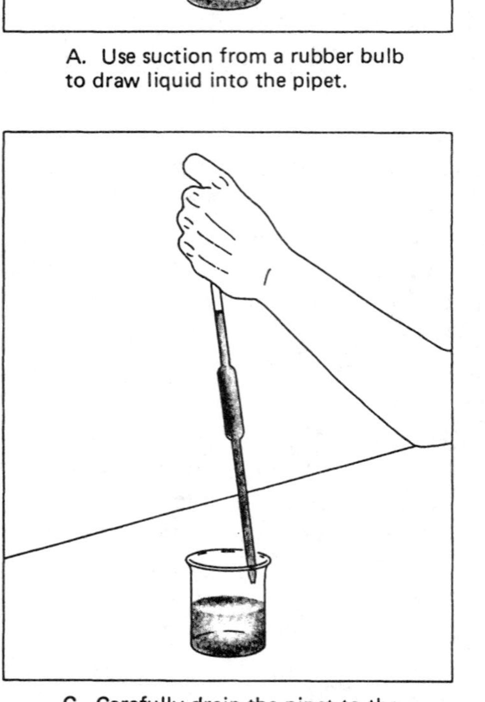

C. Carefully drain the pipet to the calibration mark.

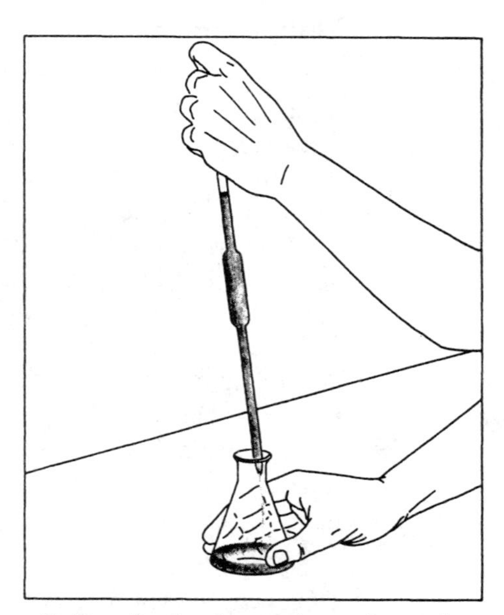

D. Transfer the pipet to the receiver and let the liquid drain out under the influence of gravity.

FIGURE I.4 Pipet technique.

Use of the Pipet

The procedure for filling a pipet by means of a rubber bulb is illustrated in Figure I.4. Hold the pipet near its upper end with your index finger in position to be moved easily to cover the open upper end of the pipet after it has been filled with liquid and the bulb has been removed. By varying the pressure of your finger against the opening, you can either hold the liquid in the pipet indefinitely or dispense it at a controlled rate. Hold the rubber bulb firmly against the end of the pipet when drawing up the liquid by suction. Do not force the bulb over the end of the pipet, as this enlarges the hole in the bulb and tends to make it unusable.

To pipet a liquid sample accurately, without diluting it or contaminating it, you should first rinse the pipet once or twice with small portions of the liquid. To do this, partially fill the pipet with the sample, then tilt and turn the pipet until the inner walls have been well rinsed with the liquid. Next drain and discard the contents of the pipet. After you have rinsed the pipet, fill it to a level approximately $^1/_2$ inch above the calibration mark, then seal the pipet with your finger. Remove the pipet tip from the beaker or bottle containing the sample liquid; then dry off the outside of the pipet using a disposable paper tissue. Next touch the pipet tip against the side of a beaker or other waste receptacle and carefully drain out the excess liquid, stopping when the bottom of the curved liquid surface just rests on the calibration mark. Finally, move the pipet to the beaker or flask provided for collecting the sample and, once more, hold its tip against the walls while the liquid drains out under gravity. When using a measuring pipet, stop the flow when the desired calibration mark is reached. When using a volumetric pipet, let the liquid drain out naturally. A drop may remain in the tip. Do not blow this out; it has been accounted for in calibrating the pipet.

Use of the **Buret**

Care of the stopcock. After cleaning the buret, clean and lubricate the stopcock. (Teflon stoppers do not require lubrication.) Wipe both the barrel and stopcock plug with a paper tissue, taking pains to remove all grease from the capillary through which the liquid is dispensed. A corner of the tissue or a pipe cleaner may be used for this purpose. Next lightly grease the stopcock, taking care not to plug the bore with excess grease.

Although Teflon stopcocks do no require lubrication, there are precautions to be taken to ensure that the stopcock continues to work properly. First, note the arrangement of the nut, washer, O-ring, and so on, so that they may be put back in the proper order after the stopcock has been disassembled. Next, use the adjustment nut to control the fit of the plug in the barrel. It should turn easily but should not leak. Finally, when you are finished with the buret for the day, loosen the stopcock nut before putting the buret away. If you do not loosen the stopcock, the plastic may become deformed and acquire a permanent leak.

Filling the buret. Before filling the buret completely, rinse it with two 5-mL portions of the titrant solution. Then fill the buret almost to the top and turn the stopcock briefly to dispense a vigorous stream of liquid. This procedure should suffice to sweep the air out of the stopcock and dispensing tip, but it may be repeated if necessary. Finally, refill the buret and then use the stopcock to adjust the liquid level to the zero mark. Practice controlling the stopcock with your left hand (if you are right-handed) or with your right hand (if you are left-handed). Although this might feel awkward at first, this procedure leaves your more skillful hand free for swirling the titration flask.

Reading the buret. Figure I.5 illustrates the procedure for reading the liquid level in the buret. The bottom of the curved liquid surface, or meniscus, is compared with the marked scale by sighting with the eye at the same elevation as the meniscus. (You can hold a small black rectangle behind and just below the meniscus to darken it and make it more distinct.) Be sure that your eye level is

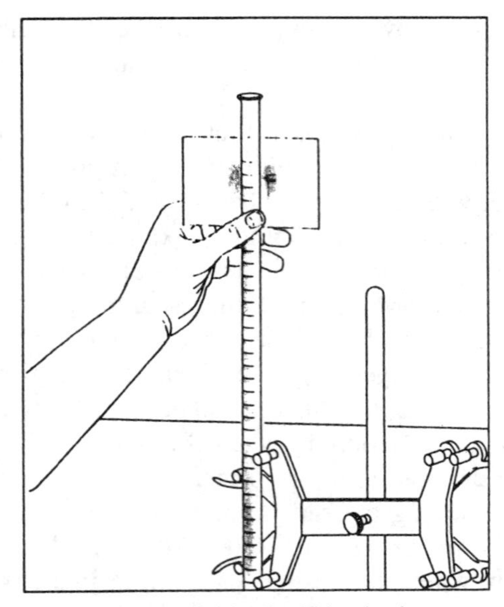

A. Read the initial liquid level using a marked card to darken the meniscus.

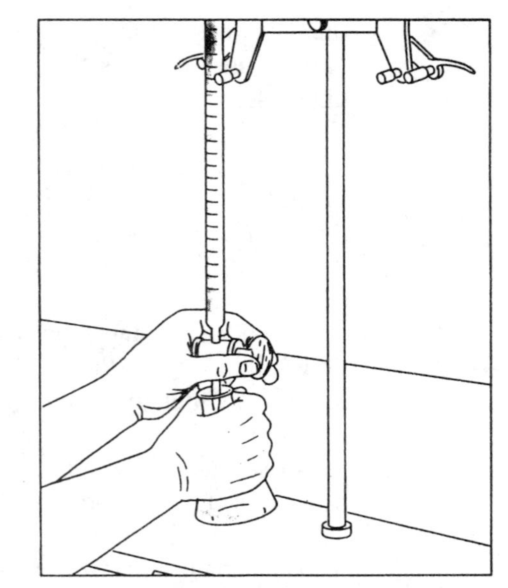

B. Control the stopcock with your left hand while swirling the flask with your right hand.

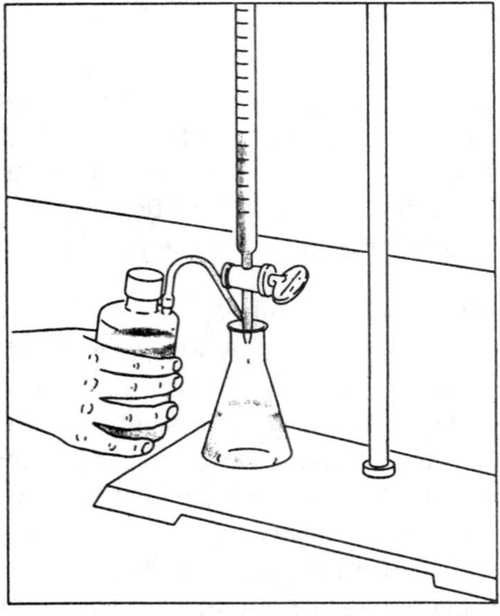

C. Rinse the walls of the flask and the buret tip when the end point is near.

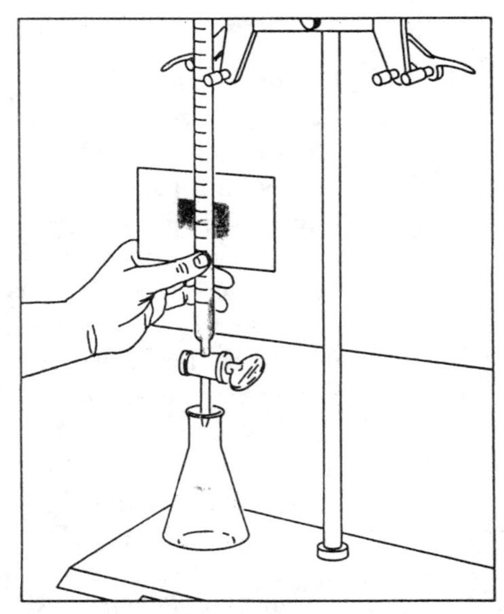

D. At the end point, read the final liquid level using the marked card to darken the meniscus.

FIGURE I.5 Titration technique.

not appreciably above or below the meniscus, or an error due to parallax will result. Read the volume to the smallest marked scale subdivision and obtain an additional figure by estimating the position of the meniscus between subdivisions. For example, when you use a buret marked at 0.1-mL intervals, you can estimate the liquid level to ±0.01 mL. Always be certain to record both the initial and the final buret readings because the difference between these figures is the volume of liquid dispensed.

REPORTING DATA ACCURATELY

Significant Figures

Significant figures are all the digits in a number except for any zeros written only for the purpose of locating the decimal point. The numbers 30.2, 0.00357, and 35.8 all have three significant figures. The number 35 has only two significant figures, but the number 35.0 has three because the final zero is not necessary for establishing the decimal point, and it must therefore be significant. Accepted practice is to report all accurately known digits plus one that has been estimated or is otherwise subject to uncertainty. One generally assumes an uncertainty of at least ±1 in the last digit reported. Reporting too few or too many significant figures gives a false picture of the accuracy of the measurement.

When you combine numbers, your answer should retain only one digit that is uncertain, in addition to all accurately known digits. This may require "rounding off" to the correct number of significant figures. The rules for rounding off are illustrated below.

1. *Rule*: If the first dropped figure (underlined in the example) is less than 5, drop excess figures without changing previous figures.
   Example: Round off 26.8037 to four significant figures.
   Answer: 26.80.
2. *Rule*: If the first dropped figure (underlined) is greater than 5, increase the previous digit by 1 unit, then drop the excess figures.
   Example: Round off 26.8037 to five significant figures.
   Answer: 26.804.
3. *Rule*: If the dropped digit equals 5, round off the previous digit to the nearest even number.
   Example: Round off 26.825 and 26.815 to four significant figures.
   Answer: 26.82 in both cases.
4. *Rule*: When adding or subtracting, round off the result to the same number of decimal places as your least-precise quantity.
   Example:
   15.806
   7.21
   5.1
   28.116 — round off to 28.1
   The uncertain digit in each number has been underlined. Note that 28.116 incorrectly reports three uncertain figures, but only one uncertain figure is retained when we round off to 28.1 in accordance with this rule.
5. *Rule*: When multiplying or dividing numbers, report the answer to the same number of significant figures as the least precise number used.
   Example: The product of 4.27 and 0.61 is 2.6657. Because the least-precise number, 0.61, has two significant figures, the answer should be rounded off to 2.7.

Experimental Uncertainties

As noted above, scientific data are reported as numbers consisting of accurately known digits plus one digit that is associated with some degree of uncertainty. A mass of 6.20 g ± 0.03 g may have a true value between 6.17 g and 6.23 g. A volume reported as 10.00 mL would have a minimum true value of 9.99 mL and a maximum true value of 10.01 mL. Any result derived from experimental data, such as the density of an object with a mass of 6.20 g ± 0.03 g and a volume of 10.00 mL ± 0.01 mL, must contain at least as high a degree of uncertainty as the original data. This section will deal with ways of determining the uncertainties to be associated with results obtained from data.

In the case of our object, the reported density would be (6.20 g/10.00 mL =) 0.620 g/mL. We could calculate the uncertainty associated with this value by considering that the density will be at its minimum value when the true value of the mass is at a minimum and the true value of the volume is at a maximum. Thus, the minimum value of the density is (6.17 g/10.01 mL =)0.616 g/mL. Likewise, the density will be at its maximum value when the true value of the mass is at a maximum and the true value of the volume is at a minimum. Thus, the maximum value of the density is (6.23 g/9.99 mL =)0.624 g/mL. Therefore, the density of the object should be reported as 0.620 g/mL ± 0.004 g/mL.

In order to avoid doing three calculations instead of one, whenever uncertainties are to be reported along with results, you may use the following rules.

1. *Rule*: When data are to be added or subtracted, add the uncertainties.
   Example 1: Find the length of the perimeter of a square that measures 2.52 cm ± 0.02 cm on a side.
   Answer: 10.08 cm ± 0.08 cm.
   Example 2: Find the mass of a sample placed in a flask, if the empty flask weighed 23.6154 g ± 0.0001 g and the flask plus sample weighed 26.2198 g ± 0.0001 g.
   Answer: 2.6044 g ± 0.0002 g
2. *Rule*: When data are to be multiplied or divided, add the relative uncertainties.
   Example 3: Find the density of an object that has a mass of 6.20 g ± 0.03 g and a volume of 10.00 mL ± 0.01 mL.
   Answer: $0.620\text{ g/mL} \pm 0.620\text{ g/mL}\left(\frac{0.03\text{ g}}{6.20\text{ g}} + \frac{0.01\text{ mL}}{10.00\text{ mL}}\right) =$

   $$0.620\text{ g/mL} \pm 0.004\text{ g/mL}$$

   Example 4: Find the volume of a rectangular solid that measures 25 mm × 45 mm × 65 mm. Assume the uncertainty in each dimension is ± 3 mm.

   Answer:
   $7.3 \times 10^4\text{ mm}^3 \pm 7.3 \times 10^4\text{ mm}^3\left(\frac{3\text{ mm}}{25\text{ mm}} + \frac{3\text{ mm}}{45\text{ mm}} + \frac{3\text{ mm}}{65\text{ mm}}\right) =$

   $$7.3 \times 10^4\text{ mm}^3 \pm 1.7 \times 10^4\text{ mm}^3 = 7 \times 10^4\text{ mm}^3 \pm 2 \times 10^4\text{ mm}^3$$

Note that, in the examples given under Rule 2, the relative uncertainty is obtained by dividing the uncertainty (e.g., ±0.01 mL) by the quantity (10.00 mL) with which it is associated. Relative uncertainties are unitless quantities. Thus, relative uncertainties for different types of data (e.g., mass and volume measurements) can be added, although absolute uncertainties with different units (± 0.03 g and ± 0.01 mL) cannot be combined. One could report the final result along with a relative uncertainty (e.g., density = 0.620 g/mL ± 6 parts per thousand). However, the more common practice is to multiply the relative uncertainty by the result to obtain an absolute uncertainty that has the same units as the result (0.620 g/mL ± 0.004 g/mL).

The rules for determining the number of significant figures in a derived result were set up so that the result would seem no less uncertain than the data from which it was derived. When actual uncertainties are computed, the number of digits known with absolute certainty may differ from the number predicted by the significant figures rules. In such a case, the computed uncertainties take precedence over the somewhat arbitrary significant figure rules. Thus, the volume considered in Example 4 should be reported as 7×10^4 mm^3 ± 2×10^4 mm^3.

DEALING WITH "BAD" DATA

A question of concern to novice experimenters is "Must I report this measurement that totally disagrees with the others" in a series of replicates (data obtained in a group of trials to determine a single value)? Of course, any data that you know to be erroneous (such as a buret reading made when the stopcock was leaking or a mass of a sample that was spilled) *should* be rejected, with a brief notation made in the original data sheet explaining the reason for the deletion. But what if a measurement merely looks strange compared with other values? Then, you must use a mathematical test such as the one described below for guidance. In order to perform this test, you should identify the suspected bad datum and temporarily ignore it. Obtain the average of the remaining data. Compare each value used to obtain the average with the average itself and then calculate the average of these deviations. Next, compute the deviation of the suspect datum from the average of the other data. If the deviation of the suspect value from the average is more than four times the average deviation for the other data, the suspect value should be rejected. The use of this method is illustrated in the examples given below.

Example 1

The measured values are volumes of base used in a titration. The values are 25.19 mL, 26.25 mL, 26.01 mL, 26.17 mL, and 26.32 mL.

Suspect datum:	25.19 mL
Remaining data:	26.25 mL, 26.01 mL, 26.17 mL, 26.32 mL
Average of remaining data:	26.19 mL
Deviations:	0.06 mL, 0.18 mL, 0.02 mL, 0.13 mL
Average deviation:	0.098 mL
Deviation of suspect datum:	1.00 mL

$$1.00 > 4(0.098 \text{ mL})$$

The suspect value is rejected and the average volume is reported as 26.19 mL.

Example 2

The values are 25.79 mL, 26.01 mL, and 26.17 mL.

Suspect datum:	25.79 mL
Remaining data:	26.01 mL, 26.17 mL
Average of remaining data:	26.09 mL
Deviations:	0.08 mL, 0.08 mL
Average deviation:	0.08 mL
Deviation of suspect datum:	0.30 mL

$$0.30 < 4(0.08)$$

The suspect value is *not* rejected and the average volume is reported as 25.99 mL.

EXPERIMENT 1

Getting Started in the Laboratory

Laboratory Time Required

Three hours

Special Equipment and Supplies

Analytical balance
Gas burner
Thermometer
Wooden applicator stick
Volumetric pipet, 10-mL
Ringstand with iron ring
Glass rod
Small paper labels
Buret
Buret clamp
Barometer
Ice
Ice cream salt
NaCl solutions, 0.5% to 2.5% by mass

Safety

The gas burner, hot glass, and boiling water used in the experiment can cause painful and serious burns to the skin. Hot glass looks like cold glass, so it pays to be cautious when heating or melting glass. You should also pay close attention when lighting or adjusting the burner and when measuring the temperature of boiling water.

First Aid

Severe burns should be treated by a physician.

The chemistry laboratory of motion pictures is an exciting but scary place—dimly lit; cluttered with smoking retorts; flasks filled with boiling, brightly colored liquid; and electrodes that discharge loudly for no apparent purpose. Although items such as these are still to be found in chemical museums or research labs, today's teaching labs tend to be furnished with more basic equipment and with less noise and smoke. However, the excitement is still there. It comes from observing an unfamiliar chemical transformation; making a set of measurements that clearly confirms a chemical law; or simply mastering a new laboratory skill.

PRINCIPLES

In this experiment, you will identify useful items of equipment and make your first measurements using four basic instruments: the balance, the pipet, the buret, and the thermometer. As will be the case for every experiment, you will be expected to follow the rules given in the Introduction pertaining to safety, chemical disposal, report preparation, significant figures, and the operation of instruments. If you are not familiar with these rules, reread the Introduction.

Our tour of the laboratory begins with an introduction to the desk equipment, which is used to contain, mix, measure, heat, stir, or separate chemicals, and for other purposes as well. Your kit of equipment will probably contain flasks and beakers of varying sizes, a wash bottle, a wire gauze, stirring rods, test tubes, and one or more graduated cylinders.

Your kit may also contain a gas burner of some sort. There is a variety of burners that are used in chemistry labs, all of which have means of containing a combustible mixture of gas and air. A mixture that has a high gas-to-air ratio will burn with a luminous yellow flame. Increasing the amount of air in the mix should produce a two-zone flame (Figure 1.1). This is a hotter, more efficient flame than the yellow flame. The blue, inner zone of the flame is the cooler of the zones. The hotter, outer zone is relatively colorless and transparent. If a glass rod is placed in the outer zone, a yellow flame may be observed above the rod.

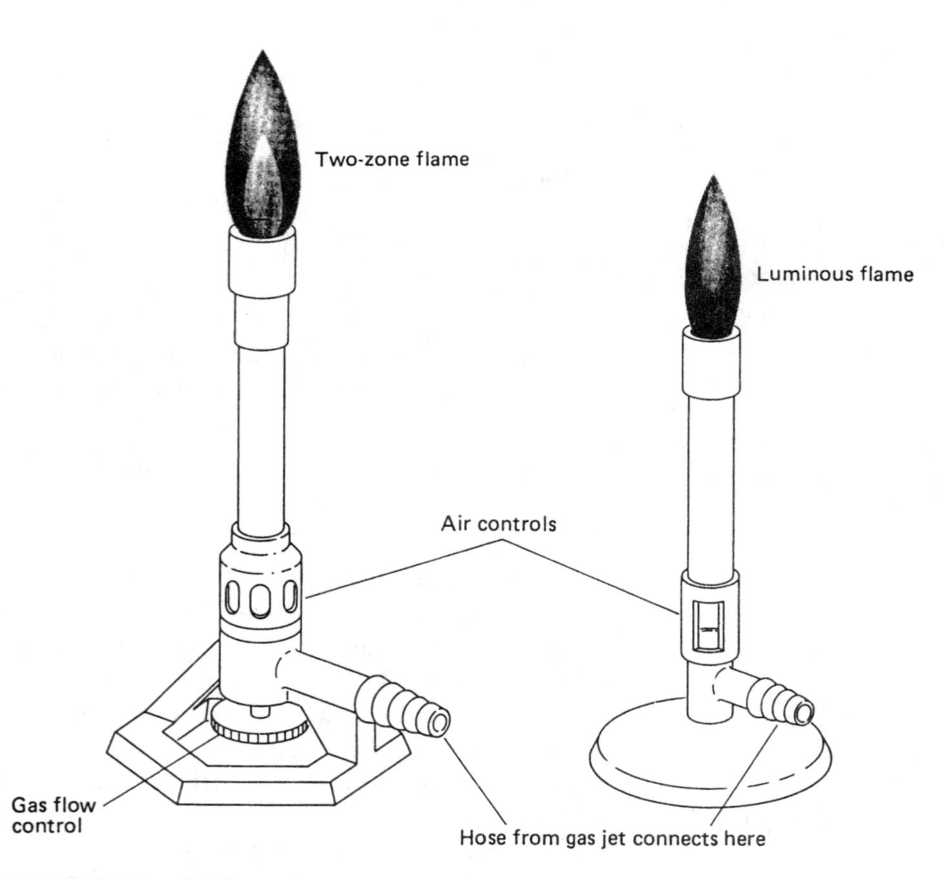

FIGURE 1.1 Gas burners.

This flame has the characteristic appearance of thermally excited sodium atoms (sodium is a component of glass).

The outer zone is hot enough to soften "soft" glass so that it can be bent, stretched, blown, or fire polished to fabricate custom laboratory glassware. Borosilicate glasses, such as Pyrex®, however, have too high a melting point to be worked in this way, although you can still use the Bunsen flame to round sharp edges slightly. Figure 1.2 shows soft glass and Pyrex rods whose ends have been rounded or "fire polished."

Although it is unlikely to be a part of your laboratory kit, the thermometer is an instrument which you will use frequently in the course of your lab work. The thermometer is used to measure the temperature of an object on the Fahrenheit (F), Celsius (C), or Kelvin (K) scale. The latter two are most useful to chemists, and are easily interconverted using the formula shown in Equation 1.1.

$$t_{\text{Celsius}} = T_{\text{Kelvin}} - 273.16 \qquad (1.1)$$

The Celsius thermometer is easily calibrated by comparing its readings with the known freezing point and boiling point of water, which are 0°C and 100°C exactly *at 760 torr atmospheric pressure*. Because the boiling point of water varies slightly with pressure, you will have to take this into account. To make this correction, you must know the barometric pressure in the laboratory. You may then look up the actual boiling point of water at that pressure in the *Handbook of Chemistry and Physics*, or, if you wish, you can calculate a value for the boiling point using the expression given in Equation 1.2, where P is the barometric pressure in torr and T_b is the boiling point of water on the Kelvin scale.

$$T_b = \frac{2124}{8.573 - \log P} \qquad (1.2)$$

Another instrument you will frequently use in the general chemistry laboratory is the balance. You will probably need to be familiar with only two types of balance: a triple-beam balance for weighings in which a tolerance of ± 0.1 g is acceptable, and a top-loading or analytical balance for weighings that require a tolerance of ± 0.001 g or smaller. General instructions concerning the use of balances are given in the Introduction. Your instructor will give you more specific directions for using the type of balance found in your lab.

In this experiment, you will use a top-loading or analytical balance to weigh samples of water that have been dispensed by a pipet and a buret. The data on the

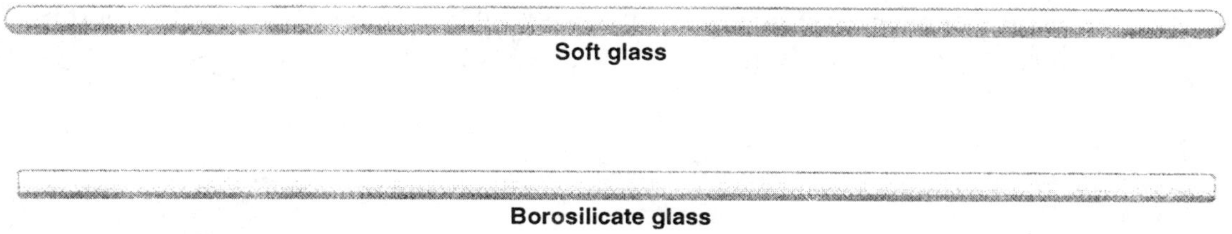

Soft glass

Borosilicate glass

FIGURE 1.2 Fire-polished glass rods.

TABLE 1.1 Specifications of 10-mL Pipets

Type	Tolerance	Cost
Class A	± 0.02 mL	$6–$10
Class B	± 0.04 mL	$4–$5
Polypropylene	± 0.06 mL	$16

masses of the water samples will be used to **calibrate** the pipet and buret (determine, accurately, what volume of water the pipet and buret actually dispensed).

The volumetric pipet (see Figures I.3 and I.4) is used to dispense a known volume of liquid quickly and precisely. A clean 10-mL pipet is probably precise (or reproducible), but may not be completely accurate because a small tolerance in volume dispensed is deemed acceptable. For three grades of 10-mL pipet, the tolerances and approximate cost per pipet are shown in Table 1.1. Although the polypropylene pipet has a higher cost and poorer precision than the glass pipets, it has one great advantage: it is unbreakable.

Like the pipet, the buret (see Figure I.3) is used to dispense liquid samples of known volumes. Burets are engraved with a scale reading downward, from 0 mL to 50 mL, and measure *volume dispensed.* A buret would normally be used when a variable amount of liquid is needed in an experiment, as in a titration. The buret used in introductory laboratory courses is usually accurate to ±0.03 mL.

You can improve the accuracy of experiments that employ pipets or burets by calibrating the instruments. This is done by weighing samples of water dispensed by the pipet or by the buret and recording the water temperature. You can then use the mass and temperature data to calculate the volume of the water dispensed by the pipet or by the buret by making use of one of water's fundamental properties—its density.

Density is defined as the ratio of an object's mass to its volume (see Equation 1.3). Densities for solids and liquids typically have units of g/mL.

$$\text{Density} = \frac{\text{mass}}{\text{volume}} \tag{1.3}$$

The density of water depends, to some extent, on its temperature. A table of the density of water at various temperatures is found in Appendix B.

Once you have determined the mass of the water delivered by your pipet or buret and looked up its density, you will use Equation 1.3 to calculate the actual volume of water dispensed in each of your calibration trials. Then, you will use your calibrated pipet to determine the density of a solution of sodium chloride of unknown concentration.

PROCEDURES

Desk Equipment

Using Figures I.1 through I.3 in the Introduction as a guide, try to identify each item that is on your equipment list. Note the names of unfamiliar items so that you will recognize them when they are called for in subsequent experiments. If any items are missing, broken, or inoperable, follow your instructor's directions to obtain replacements.

The Gas Burner

Connect your gas burner (see Figure 1.1) to the gas source using rubber tubing and check the operation of the air and gas controls before lighting the burner. Set the air control to partially open, gradually turn on the gas, and cautiously ignite the burner using a match or igniter. Note the effect of changing the gas flow, and of opening and closing the air control.

Adjust the burner to obtain a luminous yellow flame, and hold a clean glass rod in the flame briefly. Answer the questions on the Report Sheet.

Adjust the burner to give a two-zone flame, as in Figure 1.1. Place a wooden applicator stick so that it passes through the inner zone as well as the outer zone. Note your observations on the Report Sheet.

Place the glass rod briefly in the outer zone, and note whether a deposit forms or not. Obtain a glass rod from your instructor and fire polish both ends to make a stirring rod. Round one end at a time, laying the hot end on a wire gauze until cool before rounding the second end. Answer the questions on the Report Sheet.

CAUTION DO NOT TOUCH THE HOT GLASS! IT LOOKS THE SAME AS COOL GLASS.

The Thermometer

Fill a 100-mL beaker about half full with ice and add just enough water to cover the ice. Stir the mixture gently with the thermometer at intervals until a constant reading is observed. Record this melting point temperature (to ± 0.1°C) on the Report Sheet. (We use the terms *melting point* and *freezing point* interchangeably; they are the same.)

Fill a 250-mL beaker two-thirds full with water, add a few boiling chips, and heat the water to boiling over a Bunsen burner. The proper set up is shown in Figure 1.3. When the water is boiling smoothly, immerse the thermometer to the immersion mark (if possible) and cautiously read the boiling temperature to ± 0.1°C. Repeat a few minutes later to verify the accuracy of your measurement. Avoid touching the bottom of the beaker with the thermometer bulb.

Label your thermometer with the freezing point and boiling point that you have found by using it. If your freezing point and boiling point readings are too low or too high by an identical amount, you may use that amount as your thermometer correction. If the two deviations are not identical, you should plot the observed freezing and boiling temperatures versus the theoretical values for the freezing point and boiling point, and draw a straight line through the experimental points. This graph will permit you to report the corrected temperature for any thermometer reading.

The Pipet

Clean and dry a small Erlenmeyer flask and close the flask with a cork or rubber stopper of appropriate size. Take the stoppered flask to your assigned balance

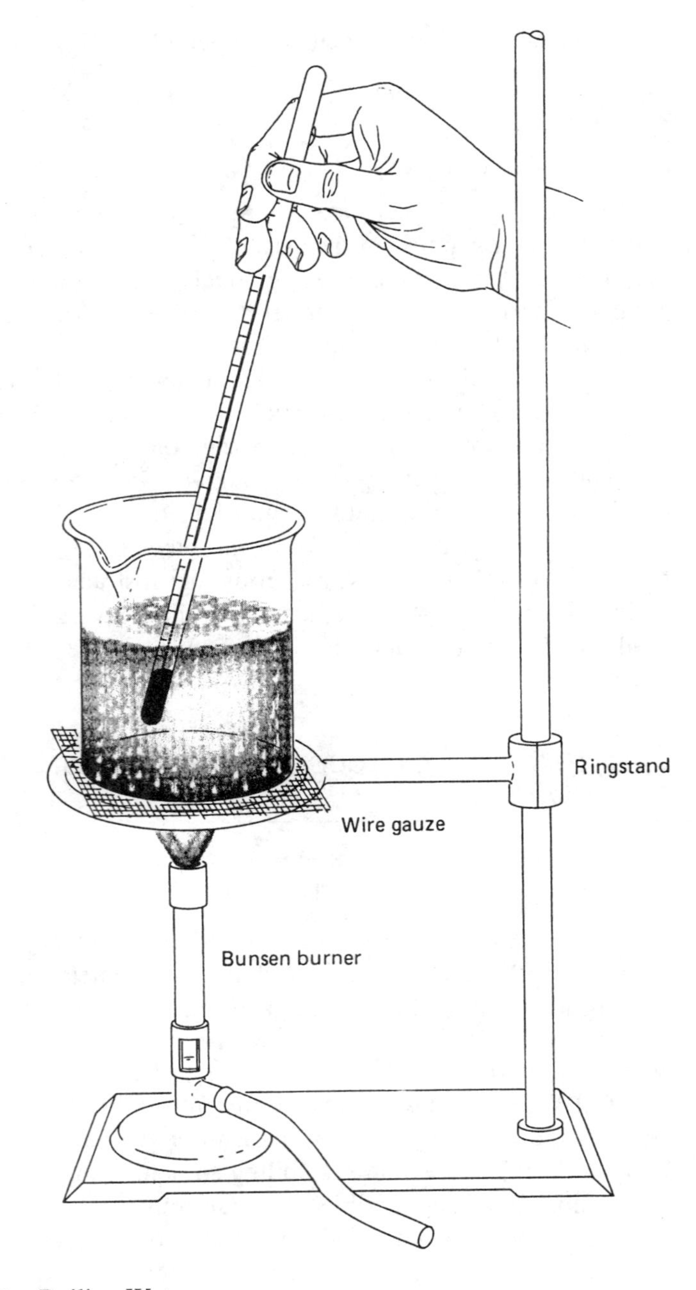

FIGURE 1.3 Boiling Water.

and determine its mass, following the weighing instructions given in the Introduction.

Practice filling your pipet using the pipet bulb until you have mastered this important skill. The steps are described in the Introduction and are illustrated in Figure I.4. When you are satisfied with your technique, use the pipet to dispense a 10-mL volume of water into the previously weighed flask. Replace the stopper and reweigh the flask, using the same balance. Record the mass on your Report Sheet. Dispense and weigh a second volume of water using the same procedure. Before discarding the excess water, measure its temperature using your calibrated thermometer. Record the corrected temperature.

Use Equation 1.3 and your data to find the volumes of the water samples delivered by your pipet. Report the average of these values, along with the average deviation. Label the pipet you used with the volume that it delivered. If there are previous calibrations noted on the pipet, comment on their consistency.

The Buret

Prepare and weigh a small Erlenmyer flask as you did when you were calibrating the pipet. Clean a buret, following the steps described in the Introduction. Fill the buret with distilled water (the water level should be just below at the 0 mL mark). Record the initial buret reading. (The procedure for reading the buret is described in Figure I.5; the procedure for reading the meniscus is shown in Figure 1.4). Use the buret to dispense a 20 mL volume of water into the previously weighed flask. Record the final buret reading. Replace the stopper and reweigh the flask, using the same balance. Record the mass on your Report Sheet. Dispense and weigh an additional 20 mL volume of water using the same procedure, and without emptying the Erlenmyer flask.

Use Equation 1.3 and your data to find the actual volumes of the (approximately 20 mL and 40 mL total) water samples delivered by your buret. Report the error resulting from dispensing a 20 mL volume and a 40 mL volume from your buret. If these errors are large, prepare a graph of apparent volume versus actual volume. Label the buret you used with the volumes that it delivered. If there are previous calibrations noted on the buret, comment on their consistency.

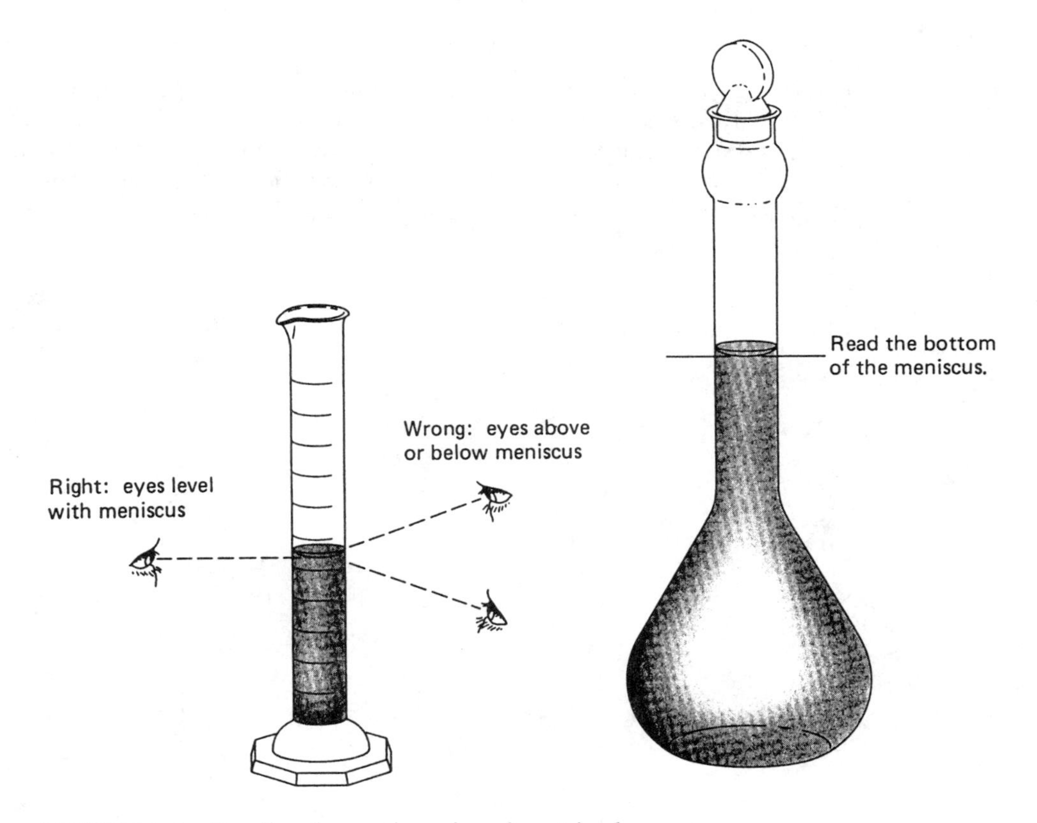

FIGURE 1.4 Reading the meniscus in volumetric glassware.

Properties of an Unknown Solution

Obtain about 25 mL of an unknown sodium chloride solution and determine its density by a procedure similar to that used to calibrate a pipet. First, weigh a clean, dry, stoppered flask. Next, use your calibrated pipet to dispense a precisely known volume of sodium chloride solution into the flask. Finally, reweigh the stoppered flask. Calculate the density of the solution.

You should also determine the freezing point as follows:

1. Fill a 6-inch test tube about two-thirds full with the solution and place the test tube in a 250 mL beaker almost filled with ice.
2. Add rock salt to the ice and stir gently with the test tube. The temperature should slowly decrease below 0°C.
3. When the unknown solution just starts to freeze, measure its temperature with your calibrated thermometer. DO NOT ALLOW YOUR UNKNOWN TO FREEZE SOLID. The temperature of the solution will decrease steadily as the ice forms. When you see the first ice crystals forming measure the temperature. Stir the mixture gently to ensure that you are observing the equilibrium temperature. Report the corrected temperature as the freezing point of your solution.

Disposal of Reagents

All chemicals used in this experiment may be diluted with water and flushed down the sink.

Questions

1. Using your data for the freezing point and boiling point of water, plot the actual Celsius temperature versus the observed (thermometer) temperature and draw a straight line through the points. Suppose your thermometer indicates that a liquid boils at 65.3°C. What would be its actual boiling point?
2. Why should you always be careful to use the same balance throughout an experiment?

Date ______ Name ______________________________ Section ______ Desk Number ______

PRE-LAB EXERCISES FOR EXPERIMENT 1

These exercises are to be performed after you have read the experiment but before you come to the laboratory to perform it.

1. Without looking at the figures, sketch the laboratory equipment specified.

   beaker pipet Erlenmeyer flask spatula

2. A thermometer indicates that water freezes at 0.3°C and boils at 100.3°C at a barometric pressure of 760 torr. The observed freezing point of an unknown liquid is –1.3°C.

   a. What is the corrected freezing point of the unknown, in °C?

   b. What is this freezing point on the Kelvin scale?

3. Suppose a 10-mL pipet dispenses 10.02 g of water and the density of water at that temperature is 0.9985 g/mL. What volume of water was dispensed by the pipet? Show your calculations.

   Is the pipet within tolerance limits for a Class A pipet? for a Class B pipet?

4. Suppose that you dispense water from a buret into a flask as described in this experiment. The initial and final buret readings are 0.13 mL and 19.67 mL, repsectively, and the flask is observed to weigh 47.315 g when empty, and 66.744 g after the water sample has been dispensed into it. Assume that the density of water at the temperature of the experiment is 0.9985 g/mL. Calculate:

   a. the apparent volume of the water dispensed;

   b. the actual volume of the water dispensed;

   c. the % error.

Date ______ Name ______________________________ Section ______ Desk Number ______

SUMMARY REPORT ON EXPERIMENT 1

The Gas Burner

1. The yellow luminous flame
   What deposits on the glass rod?

   What is the origin of this deposit?

   How is this deposit related to the yellow flame color?

2. The two-zone flame
   In which zone does the wooden applicator burn first?

   What do you conclude from this observation?

   Did you see a yellow flame above the glass rod?

   Does your glass rod appear to be soft glass or borosilicate glass? What proof do you have?

Thermometer Calibration

Observed melting point of ice	______________
Thermometer correction	______________
Observed boiling point of water	______________
Barometric pressure	______________
Actual boiling point of water	______________
Thermometer correction	______________

Pipet Calibration

	Trial 1	*Trial 2*
Mass of stoppered flask plus 10 mL water	__________	__________
Mass of empty stoppered flask	__________	__________
Mass of water dispensed	__________	__________
Temperature of water	__________	__________
Density of water at observed temperature	__________	__________

Calculated volume of water (show calculations):

Calculated volume Trial 1 __________ Trial 2 __________

Average volume ____________ ± Average deviation __________

Date ______ Name ______________________________ Section ______ Desk Number ______

Buret Calibration

Mass of empty stoppered flask ______________

Initial buret reading ______________

Buret reading, after first sample has been dispensed ______________

Apparent volume of water dispensed ______________

Mass of stoppered flask and first water sample ______________

Mass of first water sample ______________

Buret reading, after second sample has been dispensed ______________

Apparent volume of combined water sample ______________

Mass of stoppered flask and combined water sample ______________

Mass of combined water sample ______________

Temperature of water ______________

Density of water at observed temperature ______________

Calculated volumes of water dispensed (show calculations):

First water sample ______________ Combined water samples ______________

% error ______________ ______________

Density of Unknown Solution

Unknown number ______________

Actual volume dispensed ______________

Mass of stoppered flask plus unknown solution ______________

Mass of empty stoppered flask ______________

Mass of unknown solution ______________

Calculation of the density:

Observed freezing point ______________

Corrected freezing point ______________

EXPERIMENT 2 On the Nature of Pennies

Laboratory Time Required

Three hours

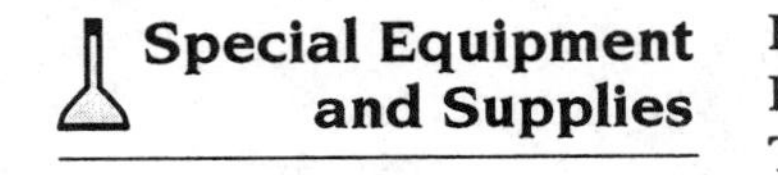

Special Equipment and Supplies

Balance
Burner
Tongs
Crucible
Triangle
Ringstand
Iron ring
Coffee cup calorimeter
Thermometer
Graduated cylinder, 100-mL
Calipers
Ruler
Iron file
Magnet
Pennies
Vinegar
Table salt
Hydrochloric acid

Safety

The gas burner can cause painful and serious burns to the skin. Be careful when lighting or adjusting the gas burner and when handling hot metals. Do not heat metals over an open flame. Use a crucible. Hydrochloric acid is corrosive and may cause skin burns or irritation.

First Aid

Severe burns should be treated by a physician. Following skin contact with hydrochloric acid, rinse the affected area thoroughly with water. If hydrochloric acid enters the eyes, rinse them in the eyewash (at least 20 minutes of flushing with water is recommended) and seek medical treatment.

One-cent pieces currently in circulation exhibit a number of different masses. In this experiment, students are asked to determine whether or not variations in the mass of a penny are significant and to develop and test hypotheses that explain both the major and minor variations.

PRINCIPLES

Hundreds of years ago, the value of a coin was determined by its mass and the worth of the metal it contained. It was common for merchants to weigh coins before making change to ensure that valuable metal had not been "shaved" from the coins. Today, the value of our currency is determined by the pledge of the federal government. Thus, damaged coins may remain in circulation, as long as their denomination is recognized.

Coins are frequently made of silver, gold, or copper. These coinage metals have low chemical activity,* with good reason. You would not want to burn your hands when handling money, as would happen if coins were made of sodium and potassium and you would not want your money to crumble, as rusting iron coins would. A common test of metal activity involves treating metal samples with hydrochloric acid. (Although hydrogen is not a metal, it does lose electrons when it combines with nonmetals to form acids.) Metals that are more active than hydrogen will be oxidized (i.e., will lose electrons) causing the protons in the acid to be reduced to hydrogen gas, which will bubble out of solution. Metals that exhibit this behavior when treated with hydrochloric acid are said to be "replacing" hydrogen in the acid. Neither gold nor silver nor copper is active enough to replace hydrogen from acid. In contrast, hydrogen gas bubbles out of solution when more active metals—such as iron or zinc—are treated with dilute hydrochloric acid. Sodium and potassium are active enough to replace hydrogen from water.

It is sometimes desirable to use an active metal in situations where its reactivity might be a problem. Iron is used in construction because of its strength. However, the iron is often alloyed with other elements, producing steel, which is more resistant to corrosion than is pure iron. Chromium and tin are often used to plate steel because they oxidize to form durable oxide coatings that prevent the interaction of oxygen and the iron beneath the surface layer.

Although copper is not a very active metal, it too becomes coated when exposed to air. Copper metal exposed to the atmosphere for a long period of time changes from its bright reddish-brown metallic luster to an almost black color and then may go on to develop a green patina. The initial blackening results from the formation of copper oxide and copper sulfide. These salts can be removed from the surface of pennies by placing the coins in vinegar to which a bit of table salt has been added. Interestingly, leaving the pennies soaking in vinegar overnight results in the formation of green deposits on the surface of coins.

Neither the mass nor the volume of a metal sample is a characteristic of the metal itself. However, the ratio of the mass to the volume (the density) is a characteristic of the metal. When attempting to determine the identity of a metal, one frequently weighs a sample (to determine its mass) and then finds the volume by direct measurement of dimensions (if the sample has a regular shape) or by measuring the volume of water displaced when the sample is submerged. The unit for density is generally g/mL, which is equivalent to g/cm^3, g/cc, and kg/dm^3.

Other characteristics that might be used to identify metals are specific heat, melting point, and appearance. The specific heat of a substance is a measure of

*This means the metals are not easily oxidized. It also means that it is relatively easy to decompose compounds of the metal so that the free metal can be obtained in element form.

the amount of energy needed to raise the temperature of one gram of the substance by one degree. A typical unit for specific heat is J/g °C. One method of determining the specific heat of metal involves placing a weighed sample of the metal in a bath of boiling water and allowing the metal to be warmed to the temperature of the boiling water. The metal sample is then transferred quickly to an insulated container which holds a sample of cool water for which the mass and temperature are known. The specific heat (C_M) of the metal can be calculated from the relation shown below, where g_M and g_{H_2O} represent the mass of the metal and the mass of the cool water, respectively, and t_b, t_c, and t_f represent the temperatures of the boiling water, the cool water, and the final system (metal plus cool water), respectively. The specific heat of water (C_{H_2O}) has a value of 4.18 J/g °C. The relationships between these variables is shown in Equations 2.1 and 2.2.

$$\text{Energy Change for Water} = -(\text{Energy Change for Metal}) \quad (2.1)$$

$$g_{H_2O}\, C_{H_2O}\, (t_f - t_c) = -g_M\, C_M\, (t_f - t_b) \quad (2.2)$$

The melting point of a substance is the temperature at which the solid and liquid forms of the substance are in equilibrium. A temperature of approximately 650°C can be achieved with a typical laboratory burner. Therefore, metals with melting points of 650°C or below will melt when held in a burner flame.

The appearance of a material might be characterized by such properties as color, shininess, and type of crystal structure. Table 2.1 lists several metals and their densities, specific heats, and melting points, along with some information about each metal's appearance.

PROCEDURES

Obtain a collection of pennies from your instructor. Weigh the pennies individually on the analytical balance. Record the data called for (mass of penny, date of mintage, description) on the Summary Report Sheet. Determine how many sets of pennies you have in your collection. Members of a single set of pennies may exhibit small variations about an average mass, which is significantly different from the average mass of pennies in any other set.

Consider each of the hypotheses listed in your Pre-Lab Exercises to explain

TABLE 2.1 Characteristics of Some Metals

	Density at 20°C	Specific Heat	Melting Point	Appearance
Aluminum	2.70 g/mL	0.89 J/g °C	660°C	silvery white, soft
Zinc	7.13	0.39	420	bluish-white, lustrous
Iron	7.87	0.45	1535	hard, brittle
Copper	8.96	0.39	1083	reddish, lustrous
Silver	10.5	0.056	962	brilliant white, lustrous
Lead	11.34	0.13	328	bluish-white, very soft
Gold	19.32	0.13	1064	yellow, soft

why the pennies do not all have the same mass. Decide whether your hypotheses are best suited to explaining the minor variations of the masses of the pennies in a single set or the major variation between the different sets. Analyze your initial data to determine if they are sufficient to confirm or refute each hypothesis. If they are not sufficient in a given case, develop and perform additional experiments aimed at confirming or refuting that hypothesis. Describe the experiments and record the data obtained from them on the appropriate spaces in the Summary Report Sheet.

Your discussion should consider how successfully your hypotheses explained the variations in mass. Estimate uncertainties in your numerical results and discuss possible improvements to your experimental design and/or further experimentation which might be needed.

Date ______ Name ______________________ Section ______ Desk Number ______

PRE-LAB EXERCISES FOR EXPERIMENT 2

1. List five hypotheses (to explain the difference of the pennies' masses) either suggested in the Principles section, or developed from some other source. State whether the hypothesis (if correct) would explain major or minor variations of the masses of the pennies.

2. Explain briefly how you might refute or obtain support for each of the hypotheses listed above.

Date ______ Name ______________________________ Section ______ Desk Number ______

SUMMARY REPORT ON EXPERIMENT 2

Mass of Penny	Mintage Date	Description of Penny	Set

Date ______ Name ______________________________ Section ______ Desk Number ______

Hypothesis ____

____

____

____

Would explain: major minor variation

Further Experimentation

Procedure ____

____

____

____

____

Data ____

____

____

____

____

Conclusions ____

____

____

____

____

Date ______ Name ______________________________ Section ______ Desk Number ______

Hypothesis ____

____

____

____

Would explain: major minor variation

Further Experimentation

Procedure ____

____

____

____

____

Data ____

____

____

____

____

Conclusions ____

____

____

____

____

Date ______ Name ______________________________ Section ______ Desk Number ______

Hypothesis ____

____

____

____

Would explain: major minor variation

Further Experimentation

Procedure ____

____

____

____

____

Data ____

____

____

____

____

Conclusions ____

____

____

____

____

Date ______ Name ______________________________ Section ______ Desk Number ______

Hypothesis ____

____

____

____

Would explain: major minor variation

Further Experimentation

Procedure ____

____

____

____

____

Data ____

____

____

____

____

Conclusions ____

____

____

____

____

Date ______ Name ________________________________ Section ______ **Desk Number** ______

Hypothesis ____

____

____

____

Would explain: major minor **variation**

Further Experimentation

Procedure ____

____

____

____

____

Data ____

____

____

____

____

Conclusions ____

____

____

____

____

EXPERIMENT 3

Identification of an Unknown Metal by the Determination of Its Density

Laboratory Time Required

One hour for well-prepared students. May be combined with Experiment 7, if desired.

Special Equipment and Supplies

Analytical balance
Graduated cylinder, 50 or 100 mL
Vernier calipers
Metric ruler or meter stick
Metal unknowns
Water

Safety

Although metal samples are not hazardous, they should not be exposed to acids with which they might react. Magnesium, aluminum, and zinc react rapidly with acids, liberating explosive hydrogen gas.

Replace any broken glassware to avoid cuts.

First Aid

For serious cuts, apply direct pressure if necessary to control bleeding and summon medical help.

Weighing using the analytical balance and measuring volumes of liquids using volumetric glassware are two of the most fundamental operations performed by a chemist. Because any quantitative experiment will most likely require one or both of these operations, it is important that you become proficient in weighing and measuring early in your laboratory training. In this experiment, you will measure both the mass and the volume of an unknown metal sample and then attempt to identify the metal using its density. Knowing the density, identity, and crystal structure of the unknown, you should be able to calculate its metallic radius.

PRINCIPLES

Density

The density of a substance is defined as its ratio of mass to volume. Consequently, density has such dimensions as g/cm^3 or lb/ft^3. The density of an element or compound is a characteristic physical property at a specified temperature and atmospheric pressure and is therefore useful in identifying the substance. Unlike liquids and, especially, gases, most solids have densities that are not greatly influenced by small changes in temperature or pressure. However, the density of a solid may depend somewhat on the method of sample preparation (i.e., whether it has been rolled, cast, or powdered). Typical densities of some common metals under normal laboratory conditions are shown in Table 3.1.

Atomic Masses and the Mole Concept

The atomic masses of the elements are all defined relative to the mass of one isotope of carbon, ^{12}C. On this scale, the atomic mass of ^{12}C is exactly 12. The atomic mass of magnesium is 24.31. This means that, on average, the mass of a magnesium atom is slightly more than twice the mass of an atom of ^{12}C.

Individual atoms are too small to be worked with in standard laboratories. The mole is a more convenient unit for measuring the chemical amount of substances. One mole of any substance contains 6.022×10^{23} particles of that substance (6.022×10^{23} is the number of atoms in 12 g of ^{12}C). One mole of magnesium contains 6.022×10^{23} atoms and has a mass of 24.31 g. (The mass of one mole of any element is simply the atomic mass of the element, expressed in grams.)

A mole of water contains a mole of oxygen atoms combined with two moles of hydrogen atoms. Thus, a mole of H_2O contains 6.022×10^{23} molecules and has a mass of 18.02 g.

When we know the density and identity (and hence the atomic mass) of an elemental metal sample, we can calculate the volume that would be occupied by one mole of the element. You might think that combining this molar volume with Avogadro's number would yield the volume of a single atom of the element (see Equation 3.1).

TABLE 3.1 Densities, Crystal Structures, and Metallic Radii for Selected Metals

Metal	Density, g/cm³	Crystal Structure*	Metallic Radius, pm†
Mg	1.74	HCP	160
Al	2.70	CCP	143
Cr	7.20	BCC	125
Fe	7.86	BCC	124
Ni	8.90	CCP	125
Zn	7.14	HCP (distorted)	139 (average)
Pb	11.34	CCP	175

*HCP = hexagonal closest-packed lattice; CCP = cubic closest-packed lattice; BCC = body-centered cubic lattice.

†Metallic radii are given in units of picometers (1 pm = 10^{-12} m).

$$\frac{\text{Atomic Mass}}{\text{Density} \times \text{Avogadro's Number}} = \frac{\text{Volume}}{\text{Atom}} \tag{3.1}$$

$$\frac{\text{g/mol}}{(\text{g/cm}^3)\ \dfrac{6.022 \times 10^{23}\ \text{atoms}}{\text{mol}}} = \frac{\text{cm}^3}{\text{Atom}}$$

However, because atoms are approximately spherical, part of the volume occupied by a sample is actually empty space. The amount of empty space in a given sample depends on the type of packing employed when the atoms come together to form the solid crystal. The nature of this packing correction is considered below.

Packing in Metals

In metals, the atoms are usually packed together in either a cubic closest-packed lattice, a hexagonal closest-packed lattice, or a body-centered cubic lattice. Figure 3.1 shows how spherical atoms can be packed together to form these structures. The two closest-packed structures are characterized by a layer arrangement in which each atom in one layer fits into the depression between three touching atoms in the layer below. These structures are packed quite efficiently, with very little unoccupied space. It can be shown that, for spherical atoms in either of the closest-packed structures, 74.0 percent of the available space is occupied by atoms. The body-centered cubic structure, however, is packed less efficiently, with only 68.0 percent of the available space occupied by atoms.

Determination of Volume

There are several methods for determining the volume of a solid. You might care-

Cubic closest-packed lattice

Hexagonal closest-packed lattice

Body-centered cubic lattice

FIGURE 3.1 Crystal structures commonly found in metals.

fully immerse the sample in a known volume of water in a graduated cylinder and note the volume of water displaced by the object. The volume of water displaced is equal to the volume of the object itself. Or you might use a ruler or calipers to measure the dimensions of the sample and calculate its volume. A clever student might find a way to use Archimedes' buoyancy principle to determine the volume.

PROCEDURE

Obtain an unknown metal sample from your instructor and record its number in your notebook or on the Summary Report Sheet. Weigh the unknown to the nearest 0.1 mg, using the analytical balance, and record its mass.

Measure the volume of the unknown as accurately as possible by using two different methods. Briefly describe each procedure followed. Record your observations and calculate the volume of the sample. Estimate the uncertainty in the volume by each method used and record the uncertainty after the volume.

Using the mass of your sample and the most accurate volume measurement, calculate the density of the metal. Estimate the uncertainty in the calculated density. Identify the metal using Table 3.1.

Calculate the apparent radius of the metal atom, showing your method of calculation as clearly as possible. Using the value of the radius given in Table 3.1, calculate your percent error.

Disposal of Reagents

The metal unknowns are to be returned to your instructor.

Questions

1. Consider spherical atoms of radius R arranged in a cubic closest-packed lattice, as shown in Figure 3.1. If a small cube is constructed so that its corners are at the centers of the corner atoms, then one-eighth of each corner atom and one-half of each face-centered atom lies within the small cube. What is the total number of atoms within the small cube? Calculate the volume occupied by these atoms, the volume of the cube, and the fraction of the available space that is occupied by atoms.

2. Show that for spherical atoms in a body-centered cubic structure exactly 68.0 percent of the available space is occupied by atoms. (Hint: Begin by constructing a cube like the one described in Question 1.)

Date ______ Name ____________________________ Section ______ Desk Number ______

PRE-LAB EXERCISES FOR EXPERIMENT 3

These exercises are to be performed after you have read the experiment but before you come to the laboratory to perform it.

1. The density of methanol, CH_3OH, is 0.792 g/cm^3 at 20°C. What volume of methanol contains Avogadro's number of molecules? How many molecules are in exactly 1 g of methanol? How many hydrogen atoms?

2. A rectangular metal sample has a mass of 7.1774 g and is 2.5 mm thick, 12.7 mm wide, and 25.4 mm long. What is its density? Could the sample be one of the pure metals listed in Table 3.1? If so, which one? If not, why not?

3. If the density of magnesium is 1.74 g/cm^3, what is the volume occupied by exactly one mol of magnesium atoms? What is the total volume per magnesium atom? If 68.0 percent of this volume is occupied, what is the actual volume of a single magnesium atom? What is the radius, in pm, of a single, spherical atom? The volume of a sphere is given by the formula $V = (4/3)\,\pi\, r^3$.

Date ______ Name ________________________________ Section ______ Desk Number ______

SUMMARY REPORT ON EXPERIMENT 3

Number and description of unknown

Mass of unknown

First weighing ________________ g

Second weighing ________________ g

Average ________________ g

Volume of unknown

1. Brief outline of first procedure followed

Experimental data

First volume ________________ cm^3 ± ________________ cm^3

Date ______ Name ______________________________ Section ______ Desk Number ______

2. Brief outline of second procedure followed

Experimental data

Second volume ________________ cm^3 ± ________________ cm^3

Density of unknown

Calculated density ________________ cm^3 ± ________________ cm^3

Identification of unknown and calculation of metallic radius

Identity ________________

Calculated radius ________________ pm

Actual radius ________________ pm

Percent error ________________ %

EXPERIMENT 4

The Strange Case of the Floating Cans

Laboratory Time Required

Three hours

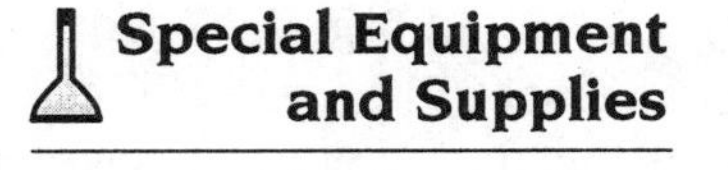

Special Equipment and Supplies

Balance
Buret
Buret clamp
Pipet
Pipet bulb
Volumetric flask, 100-mL
Graduated cylinder, 10-mL
Graduated cylinder, 1000-mL
Tape measure
Vernier calipers
Ruler
Diet and regular soft drinks
Sugar
Artificial sweetener

Safety

This experiment does not expose students to chemical hazards. However, soft drinks and aqueous solutions that have been opened or used in the lab must not be consumed.

First Aid

It is not expected that any injuries could result from the performance of this experiment.

An experimental determination of density is often a student's first experience in the laboratory. In this experiment, students become familiar with a variety of instruments for the measurement of mass, volume, and density, while exploring an intriguing, but commonplace, phenomenon.

PRINCIPLES

Some cans of soda have fallen off a truck and landed in a shallow creek. The cans of regular soda (sweetened with sugar or corn syrup) sink quickly to the bottom of the creek, but the cans of diet soda (sweetened with aspartame) float! Your job is to explain this phenomenon. To assist you in planning your experiments, we review the story of Archimedes, the crown, and the bathtub.

Archimedes was a Greek scientist who lived over two thousand years ago. He was commissioned by his king to determine whether a goldsmith had surreptitiously replaced gold with a less-precious metal in making a crown for the king. The king suspected that he had done so; however, the finished crown looked like it was made of gold and it had the same weight as the gold that had been given to the smith. Thus, the king could not prove his suspicions. Archimedes could not find an answer for the king quickly. However, when Archimedes was settling into his tub for a bath one day, he noticed that some water spilled out of the tub. He shouted "Eureka" (I've found it!) and proceeded to perform the experiment that proved the goldsmith guilty.

What Archimedes realized was that an object submerged in a fluid displaces a volume of fluid equal to its own volume. Therefore, he could find the volume of the crown by catching the water it displaced from a vessel full of water. Gold had the highest density (mass to volume ratio) of all metals known in ancient times. A crown that had been adulterated with a base metal would have a volume larger than an equal weight of gold.

What does this have to do with floating cans? All fluids exert a buoyant force on objects immersed in them. This force is a measure of the difference between the mass of the object and the mass of the fluid it displaces. If the mass of the object is less than the mass of the displaced fluid, the object will float in the fluid. Thus, objects that float in a fluid have densities less than the densities of the fluids in which they float.

Your starting point in investigating the mystery of the floating cans should be to determine the densities of a can of regular soda and a can of the corresponding diet soft drink. Your investigation should be expanded from there to answer some or all of the following questions: Do containers of noncarbonated diet soft drinks float in water whereas containers of the corresponding "regular" soft drink sink? Is the density of the diet soda less than that of pure water? Are containers for diet soda and the corresponding "regular" soda the same size? Is there some reason for diet drinks to be packed with more empty space than is needed for "regular" drinks? Does this need for more space—if there is a need—exist in bottled soda as well as in soda packaged in cans? What is the significance of the presence or absence of sugar in determining the density of the drink itself and of the container filled with the drink?

PROCEDURE

Obtain a can, glass bottle, or plastic bottle of any diet drink—carbonated or noncarbonated—and another can, glass bottle, or plastic bottle of the corresponding "regular" drink. Determine whether either or both of the drinks floats in water. Using the equipment provided (and any other materials your instructor can provide upon request), determine whether the density of your unopened container is what is expected (less than the density of water if the container floats; greater than the density of water if it sinks). You might do this by weighing the unopened containers and then measuring the volume of water the containers displace when they are submerged (how will you do that for a container that floats?). Or you might measure the dimensions of the container and calculate its volume. Or you might empty the container (be sure to save its contents) and see how much water is needed to refill the container.

Next, determine the densities of the contents of your containers. You may use any procedure you wish. Use the information you have about the container and about the drink itself to determine if each container does contain at least the volume of liquid specified on the container. Also, determine the amount of empty space in the container.

Use the nutritional information printed on the container to determine how much sugar should be dissolved in water to prepare a solution comparable to a "regular" drink. Prepare an equal volume of an equally sweet solution using artificial sweetener. Measure the densities of the sweetened water and compare these densities to those of the corresponding commercial soft drinks.

Be sure to record all necessary data on the Summary Report Sheets. Share your findings with other students who worked with different drinks and obtain their results. Prepare a table of results, including your own work and the information provided by classmates. Give a full discussion of your answers to the questions posed above and to any other questions which may have arisen in class as the experiment was being conducted.

Disposal of Reagents

All soft drinks and sweetened waters can be flushed down the drain.

Date ______ Name ______________________ Section ______ Desk Number ______

PRE-LAB EXERCISES FOR EXPERIMENT 4

These exercises are to be completed after you have read the experiment but before you come to the laboratory to perform it.

An unopened can of soda weighed 410.12 g. The length of the can was 12.6 cm. The circumference of the can (at its widest) was 20.1 cm.

1. Calculate the volume of the can.

2. Determine the density of the unopened (full) can and predict whether the can would float in water.

3. A volume of 10.0 mL of the contents of the can weighed 10.04 mL. Determine the density of the soda contained in the can.

4. The empty can weighed 13.2 g. Did the unopened can contain the 355 mL of soda promised on the label?

5. List any errors that might be present in the data cited above. Give suggestions on how you will minimize these errors when you perform the experiment.

Date ______ Name ______________________________ Section ______ Desk Number ______

SUMMARY REPORT ON EXPERIMENT 4

Soft drinks you are studying ______________________________

Data you are seeking ______________________________

Procedure ______________________________

____

____

____

____

____

Data ______________________________

____

____

____

____

____

____

____

____

____

____

____

____

____

____

____

____

Date ______ Name ______________________________ Section ______ Desk Number ______

Results ____

____

____

____

____

____

____

____

____

____

____

Date ______ Name ______________________________ Section ______ Desk Number ______

Soft drinks you are studying ______________________________

Data you are seeking ______________________________

Procedure ______________________________

____

____

____

____

____

Data ______________________________

____

____

____

____

____

____

____

____

____

____

____

____

____

____

____

____

____

Date ______ Name ______________________________ Section ______ Desk Number ______

Results ____

____

____

____

____

____

____

____

____

____

____

Date ______ Name ______________________________ Section ______ Desk Number ______

Soft drinks you are studying ______________________________

Data you are seeking ______________________________

Procedure ______________________________

Data ______________________________

Date ______ Name ______________________________ Section ________ Desk Number ______

Results ____

____

____

____

____

____

____

____

____

____

____

Date ______ Name ______________________ Section ______ Desk Number ______

Soft drinks you are studying ______________________

Data you are seeking ______________________

Procedure ______________________

Data ______________________

Date ______ Name ______ Section ______ Desk Number ______

Results ______

Date ______ Name ________________________________ Section ______ Desk Number ______

Tabluated Results

Product	Studied By	Does it float?	Density of drink	Density of full container

EXPERIMENT 5

Separation of a Mixture into Its Components by Fractional Crystallization

Laboratory Time Required

Three hours

Special Equipment and Supplies

Hot plate
Buchner funnel and vacuum flask
Ice bath
3 M H_2SO_4
Salicylic acid/ $CuSO_4 \cdot 5H_2O$ mixture
Boiling chips
95% Ethanol

Safety

The usual precautions should be taken to prevent exposure of the skin or eyes to chemicals. Boiling solutions may bump suddenly, splashing hot liquid over a wide area. Add boiling chips to solutions *before heating is begun* to prevent overheating and the resulting bumping. Beware of hot liquids that are still; they may be superheated.

First Aid

Rinse chemicals from the skin with water. For treatment of severe burns, see a doctor.

The resolution of mixtures into their components is an important part of experiments concerned with analysis or synthesis. **Analysis** is concerned with finding the quantitative composition of a mixture. **Synthesis** is the formation of a substance from simpler substances. Synthesis usually includes a separation to isolate the desired product from large amounts of solvent and by-products. This separation is often followed by a purification step that removes trace amounts of other substances from the final product.

Some of the experimental techniques that may be used to separate or purify mixtures are: fractional crystallization, fractional distillation, chromatography, and zone refining. Each of these makes use of small differences in a physical property such as solubility, vapor pressure, melting point, or tendency to be adsorbed on a surface, to separate the components of a mixture. In this experiment, you will use fractional crystallization to resolve a mixture of salicylic acid and $CuSO_4{\cdot}5H_2O$ into its components.

PRINCIPLES

The oft-quoted rule of thumb "like dissolves like" suggests that the ionic salt $CuSO_4{\cdot}5H_2O$ would be more soluble in the highly polar solvent H_2O than in a solvent of lower polarity, such as ethanol. For the organic solute salicylic acid (2-hydroxybenzoic acid), these solubility tendencies might well be reversed. Such is indeed the case.

Figure 5.1 shows the solubility (in g solute/100 g of H_2O) of $CuSO_4{\cdot}5H_2O$ to be one to two orders of magnitude higher than that of salicylic acid. Further, the solubilities of both solutes are seen to increase with temperature. A mixture of these two solutes could be dissolved in 100 mL of water at 75°C to 100°C. Most of the salicylic acid could then be precipitated as fine, off-white needles when the solution is cooled to 0°C. Up to 25 or 30 g of $CuSO_4{\cdot}5H_2O$ could remain in the cooled solution without precipitating. This procedure will be used in this experiment to separate the salicylic acid in good yield and purity.

It is harder to crystallize $CuSO_4{\cdot}5H_2O$ from water, especially if we wish to prevent the precipitation of the small amount of salicylic acid left in solution from the first crystallization. We can obtain pure $CuSO_4{\cdot}5H_2O$ by carefully evaporating the solution to about 25 mL and then adding an equal volume of 95% ethanol. This lowers the dielectric constant of the solvent medium so that

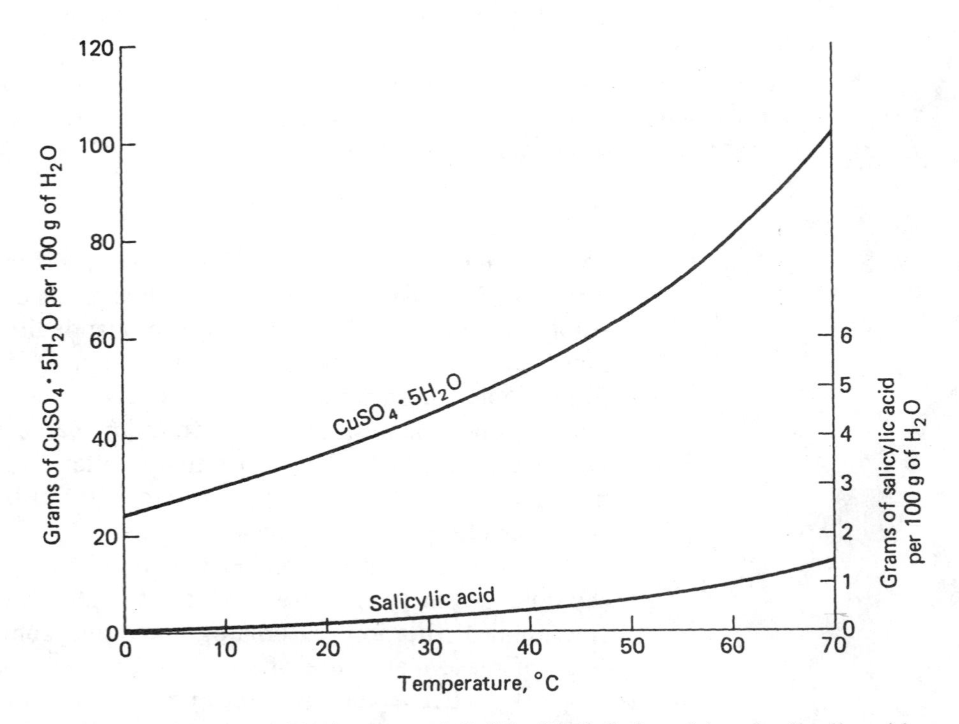

FIGURE 5.1 The solubility in water of $CuSO_4{\cdot}5H_2O$ (left scale) and salicylic acid (right scale).

essentially all of the $CuSO_4{\cdot}5H_2O$ precipitates as fine blue crystals when the solution is cooled to room temperature or lower. At the same time, the salicylic acid will remain in solution.

Although both precipitates may appear pure, an analysis would no doubt show that the salicylic acid is contaminated by trace amounts of $CuSO_4{\cdot}5H_2O$, and vice versa. If, for some reason, we need ultrapure salicylic acid or $CuSO_4{\cdot}5H_2O$, we obviously could repeat the dissolution and crystallization procedure to obtain recrystallized products. That will not be necessary for this experiment.

PROCEDURE

Obtain an unknown mixture and record its number on the Summary Report Sheet. Add about 5 grams of the unknown mixture to a previously weighed, clean and dry, 250-mL beaker. Weigh the beaker containing the mixture and calculate the mass of the unknown.

Add 100 mL of water containing about 1 mL of 3 M H_2SO_4 solution to the beaker. The acid will prevent any possible chemical reaction between the copper sulfate and the salicylic acid. Heat the mixture on a hot plate, constantly stirring until all of the solid material has dissolved. Avoid boiling the solution. Remove the beaker from the heat, cover it with a watch glass, and allow it to cool enough to be comfortable to the touch.

CAUTION WAIT A FEW MINUTES BEFORE TRYING TO TOUCH THE BEAKER.

Place the beaker in an ice bath. A large quantity of salicylic acid needles should settle to the bottom of the beaker. Keep the beaker in the ice bath until crystallization is complete.

Set up a vacuum filtration apparatus as shown in Figure 5.2 and place a piece of filter paper in the Buchner funnel. Swirl the beaker and pour its contents into the Buchner funnel. Transfer the salicylic acid crystals in this way as completely as possible to the filter paper. A minimum amount of cold rinse water may be used to complete the transfer. Pour the blue solution collected in the filtration flask into a second clean beaker and reserve it for the crystallization of $CuSO_4{\cdot}5H_2O$. Draw air through the Buchner funnel for several minutes to remove as much water as possible; then remove the filter paper and its mat of salicylic acid crystals. Carefully transfer the crystals to a dry, preweighed filter paper to finish drying. At the end of the lab period, weigh the paper and its crop of crystals and calculate the mass of salicylic acid recovered. Also report the percent by mass of salicylic acid in the original mixture.

Place the beaker containing the blue copper sulfate solution on a hot plate, add a stirring rod and one or two boiling chips to the beaker, and boil the solution gently until its volume has been reduced to about 25 mL.

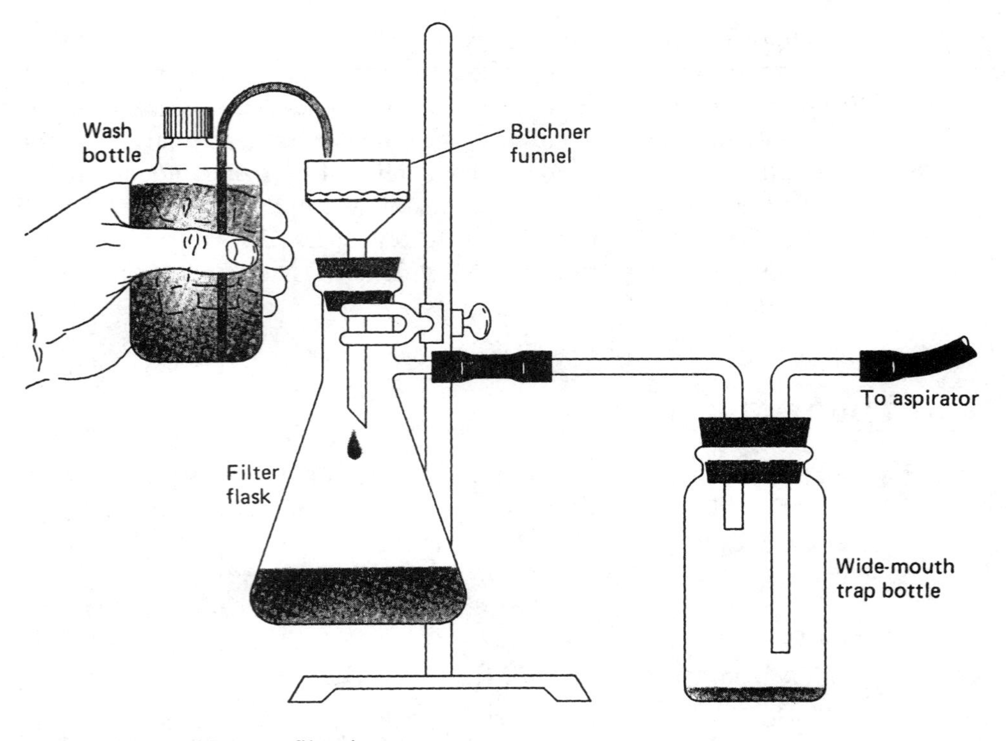

FIGURE 5.2 Vacuum filtration apparatus.

CAUTION THE STIRRING ROD AND THE BOILING CHIPS ARE INTENDED TO BE A SOURCE OF BUBBLES THAT WILL PREVENT SUPERHEATING, BUT BE VERY CAUTIOUS OF THE HOT SOLUTION ANYWAY, ESPECIALLY IF NO BUBBLES ARE OBSERVED. IF THE SOLUTION DOES BECOME SUPERHEATED, ANY DISTURBANCE COULD CAUSE IT TO BOIL OVER VIOLENTLY, RISKING INJURY TO PERSONS NEARBY.

Remove the beaker from the hot plate and, while stirring, slowly add about 25 mL of ethanol, or a sufficient volume to just cause the solution to become permanently cloudy. Cool the beaker to room temperature, then further cool it in an ice bath. A nice crop of blue $CuSO_4{\cdot}5H_2O$ crystals should settle to the bottom of the beaker. Collect the copper sulfate crystals on a preweighed filter paper as before, using the Buchner funnel. Rinse the product with a small amount of ethanol and draw air through it to dry it, using the vacuum filtration apparatus. Weigh the dry product and report the mass of $CuSO_4{\cdot}5H_2O$ recovered, as well as its percentage in the original mixture. (If the mass of the crystals seems to fall constantly, the $CuSO_4{\cdot}5H_2O$ may be contaminated with ethanol. Allow additional time for the ethanol to evaporate before attempting to weigh the crystals again.)

Comment briefly on the appearance of the products. Is there any visual evidence that they are not pure?

Disposal of Reagents

Place the $CuSO_4{\cdot}5H_2O$ and the salicylic acid, respectively, in the labeled collection bottles. All other chemicals, including the filtrate, may be rinsed down the drain.

Date ______ Name ______________________________ Section ______ Desk Number ______

PRE-LAB EXERCISES FOR EXPERIMENT 5

These exercises are to be completed after you have read the experiment but before you come to the laboratory to perform it.

1. Explain briefly why $CuSO_4$ is more soluble in water than in an organic (nonpolar) solvent. Why is the reverse true for salicylic acid?

2. Suggest reasons why the solubility of most solid substances in liquid solvents increases with temperature, whereas the solubility of gases decreases with temperature.

3. E.D. Student used 5.124 g of a mixture of copper sulfate pentahydrate and salicylic acid in the performance of this experiment. E.D. recovered 3.062 g of salicylic acid and 2.139 g of copper sulfate pentahydrate. What is wrong with E.D.'s results? How might this error have been avoided?

Date ______ Name ______________________ Section ______ Desk Number ______

SUMMARY REPORT ON EXPERIMENT 5

Unknown number ______________

Mass of unknown + beaker ______________

Mass of beaker ______________

Mass of unknown mixture ______________

Salicylic Acid Crystallization

Mass of weighing paper + salicylic acid ______________

Mass of weighing paper ______________

Mass of salicylic acid ______________

Calculation of percent salicylic acid in original mixture:

$CuSO_4 \cdot 5H_2O$ Crystallization

Mass of weighing paper + $CuSO_4 \cdot 5H_2O$ ______________

Mass of weighing paper ______________

Mass of $CuSO_4 \cdot 5H_2O$ ______________

Calculation of percent $CuSO_4 \cdot 5H_2O$ in original mixture:

Percent of original sample not recovered ______________

Comments concerning appearance and purity:

Fun With Solutions

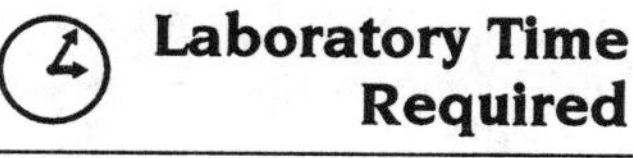

Laboratory Time Required

Three hours

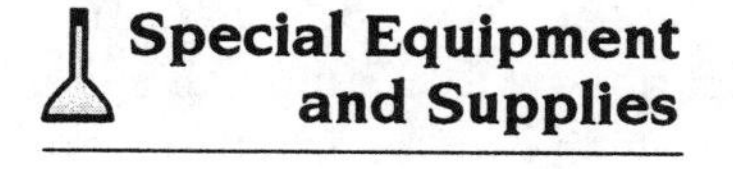

Special Equipment and Supplies

Paper plates
Cotton swabs
Plastic sandwich bags
Conductivity tester
Spray bottles
Labels
Dropper bottles

Phenolphthalein (aq)
$Na_2S_2O_3$ (aq)
$CuSO_4$ (aq)
NaOH (aq)
I_2 (aq)
NH_3 (aq)
$MgSO_4$ (s)
$MgSO_4 \cdot 7H_2O$ (s)
1 M $BaCl_2$
6 M NH_3
0.1 M $Pb(NO_3)_2$
0.05 M $CoCl_2$
0.05 M $CuSO_4$
0.05 M $Ni(NO_3)_2$
Table salt
Table sugar

Safety

Avoid getting acids or bases in your eyes or on your skin. After completing the experiment, wash your hands with soap and water to remove any traces of toxic chemicals, such as the salts of lead, cobalt, or nickel.

First Aid

Following skin contact with any of the reagents, wash the area with copious amounts of water. If acid or base enters your eyes, use the eyewash fountain to flush away the chemical. Then, consult a physician.

Two of the hardest skills for students to master in introductory chemistry classes are the ability to determine which chemical changes account for the observations they make when a reaction occurs, and the ability to represent those reactions with balanced chemical equations. This experiment allows students to explore the properties of solutions in an entertaining setting, provides support for distinguishing between physical changes and chemical changes, and introduces students to the processes of writing equations and designing experiments.

PRINCIPLES

A magic show can be appreciated on several levels. A young child may believe that a magician has really pulled a rabbit out of an empty hat. An older child may sense that the magician probably has not really managed to change a scarf into a dove, but may not be able to offer a plausible explanation for the magician's illusion. Someone who has practiced sleight-of-hand may indeed know how the trick is done and still be able to enjoy seeing a talented practitioner perform the trick skillfully. In this experiment, you will have the opportunity to learn the chemistry behind some things that might appear in a "chemical magic show." You will prepare "secret" messages by writing in "invisible" ink. You will make mixtures that get hot or cold. You will learn how to distinguish sugar from salt without tasting either. You will find a liquid than can "pour out" a variety of colors, and mix colors that will disappear before your very eyes. Some of these "tricks" will involve physical changes alone; others will involve chemical reactions. Part of your job will be to determine which is which. In the course of finding the answers, you will learn to use this manual and your text as research tools and, ultimately, be able to design some "tricks" of your own.

"Invisible" inks can be prepared from a variety of substances. You will use three different kinds. One ink is an indicator, a substance that changes color as the acidity of the medium in which it is dissolved changes. You can read about indicators in Experiments 7, 8, 13, 21, 30, and 39 of this manual. You could also look into your textbook's chapters on acids and bases and on titrations or look up indicators in the text's index.

Another "invisible" ink works because of an oxidation/reduction reaction in which iodine (or its aqueous form, the triiodide ion) is reduced by the action of sodium thiosulfate. References to this reaction are found in Experiments 26, 27, 28, and 33. You might find further information by checking your text's chapter on redox reactions or looking for "thiosulfate" or "iodometric titrations" in the index.

The "magic" of the third "invisible" ink uses the reaction between copper ions and ammonia to form a complex ion. This phenomenon is mentioned in Experiments 11, 19, 28, 41, and 43. You could also find information by checking your text's sections on coordination compounds or complex ions. Complex ions are sometimes discussed in connection with the solubility of salts in water or in connection with transition metal chemistry.

The process of dissolving a salt (an ionic compound) in water is sometimes accompanied by the generation of a significant quantity of heat. Such a process is said to be "exothermic" and may be used to make chemical hot packs. Some salts dissolve in an "endothermic" manner, meaning that energy is absorbed from the surroundings as the salt dissolves. An endothermic dissolution process could be used to prepare a chemical cold pack. The generation of heat when a salt is dissolved is used for the identification of an unknown in Experiment 39. The amount of heat absorbed when another salt dissolves is measured in Experiment 30. For a more complete discussion, consult the sections of your textbook that deal with preparing solutions or with thermochemistry.

To illustrate that creating the hot or cold pack has not altered the salts, they are subjected to identical chemical reactions. The reaction of ammonia with the dissolved salts depends on the ability of ammonia to act as a base in aqueous solution. This reaction is used in Experiment 39. Ammonia's ability to act as a base is discussed in your textbook in the sections on acid/base chemistry and/or on the Bronsted/Lowry model. The reaction with barium chloride involves the formation of an insoluble barium compound. This compound is formed in Experiments 41 and 42. You should be able to find further information in your text's chapter on solubility or in a section on precipitation or metathesis (double displacement) reactions.

It would be easy to build a machine that could tell sugar from salt. Table sugar is a nonelectrolyte, which has a large molar mass. Table salt is an electrolyte, which has a much lower molar mass than table sugar. Thus, one could distinguish sugar from salt by measuring the conductivity of an aqueous solution of an unknown crystal. Or, one could check the freezing point of solutions of equal masses of crystals, each dissolved in the same mass of water. The salt, because it dissociates and because it has a lower molar mass, would have the lower freezing point. The phenomenon of freezing point depression is discussed in Experiment 17 and the tests to distinguish sugar from salt are employed in Experiment 39. Your text has more information in the sections on electrolytic solutions, freezing point depression, or colligative properties.

Ammonia changes the colors of solutions by acting as a base (generating hydroxide ions that affect the colors of acid/base indicators or forming insoluble hydroxide compounds with metal ions) or by acting as a complexing agent (as it does with copper(II) ions). Sources of information on these phenomena have been cited above.

Many transition metal ions are highly colored. Mixing the appropriate amounts of solutions containing cobalt(II) ions, copper(II) ions, and nickel(II), can produce a startling result as the colors seem to disappear. The absorbance of light by colored solutions is discussed in Experiment 19. Your text may discuss color in the sections on the beginnings of quantum mechanics, on spectroscopy, or on transition metal chemistry.

PROCEDURE

A. Invisible Inks

Using a pen or pencil, write "1," "2," and "3," respectively, on three paper plates. Also label each plate with your name. Dip a cotton swab into phenolphthalein solution. Write a secret message on plate #1 with the swab. Dip the swab in phenolphthalein as many times as is necessary to complete the writing. Set the plate aside to dry.

Repeat the procedure described above, using a new cotton swab, dipped in thiosulfate solution, to draw a picture on plate #2.

Repeat the procedure described above, using a new cotton swab, dipped in copper sulfate solution, to draw a picture on plate #3.

When the plates are dry, lightly spray the first with sodium hydroxide solution, the second with iodine solution, and the third with ammonia solution. Describe the appearance of each plate on the Summary Report Sheet.

B. Hot and Cold

Fill the bottom of a plastic sandwich bag with a thin layer of epsom salt (magnesium sulfate heptahydrate, $MgSO_4 \cdot 7H_2O$). Add 10 mL of water. Note whether the bag gets hot or cold.

Fill the bottom of a second sandwich bag with a thin layer of anhydrous magnesium sulfate, $MgSO_4$. Add 10 mL of water. Note whether the bag gets hot or cold.

Place a pea-sized amount of epsom salt in each of two clean test tubes. Add 5 mL of water to each tube and stir the contents until the solid dissolves. Add one dropperful of 6 M ammonia to one tube and one dropperful of 1 M barium chloride to the other tube. Note your observations on the Summary Report Sheet.

Place a pea-sized amount of anhydrous magnesium sulfate in each of two clean test tubes. Add 5 mL of water to each tube and stir the contents until the solid dissolves. Add one dropperful of 6 M ammonia to one tube and one dropperful of 1 M barium chloride to the other tube. Note your observations on the Summary Report Sheet.

C. Is It Sugar or Is It Salt?

Weigh out 4 g of table sugar (sucrose, $C_{12}H_{22}O_{11}$) and place the sample in a small beaker. Add 10 mL of water and stir the mixture to dissolve the solid. Weigh out 4 g of table salt (sodium chloride, NaCl) and place the sample in a second small beaker. Add 10 mL of water and stir the mixture to dissolve the solid.

Test the sugar and salt solutions you've just prepared for electrical conductivity. Obtain a conductivity tester and secure the battery connection. Place the copper leads in the sugar solution, and be sure that the leads do not touch each other. Note your observations on the Summary Report Sheet. Rinse the leads with distilled water and dry them. Repeat the conductivity test with the salt solution. Note your observations on the Summary Report Sheet. Clean the leads and disconnect the battery.

Next, determine whether the sugar solution or the salt solution has the lower freezing point. Prepare an ice/water/rock salt bath by placing equal amounts of each material in a large beaker. Stir the mixture well with a stirring rod or a thermometer. Be sure the temperature of the bath is –10°C or below. (If it is not, pour off some water and add more ice and rock salt; stir the mixture again before checking its temperature.) Place the beaker of sugar solution in the cold bath and stir the sugar solution with a thermometer until ice crystals begin to form or until the temperature of the solution reaches the temperature of the bath. Record your observations on the Summary Report Sheet. Rinse the thermometer and dry it. When the thermometer is again reading room temperature, repeat the procedure with the beaker containing the salt solution. Record your observations on the Summary Report Sheet.

D. Patriotic Liquid

Place a dropperful of phenolphthalein solution in a clean test tube. Add a dropperful of 6 M ammonia. Record your observations on the Summary Report Sheet.

Place a dropperful of lead nitrate solution in a clean test tube. Add a dropperful of 6 M ammonia. Record your observations on the Summary Report Sheet.

Place a dropperful of copper sulfate solution in a clean test tube. Add a dropperful of 6 M ammonia. Record your observations on the Summary Report Sheet.

E. Now You See It, Now You Don't

Place one dropperful of 0.05 M cobalt chloride solution in a clean test tube. Note its color. Add 0.05 M copper sulfate solution, dropwise, until the volume of the solution in the test tube has doubled. Note the color of the mixture. Add 0.05 M nickel nitrate solution to the mixture, one drop at a time. Stir the mixture after each addition of nickel nitrate solution. Add nickel nitrate solution until the mixture appears colorless. On the Summary Report Sheet, note whether you have succeeded in producing a colorless mixture.

F. On Your Own

Design a chemical "magic trick" to demonstrate for your classmates. You might demonstrate how to make a blue message materialize on a white background, or how to have a white message appear on a brown background. You might prepare a cold pack or a hot pack using salts you've read about while doing research for this experiment. You might produce your school colors using a single liquid and various reagents, or you might devise a way for distinguishing between cornstarch and baking soda. Whether you devise a method for performing one of the suggested transformations, or devise something totally on your own, have your instructor check your idea for safety before you proceed.

DISCUSSION

Explain each of the observations you've noted in performing this experiment with a sentence or two and/or an equation. Note which procedures involved chemical changes and which did not. Explain your reasons for classifying certain changes as physical changes and others as chemical changes.

Disposal of Reagents

Place any mixtures containing nickel, cobalt, or lead in the collection bottles provided. All other solutions and suspensions may be diluted and flushed down the drain with a good supply of water.

Date ______ Name ________________________________ Section ______ Desk Number ______

PRE-LAB EXERCISES FOR EXPERIMENT 6

These exercises are to be completed after you have read the experiment but before you come to the laboratory to perform it.

1. Find the colors of the acid and base forms of phenolphthalein and two other indicators.

2. Complete and balance the equations started below.

   a. ______ I_2 (aq) + ______ $S_2O_3^{2-}$ (aq) $\longrightarrow$

   b. ______ Cu^{2+} (aq) + ______ NH_3 (aq) $\longrightarrow$

   c. ______ Mg^{2+} (aq) + ______ OH^- (aq) $\longrightarrow$

   d. ______ Ba^{2+} (aq) + ______ SO_4^{2-} (aq) $\longrightarrow$

3. Dilute solutions have physical properties that are close to those of the solvent. However, a solution's freezing point does differ from that of the solvent. Is the freezing point of a solution higher or lower than that of the solvent? Which factors determine how close the freezing point of the solution is to the freezing point of the solvent?

4. What are the "primary" colors?

Date ______ Name ______________________________ Section ______ Desk Number ______

SUMMARY REPORT ON EXPERIMENT 6

A. Invisible Inks

Mixture	Observations
phenolphthalein/NaOH	______________________
thiosulfate/iodine	______________________
copper sulfate/ammonia	______________________

B. Hot and Cold

Epsom salt plus water was hot cold.

Anhydrous $MgSO_4$ plus water was hot cold.

Mixture	Observations
NH_3/epsom salt solution	______________________
$BaCl_2$/epsom salt solution	______________________
NH_3/solution of $MgSO_4$	______________________
$BaCl_2$/solution of $MgSO_4$	______________________

C. Is It Sugar or Is It Salt?

Complete the table.

	Did bulb light up?	Did solution freeze?
Sugar solution	__________	__________
Salt solution	__________	__________

D. Patriotic Liquid

Mixture	Colors
phenolphthalein/NH_3	______________________
lead nitrate/NH_3	______________________
copper sulfate/NH_3	______________________

Date ______ Name ______________________________ Section ______ Desk Number ______

E. Now You See It, Now You Don't

What is the color of:

cobalt chloride (aq) ______________________

nickel nitrate(aq) ______________________

copper sulfate (aq) ______________________

Co^{2+}/Cu^{2+} mixture ______________________

$Co^{2+}/Cu^{2+}/Ni^{2+}$ mixture ______________________

F. On Your Own

Outline the demonstration you wish to perform.

EXPERIMENT 7

Properties of Oxides, Hydroxides, and Oxo-Acids

Laboratory Time Required

Three hours

Special Equipment and Supplies

Red and blue litmus paper
Drinking straw
Glass tubes for gas generator
Sandpaper
Labels
Phenolphthalein indicator
Limewater, saturated
Limestone chips
Na_2SO_3, crystal
Aluminum foil
Granulated zinc
Steel wool
Light copper turnings
$HC_2H_3O_2$ (aq), 3 M
NaOH (aq), 3 M
$CuSO_4$ (aq), 0.1 M
$Al_2(SO_4)_3$ (aq), 0.05 M
$Fe_2(SO_4)_3$ (aq), 0.05 M

Safety

All of the acids, bases, and oxides should be regarded as **corrosive** or **caustic** and capable of causing burns to the skin or possible blindness. Consequently, safety glasses with side shields and laboratory aprons are mandatory. Take every precaution against skin or eye exposure to chemicals.

First Aid

Rinse the exposed skin or eyes thoroughly with water. See a doctor if chemicals have entered the eye or if skin irritation is extensive.

You will examine the physical and chemical properties of a number of compounds in this experiment. You will be asked to represent the observed reactions by balanced equations and to use the Periodic Law to account for trends and properties. A listing of equations for reactions you may encounter in performing this experiment is given in Table 7.1 at the end of the experiment.

PRINCIPLES

People have classified materials as acids or bases since ancient times, long before the development of the modern Atomic Theory. Acids characteristically have a sour taste. Limestone or marble will fizz when treated with acid substances. Bubbles form when acid is poured onto samples of zinc or iron. Many natural dyes take on characteristic colors in the presence of acids. For instance, purple grape juice turns red in acid solution. Red apple skins turn orange when treated with acids. Both the grape juice and the apple skins are acting as "indicators" in these cases.

Acids react readily with bases. Bases—or alkalis—characteristically have a bitter taste and a soapy feel. Like acids, they cause indicators to take on distinctive coloration: grape juice turns green in basic solution; blueberry juice, which is dark red in the presence of acids, is blue in the presence of bases.

If base is added to an acid in the presence of an indicator, the color of the indicator will change. Red cabbage juice is orange-red in a strongly acidic environment. Its color will change to pinkish-red with the addition of small amounts of base. Continued addition of base will change the color to purple, green, yellow, and, finally, blue, in a strongly basic environment. These color changes are manifestations of the fact that an acid and a base react.

Once chemists accepted Dalton's Atomic Theory, it was natural that they attempt to discover the relationship between a substance's chemical formula and its acid or base properties. Many acids (today called "oxo-acids") contain oxygen and, in the late eighteenth century, it was thought that all acids were oxygen compounds; in fact, the German word for *oxygen* is *Sauerstoff*, meaning "acid material." The notion that all acids contain oxygen was proven wrong by Sir Humphrey Davy in 1810 when he demonstrated that "muriatic acid" (now known as hydrochloric acid) did not contain oxygen.

The first comprehensive acid/base theory in modern times was developed by Arrhenius in the late nineteenth century. The theory applies only to aqueous solutions, but is not quite useful nonetheless, since water is an excellent solvent and many reactions take place in aqueous solution. According to the Arrhenius theory, in aqueous solution, an acid acts as donor of protons (H^+ ions) and a base acts as a donor of hydroxide ions (OH^- ions). The reaction of an acid and a base is called "neutralization" and results in the formation of water and a salt. The reaction of nitric acid with sodium hydroxide is shown below in Equation 7.1; the products of this reaction are water and sodium nitrate.

$$HNO_3\,(aq) + NaOH\,(aq) \longrightarrow H_2O\,(\ell) + NaNO_3\,(aq) \qquad (7.1)$$

Oxo-acids and hydroxides are ternary compounds composed of hydrogen, oxygen, and some other element. A number of such compounds can be obtained by the reaction of soluble, binary oxides with water, as shown in Equations 7.2 and 7.3. (Soluble, binary oxides are compounds of oxygen and another element that dissolve in water.)

$$N_2O_5\,(g) + H_2O\,(\ell) \longrightarrow 2(NO_2)OH\,(aq) \qquad (7.2)$$

$$Na_2O\,(s) + H_2O\,(\ell) \longrightarrow 2NaOH\,(aq) \qquad (7.3)$$

Examination of Equation 7.3 reveals that the reaction of sodium oxide with water has produced sodium hydroxide. Because sodium hydroxide is a base, sodium oxide is said to be a basic anhydride.

You may not readily recognize the product of the reaction shown in Equation 7.2, which shows that N_2O_5 is the acid anhydride of nitric acid (HNO_3). In Equation 7.2 the formula for nitric acid is written in an unusual way. The formula, $(NO_2)OH$, is a condensed structural formula, which tells us more than the molecular formula, HNO_3. All that the molecular formula reveals is that one hydrogen atom, one nitrogen atom, and three oxygen atoms somehow combine to make one molecule of nitric acid. The condensed structural formula tells us that, in nitric acid, a nitrogen atom is bonded directly to two oxygen atoms and to one hydroxyl (OH) group. The hydroxide ion is basic, but the hydroxyl group is acidic.

All oxo-acids are in fact hydroxyl compounds. Writing condensed structural formulas for acids, such as $HC_2H_3O_2$ (acetic acid), can help us to understand why all hydrogens in oxo-acids are not necessarily acidic. The condensed structural formula for acetic acid is CH_3COOH. The central carbon in this molecule is bonded to a CH_3 group, an oxygen atom, and a hydroxyl group. The hydrogen in the hydroxyl group is the acidic hydrogen; it may separate from the molecule, as a proton, in aqueous solution. The three hydrogens of the CH_3 group do not separate from the carbon.

According to the Periodic Law, the properties of the elements vary in a periodic manner. These properties include the formulas of binary oxides, the tendency for atoms to acquire positive or negative charges when combined in molecules, and the tendency for binary oxides to act as anhydrides for acids, bases, or amphoteric substances. Amphoteric substances are substances that may act as either acids or bases. Because it is a characteristic of acids to react with bases, and vice versa, one criterion for calling a substance "amphoteric" is that the material reacts with both acids and bases. Thus, water-insoluble zinc hydroxide is amphoteric because it "dissolves" in both acidic and basic solutions. This dissolution process actually involves reaction with either the acid or the base, as is shown in Equations 7.4 and 7.5.

$$Zn(OH)_2\,(s) + 2HCl\,(aq) \longrightarrow ZnCl_2\,(aq) + 2H_2O\,(\ell) \qquad (7.4)$$

$$Zn(OH)_2\,(s) + 2NaOH\,(aq) \longrightarrow Na_2Zn(OH)_4\,(aq) \qquad (7.5)$$

The tendency for insoluble binary oxides to act as acid or base anhydrides is, likewise, determined by the ability of the insoluble oxide to dissolve in an acidic or basic solution. Silicon dioxide is shown to be an acid anhydride in Equations 7.6 and 7.7, whereas magnesium oxide is shown to be a basic anhydride in Equations 7.8 and 7.9. In Equations 7.6 and 7.9, NR signifies no reaction.

$$SiO_2\,(s) + HCl\,(aq) \longrightarrow NR \qquad (7.6)$$

$$SiO_2\,(s) + 2NaOH \longrightarrow Na_2SiO_3\,(aq) + H_2O\,(\ell) \qquad (7.7)$$

$$MgO\,(s) + 2HCl\,(aq) \longrightarrow MgCl_2\,(aq) + H_2O\,(\ell) \qquad (7.8)$$

$$MgO\,(s) + NaOH\,(aq) \longrightarrow NR \qquad (7.9)$$

PROCEDURE

A. Properties of Acids and Bases

1. *Reactions with indicators.* Prepare about 3 mL each of 1 M acetic acid and 1 M sodium hydroxide in labeled test tubes by adding (with caution) 1 mL each of the 3 M stock solutions to 2 mL of water. Use a clean stirring rod to transfer a drop of the acetic acid solution to pieces of red and blue litmus paper. In a similar manner, place 1 drop of the sodium hydroxide solution on red and blue litmus papers. Report any color changes. Pour about 1 mL each of the acetic acid and sodium hydroxide solutions into separate test tubes and add 1 drop of phenolphthalein indicator solution to each test tube. Note the color of the indicator in the acidic and basic solutions, respectively.
2. *Neutralization.* Using an eyedropper or pipet, place 5 drops of the NaOH/phenolphthalein solution in a small beaker or test tube. Using a second eyedropper, add the CH_3COOH solution, drop by drop, with mixing, until the color just disappears. Note the color after each addition of acid. Write the equation for the reaction between acetic acid and sodium hydroxide.
3. *Reactions of acid with active metals.* Place four test tubes in a test tube rack. Label one tube "Al" and place a small piece of aluminum foil in it. Label a second tube "Zn" and add one-half inch of granular zinc. Label a third tube "Cu" and add one-half inch of light copper turnings. Label a fourth tube "Fe" and put a small ball of steel wool in it. Add about 1 mL of 3 M acetic acid to each test tube. Note your observations. Write the chemical equations corresponding to the reactions observed.

B. Properties of CaO

Limewater is a solution that is prepared by dissolving 2 g of calcium oxide (called lime or quicklime) in a liter of water. Carefully pour 5 mL of the provided limewater into a 6-inch test tube without disturbing the solid at the bottom of the limewater container. If the solution obtained is not clear, filter it through filter paper as shown in Figure 7.1. Dip a clean stirring rod into the solution and touch the wet tip to pieces of red and blue litmus paper. Note your observations. Write an equation for the reaction of calcium oxide with water and state whether the product of the reaction is a (basic) hydroxide or an oxo-acid.

Put a drinking straw into the limewater that remains in the test tube Gently blow through the straw into the limewater solution until you see a definite change in the appearance of the limewater. (You may need to keep *exhaling* into the limewater for a minute or two.) Record your observations. Set aside the test tube into which you have exhaled; it will be tested again in section C.

CAUTION KEEP THE LIMEWATER OUT OF YOUR EYES.

C. Properties of CO_2

Test the acidity of aqueous carbon dioxide as follows. Moisten a strip of blue litmus paper with water and drape it over the convex side of a watch glass as shown in Figure 7.2. Place about one spatulaful of granular $CaCO_3$, or limestone chips in a clean, dry test tube, and stand the test tube upright in a test tube rack.

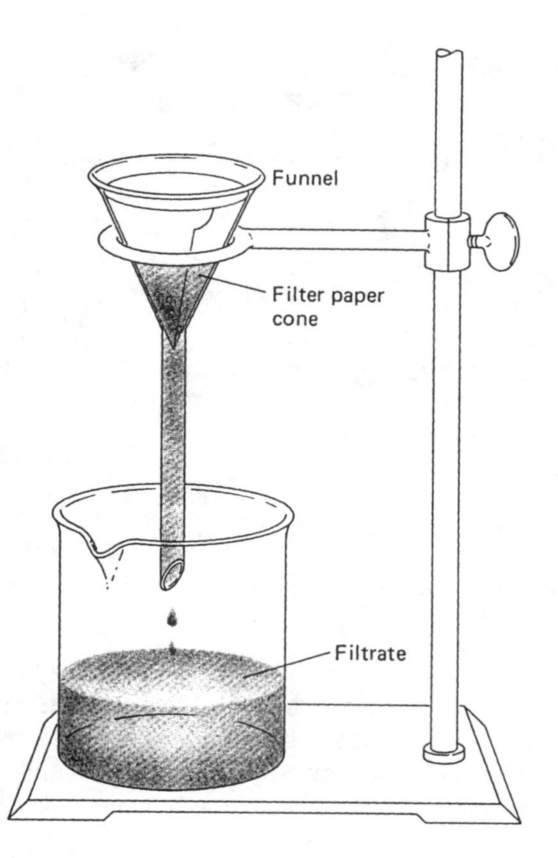

FIGURE 7.1 Gravity filtration apparatus.

Carefully pour about 2 mL of 3 M acetic acid into the test tube without wetting the test tube lip. Cover the test tube with the watch glass so that the litmus paper is exposed to any gas generated in the reaction. Note any *uniform* color change of the litmus paper covering the test tube but ignore any random specks of color caused by acid splattering onto the litmus paper during the course of the reaction. Record your observations. Write an equation for the reaction of calcium carbonate with acetic acid and an equation accounting for the fizzing that occurs when carbonates are treated with acid. Also, write an equation for the reaction of carbon dioxide with water. Is the product of this reaction a (basic) hydroxide or an oxo-acid?

Rinse the test tube containing the calcium carbonate and acetic acid mixture. Pour off the rinse water, but retain the solid carbonate sample. Obtain a bent glass tube, a one-hole stopper or cork to fit your test tube, and a piece of rubber tubing about 1 foot long. These are to be connected as shown in Figure 7.3. Add about 2 mL of 3 M acetic acid to the test tube containing the $CaCO_3$, insert the stopper and exit tube, and allow any gas produced by the reaction of the calcium carbonate and the acid to bubble through the limewater for a few minutes. Record your observations.

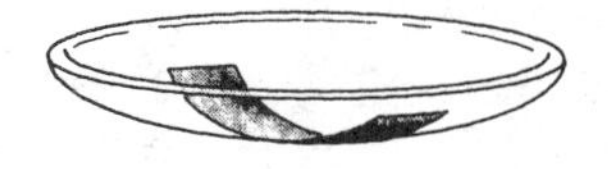

FIGURE 7.2 Litmus paper adhering to watch glass.

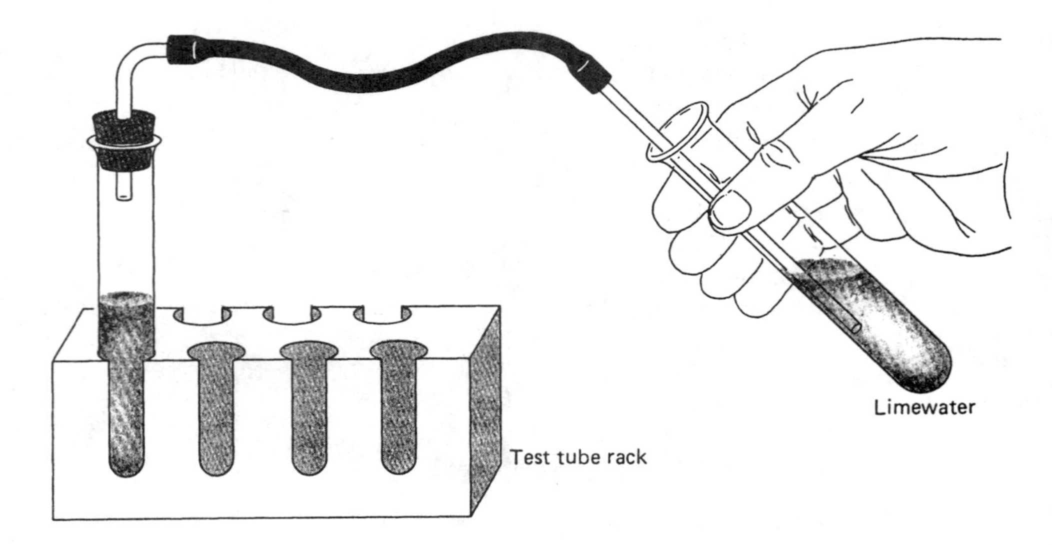

FIGURE 7.3 Limewater test for CO_2.

Add about 5 mL of 3 M acetic acid to the test tubes containing the limewater through which the bubbles have passed and into which you have exhaled. Note any observation. Identify the product of the reaction between the gas and the limewater, and write an equation showing its formation.

D. Properties of SO_2

Place a spatulaful of sodium sulfite in a clean test tube and add about 2 mL of water. Stir to dissolve the solid and use a stirring rod to place a drop of the solution on a piece of red litmus paper and on a piece of blue litmus paper. Record your observations. Stand the test tube upright in a test tube rack, and carefully pour about 3 mL of 3 M acetic acid into the test tube, taking care not to wet the test tube lip. Cover the test tube with a watch glass that has a fresh piece of moist blue litmus paper adhering to its convex side. Note any *uniform* color change of the litmus paper. Record your observations. Write an equation for the reaction of the sodium sulfite and the acetic acid, and an equation that explains the origin of the gas which arises in this reaction. Finally, write an equation showing the reaction between the gas and water. Is the product of that reaction a (basic) hydroxide or oxo-acid?

E. Properties of Metal Hydroxides

Place six clean test tubes, each containing about 1 mL of water, in a test tube rack. Add 2 drops of 0.1 M $CuSO_4$ to each of the first two tubes; add 2 drops of 0.05 M $Al_2(SO_4)_3$ to each of the next two tubes; add 2 drops of 0.05 M $Fe_2(SO_4)_3$ to each of the last two tubes. Then, add 5 drops of 3 M NaOH to the first of the six test tubes, stirring the test tubes and examining its contents for any sign of reaction after each addition of base. Record your observations and repeat the process of adding 5 drops of 3 M NaOH, stirring, and examining, until each of the six test tubes has been treated in this manner (be sure to clean your stirring rod before working with each new tube).

Add 10 drops of 3 M acetic acid to the *first* tube of each of the three test tubes, so that you will be testing the product of the reaction of each salt and sodium hydroxide with acid. As you did before, stir the contents of the test tube and examine the contents of the tube after each addition of reagent. Note your observations.

Add 10 drops of 3 M sodium hydroxide to the *second* tube of each of the three pairs of test tubes, so that you will be testing the product of the reaction of each salt and sodium hydroxide with additional sodium hydroxide. As you did before, stir the contents of the test tube and examine the contents of the tube after each addition of reagent. Note your observations.

Write equations for the reaction of each of the salts with sodium hydroxide, and for the reaction of the product of these reactions with additional acid or base. State whether the product of the initial reaction was a (basic) hydroxide, an oxo-acid, or an amphoteric material.

Disposal of Reagents

All solutions may be disposed of by being neutralized, diluted with water, and then flushed down the drain. Insoluble solids, such as limestone chips and strips of aluminum, copper, iron, and zinc should be rinsed, dried, and then placed in the labeled collection bottles for reuse.

Questions

1. The procedure of adding aqueous acid, drop by drop, to a solution of base, as was done in this experiment, is a rather crude titration. One purpose of a titration is to determine the molarity of an acid or base. Assume your CH_3COOH was 1 M. Find the molarity of the NaOH, based on your data for the neutralization portion of this experiment.
2. In this experiment, you worked directly with three anhydrides (CaO, CO_2, and SO_2). You have also been given information about the behavior of two other anhydrides, SiO_2 and MgO (see Equations 7.6 through 7.9). In addition, you performed tests on three hydroxides ($Al(OH)_3$, $Cu(OH)_2$, and $Fe(OH)_3$), for which the corresponding anhydrides are Al_2O_3, CuO, and Fe_2O_3, respectively. Based on this information, state whether metal oxides are basic anhydrides or acid anhydrides. Also, state whether nonmetal oxides are basic anhydrides or acid anhydrides. Which oxides dissolve in water to form amphoteric substances? In which part of the Periodic Table are the elements whose oxides form amphoteric substances found?
3. Acids and bases may be characterized as being strong or weak. A strong acid will succeed in donating essentially all of its protons to water, just as a strong base will succeed in donating essentially all of its hydroxide ions when in aqueous solution. Weak acids, on the other hand, donate only a fraction of their protons to the water, while weak bases also tend to cause the formation of only a few hydroxide ions in water. Dissolving sodium sulfite in water has a tendency to produce both sulfurous acid and sodium hydroxide. Based on your observation of the effect of aqueous sodium sulfite on litmus paper, which of the products of the reaction between sodium sulfite is strong and which is weak?

TABLE 7.1 Equations for Selected Reactions

Equations
$2Al\ (s) + 6CH_3COOH\ (aq) \longrightarrow 3H_2\ (g) + 2Al(CH_3COO)_3\ (aq)$
$2Al(OH)_3\ (s) + 6CH_3COOH\ (aq) \longrightarrow 6H_2O\ (\ell) + 2Al(CH_3COO)_3\ (aq)$
$Al(OH)_3\ (s) + NaOH\ (aq) \longrightarrow NaAl(OH)_4\ (aq)$
$Al_2(SO_4)_3\ (aq) + 6NaOH\ (aq) \longrightarrow 2Al(OH)_3\ (s) + 3Na_2SO_4\ (aq)$
$CaO\ (s) + H_2O\ (\ell) \longrightarrow Ca(OH)_2\ (aq)$
$Ca(OH)_2\ (aq) + CO_2\ (g) \longrightarrow CaCO_3\ (s) + H_2O\ (\ell)$
$CaCO_3\ (s) + 2CH_3COOH\ (aq) \longrightarrow Ca(CH_3COO)_2\ (aq) + CO(OH)_2\ (aq)$
$CH_3COOH\ (aq) + NaOH\ (aq) \longrightarrow H_2O\ (\ell) + NaCH_3COO\ (aq)$
$CO(OH)_2\ (aq) \longrightarrow H_2O\ (\ell) + CO_2\ (g)$
$CO_2\ (g) + H_2O\ (\ell) \longrightarrow H^+\ (aq) + HCO_3^-\ (aq)$
$Cu\ (s) + CH_3COOH\ (aq) \longrightarrow NR$
$Cu(OH)_2\ (s) + 2CH_3COOH\ (aq) \longrightarrow 2H_2O\ (\ell) + Cu(CH_3COO)_2\ (aq)$
$Cu(OH)_2\ (s) + NaOH\ (aq) \longrightarrow NR$
$CuSO_4\ (aq) + 2NaOH\ (aq) \longrightarrow Cu(OH)_2\ (s) + Na_2SO_4\ (aq)$
$2Fe\ (s) + 6CH_3COOH\ (aq) \longrightarrow 3H_2\ (g) + 2Fe(CH_3COO)_3\ (aq)$
$2Fe(OH)_3\ (s) + 6CH_3COOH\ (aq) \longrightarrow 6H_2O\ (\ell) + 2Fe(CH_3COO)_3\ (aq)$
$Fe(OH)_3\ (s) + NaOH\ (aq) \longrightarrow NR$
$Fe_2(SO_4)_3\ (aq) + 6NaOH\ (aq) \longrightarrow 2Fe(OH)_3\ (s) + 3Na_2SO_4\ (aq)$
$Na_2SO_3\ (s) + 2H_2O\ (\ell) \longrightarrow SO(OH)_2\ (aq) + 2\ NaOH\ (aq)$
$Na_2SO_3\ (aq) + 2CH_3COOH\ (aq) \longrightarrow SO(OH)_2\ (aq) + 2NaCH_3COO\ (aq)$
$SO(OH)_2\ (aq) \longrightarrow H_2O\ (\ell) + SO_2\ (g)$
$SO_2\ (g) + H_2O\ (\ell) \longrightarrow H^+\ (aq) + HSO_3^-\ (aq)$
$Zn\ (s) + 2CH_3COOH\ (aq) \longrightarrow Zn(CH_3COO)_2\ (aq) + H_2\ (g)$

Date ______ Name ______________________________ Section ________ Desk Number ________

PRE-LAB EXERCISES FOR EXPERIMENT 7

These exercises are to be completed after you have read the experiment but before you come to the laboratory to perform it.

1. Define the terms "acid" and "base." Give examples of an oxo-acid and a (basic) hydroxide.

2. List two or more characteristics properties of acids. Do the same for bases.

3. Dissolving 1 mole of hydrochloric acid, a strong acid, in a liter of water will produce 1 mole of protons. Dissolving one mole of potassium hydroxide, a strong base, in a liter of water will produce one mole of hydroxide ions. Acetic acid is a weak acid. Will dissolving 1 mole of acetic acid in a liter of water produce 1 mole of protons, less than 1 mole of protons, or more than 1 mole of protons?

4. Ammonia (NH_3) is not an Arrhenius base because it does not contain hydroxide ions. However, ammonia does have a tendency to produce ammonium ions (NH_4^+) and hydroxide ions when it is dissolved in water. Only a few of the dissolved ammonia molecules are transformed into NH_4^+ and OH^- ions. If we expand the definition of bases to include substances that produce hydroxide ions in aqueous solutions, is ammonia a weak base or a strong base?

Date ______ Name ________________________________ Section ______ Desk Number ______

SUMMARY REPORT ON EXPERIMENT 7

A. Properties of Acids and Bases

1. Indicator colors

	Red litmus	Blue litmus	Phenolphthalein
Acetic Acid	________	________	________
Sodium hydroxide	________	________	________

2. Neutralization

Drops of Acid added	Color or NaOH/Phenolphthalein
________	________
________	________
________	________
________	________
________	________
________	________
________	________

Equation:

3. Reactions of acid with active metals

Aluminum + acetic acid Observations:

Equation:

Copper + acetic acid Observations:

Equation:

Iron + acetic acid Observations:

Equation:

Zinc + acetic acid Observations:

Equation:

B. Properties of CaO

Effect of limewater on: red litmus paper ____________________

blue litmus paper ____________________

Equation for the reaction of CaO and water:

Is product a hydroxide or an oxo-acid?
Effect of exhaling into limewater.

Effect of adding acid to aerated limewater.

C. Properties of CO_2

Reaction of calcium carbonate with acid

Observations:

Equations:

Effect of CO_2 on moist litmus paper:

Equation for the reaction of CO_2 with water:

Is product a hydroxide or an oxo-acid?

Effect of bubbling CO_2 through limewater:

Effect of adding acid to limewater through which CO_2 has bubbled:

Equations:

Date ______ Name ______________________ Section ______ Desk Number ______

D. Properties of SO_2

Dissolving sodium sulfite

Observations:

Equations:

Effect of SO_2 on moist litmus paper.

Equation for the reaction of SO_2 with water:

Is product a hydroxide or an oxo-acid?

E. Properties of Metal Hydroxides

Addition of NaOH to:

	Observations	Equations
$CuSO_4$	______	______
$Al_2(SO_4)_3$	______	______
$Fe_2(SO_4)_3$	______	______

Addition of CH_3COOH to products of reaction of NaOH and:

	Observations	Equations
$CuSO_4$	______	______
$Al_2(SO_4)_3$	______	______
$Fe_2(SO_4)_3$	______	______

Addition of NaOH to products of reaction of NaOH and:

	Observations	Equations
$CuSO_4$	______	______
$Al_2(SO_4)_3$	______	______
$Fe_2(SO_4)_3$	______	______

EXPERIMENT 8

Volumetric Analysis: Acid/Base Titration Using Indicators

Laboratory Time Required

Three hours

Special Equipment and Supplies

Analytical balance
Buret
Buret clamp
0.1 M NaOH
Potassium hydrogen phthalate (KHP)
Phenolphthalein indicator
Vinegar

Safety

Bases, such as sodium hydroxide,can cause skin burns and are especially hazardous to the eyes. Although vinegar is a dilute solution of a weak acid, it is, nevertheless, advisable to avoid splashing it in the eyes.

First Aid

Following skin contact with sodium hydroxide, wash the area thoroughly with water. Should sodium hydroxide (or even vinegar) get in the eyes, rinse them thoroughly with water (at least 20 minutes of flushing with water is recommended) and seek medical attention.

In volumetric analysis, a known volume of a standard solution (one whose concentration is known) reacts with a known volume of a solution of unknown concentration. This procedure **standardizes** the latter solution, by allowing a calculation of its concentration.

The preparation and dispensing of solutions requires the use of calibrated glassware such as burets, pipets, and volumetric flasks. These items are illustrated in Figure I.3, and instructions for using the equipment are provided in the Introduction. In this experiment, you will standardize a sodium hydroxide solution. You will then use this solution to analyze a solution containing an unknown concentration of acetic acid (the ingredient that gives vinegar its sour taste), using phenolphthalein as the indicator in the titration.

PRINCIPLES

Titration

In a titration, a buret is used to dispense measured increments of one solution into a known volume of another solution. The object of the titration is the detection of the **equivalence point**, that point in the procedure where chemically equivalent amounts of the reactants have been mixed. Whether or not the equivalence point comes when equimolar amounts of reactants have been mixed depends on the stoichiometry of the reaction. In the reaction of acetic acid, $HC_2H_3O_2$, and NaOH, the equivalence point does occur when 1 mole of $HC_2H_3O_2$ has reacted with 1 mole of NaOH. However, in the reaction of H_2SO_4 and NaOH, the equivalence point occurs when 2 moles of NaOH have reacted with 1 mole of H_2SO_4.

The titration technique can be applied to many types of reactions, including oxidation-reduction, precipitation, complexation, and acid-base neutralization reactions.

Although a variety of instrumental methods for detecting equivalence points are now available, it is frequently more convenient to add an indicator to the reaction mixture. An indicator is a substance that undergoes a distinct color change at or near the equivalence point. The point in the titration at which the color change occurs is called the **end point**. Obviously, the titration will be accurate only if the end point and the equivalence point coincide fairly closely. For this reason, the indicator used in a titration must be selected carefully. Fortunately, a large number of indicators are commercially available and finding the right one for a particular titration is not a difficult task.

Acids and Bases

Although several definitions of acids and bases may be given, the classical Arrhenius concept will suffice for this experiment. According to this concept, an acid is a substance that dissociates in water to produce hydrogen ions; a base is a substance that dissociates in water to produce hydroxide ions. The classical Arrhenius acid-base reaction is one in which an acid reacts with a base to form water (from the combination of hydrogen ions and hydroxide ions) and a salt. Such a reaction is called a **neutralization** reaction. The neutralization reaction of acetic acid, the primary ingredient of vinegar, and sodium hydroxide is shown in Equation 8.1.

$$HC_2H_3O_2\,(aq) + NaOH\,(aq) \longrightarrow H_2O\,(\ell) + NaC_2H_3O_2\,(aq) \qquad (8.1)$$

Because the mole ratio of acetic acid to sodium hydroxide is 1:1, the number of moles of acid present in the sample is equal to the number of moles of base that must be added to reach the equivalence point of the titration (see Equation 8.2).

$$M \times V\,(\text{in liters}) = \text{moles (base)} = \text{moles (acid)} \qquad (8.2)$$

Indicators

Acid-base indicators are weak acids or bases that have different colors when in their dissociated and undissociated forms, respectively. The dissociation of the indicator HIn may be represented as shown in Equation 8.3.

$$HIn + H_2O \rightleftharpoons H_3O^+ + In^- \quad (8.3)$$

In acid solutions, the indicator exists predominantly as HIn. In basic solutions, it is present mainly as In^- ions, which impart a different color to the solution. Phenolphthalein is an indicator that is used frequently in acid-base titrations. It is colorless as long as the titration mixture contains an excess of acid and turns pink when the amount of base added slightly exceeds the amount of acid originally present. Thus, the end point will come shortly after the equivalence point. The conditions that cause phenolphthalein to change from its colorless (HIn) form to its colored (In^-) form are such that the difference between the end point and the equivalence point in a carefully performed titration is negligible.

Preparation of NaOH Solution

The general procedure for preparing carbonate-free NaOH solution involves decanting (pouring off) clear liquid from a 50% aqueous NaOH solution and then diluting this with freshly boiled and cooled water. This precaution is needed because atmospheric carbon dioxide tends to react with sodium hydroxide to form sodium carbonate. Because sodium carbonate has a very low solubility in 50% NaOH solution, any sodium carbonate present as an impurity in the solid NaOH used to prepare the solution (or formed by reaction of the solution of NaOH with carbon dioxide in the air) does not remain in solution. Rather, it slowly settles to the bottom of the solution as a white precipitate, from which the pure NaOH solution can be separated easily by decantation. The water used subsequently to dilute this solution should first be boiled to remove dissolved carbon dioxide.

Because of the very great hazard to the eyes presented by concentrated solutions of bases, you will not be given 50% NaOH. Rather, dilute (approximately 0.1 M), carbonate-free solution will be provided for your use. You will determine the exact concentration of NaOH in the approximately 0.1 M solution by titrating a known mass of potassium hydrogen phthalate. The reaction between these substances is shown in Equation 8.4.

$$NaOH\,(aq) + KHC_8H_4O_4\,(aq) \longrightarrow H_2O\,(\ell) + KNaC_8H_4O_4\,(aq) \quad (8.4)$$

Thus, 1 mole of NaOH reacts exactly with 1 mole (204.22 g) of potassium hydrogen phthalate (KHP).

PROCEDURE

Standardization of NaOH Solution

Clean two burets and prepare them for use in the titration according to the directions provided in the Introduction. Rinse both burets with small amounts of distilled water. Then rinse one buret twice with small portions of the NaOH solution provided. Finally, fill the buret with this solution.

Accurately weigh, to the nearest tenth of a milligram, 0.4 to 0.6 g of the dry KHP and add it to a 125-mL Erlenmeyer flask. Add 25 mL of distilled water and 2 to 3 drops of phenolphthalein indicator solution. Swirl to dissolve the solid.

Read the initial volume of NaOH solution in the buret to the nearest hundredth of a milliliter. Titrate the KHP solution with the NaOH solution until a faint pink color, which does not disappear when the solution is mixed, is obtained. Read the final volume of NaOH solution in the buret to the nearest hundredth of a milliliter. Calculate the molarity of the NaOH solution. Repeat the standardization and average the results. If time permits, you may wish to perform a third standardization and use the criteria given in the Introduction to decide whether you should average the results of two or three trials.

Titration of an Unknown

Obtain an unknown volume of vinegar in your 100-mL volumetric flask. Dilute the solution to exactly 100 mL with distilled water. Mix well and then rinse your second buret with two 5-mL portions of the diluted acid. Fill the buret with the solution, accurately read the initial volume, and then dispense 25 mL into a 125-mL Erlenmeyer flask. (Accurately measure the final volume.) Add 2 or 3 drops of phenolphthalein indicator and then titrate the acid with your 0.1 M NaOH solution until a faint, persistent pink color is obtained. Calculate the molarity of the acetic acid solution. Repeat the titration once or twice. Report the results of each titration and the average concentration of your acetic acid unknown.

Disposal of Reagents

Excess KHP should be placed in the containers used for solid waste. Solutions should be neutralized and diluted. They may then be flushed down the drain.

Questions

1. The letters **TC** and **TD** on volumetric glassware indicate whether it is calibrated to contain or to deliver a given volume. Which mode of calibration would be best for the following items as they are used in the experiment?
   (a) Buret
   (b) Volumetric flask
2. Graduated cylinders may be purchased marked either **TC** or **TD**. Briefly suggest a suitable use for each type.
3. A 50% NaOH solution has a density of 1.53 g/cm^3 and contains 50% NaOH by mass. What is the molar concentration of NaOH in this solution?

Date ______ Name ______________________________ Section ______ Desk Number ______

PRE-LAB EXERCISES FOR EXPERIMENT 8

These exercises are to be completed after you have read the experiment but before you come to the laboratory to perform it.

1. A 0.5230 g sample of KHP required 26.75 mL of sodium hydroxide to reach the phenolphthalein end point. Calculate the molarity of the NaOH solution.

2. A volume of 32.15 mL of the NaOH solution described in Question 1 was needed to titrate 25.00 mL of acetic acid to the phenolphthalein end point. Find the concentration of the acetic acid solution.

Date ______ Name ______________________________ Section ______ Desk Number ______

SUMMARY REPORT ON EXPERIMENT 8

Standardization of NaOH Solution

	Trial 1	Trial 2	Trial 3*
Mass of KHP and container	______	______	______
Mass of container	______	______	______
Mass of KHP	______	______	______
Final buret reading, NaOH	______	______	______
Initial buret reading, NaOH	______	______	______
Volume used, NaOH	______	______	______
Molarity of NaOH solution	______	______	______
Average molarity of NaOH		______	

Titration of an Unknown

Unknown number ______

	Trial 1	Trial 2	Trial 3*
Final buret reading, acetic acid	______	______	______
Initial buret reading, acetic acid	______	______	______
Volume used, acetic acid	______	______	______
Final buret reading, NaOH	______	______	______
Initial buret reading, NaOH	______	______	______
Volume used, NaOH	______	______	______
Molarity of acetic acid solution	______	______	______
Average molarity of acetic acid solution		______	

*Optional

Job's Method for Determining the Stoichiometry of a Reaction

Laboratory Time Required

Three hours

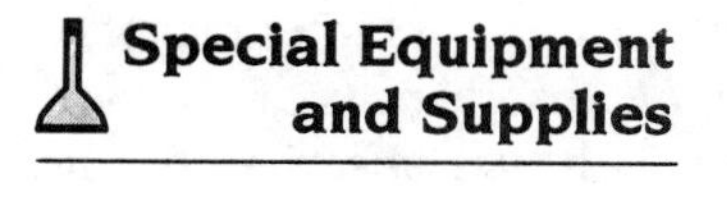

Special Equipment and Supplies

Analytical balance
Volumetric flask, 1-L
Graduated cylinders, 100-mL
Buret
Buret clamp
Water bath
Coffee cup calorimeters
Thermometer

1 M NaOH
Citric acid
Potassium hydrogen phthalate (KHP)
Phenolphthalein

Safety

Bases, such as sodium hydroxide, can cause skin burns and are especially hazardous to the eyes.

First Aid

Following skin contact with sodium hydroxide, wash the area thoroughly with water. If there has been eye contact, flush the eyes with water for at least 20 minutes and seek medical attention.

Like many organic acids, citric acid contains some hydrogens that are acidic and others that are not. In this experiment, the reaction between citric acid and sodium hydroxide is studied and the stoichiometric ratio of the reactants is determined.

PRINCIPLES

Job's method of continuous variation offers a relatively quick procedure for determining the stoichiometric ratio between two reactants. One form of Job's method involves the preparation of a series of mixtures in which the chemical amount of each reactant varies, although the total number of moles in each mixture remains the same. Thus, in the study used to prepare the graph shown in Figure 9.1, 0.100 M solutions of reagent A were mixed with 0.100 M solutions of reagent B. The volumes of reagent A used varied from 0.0 mL to 20.0 mL, while the volumes of reagent B used varied from 20.0 mL to 0.0 mL. One mixture studied contained 2.00 mmols of A; another contained 0.80 mmols of A and 1.20 mmols of B; a third contained 1.60 mmols of A and 0.40 mmols of B. In each case, the total number of mmols was 2.00. Of course, in some of the mixtures, A was in excess. In others, B was in excess and A was the limiting reagent.

Job's method utilizes some factor (color change, temperature change, volume change) which is maximized when the reactants are mixed in essentially their stoichiometric ratio (and, consequently, there is no excess of either reagent). The graph in Figure 9.1 was prepared by measuring the absorbances of the various mixtures of A and B at a wavelength of light which was absorbed by the product of their reaction. Note that the x-axis is labeled with both the amount of A and the amount of B in each mixture. The maximum in absorbance is observed to occur in a mixture containing 0.50 mmols of A and 1.50. mmols of B, indicating that the stoichiometric ratio is 1A:3B.

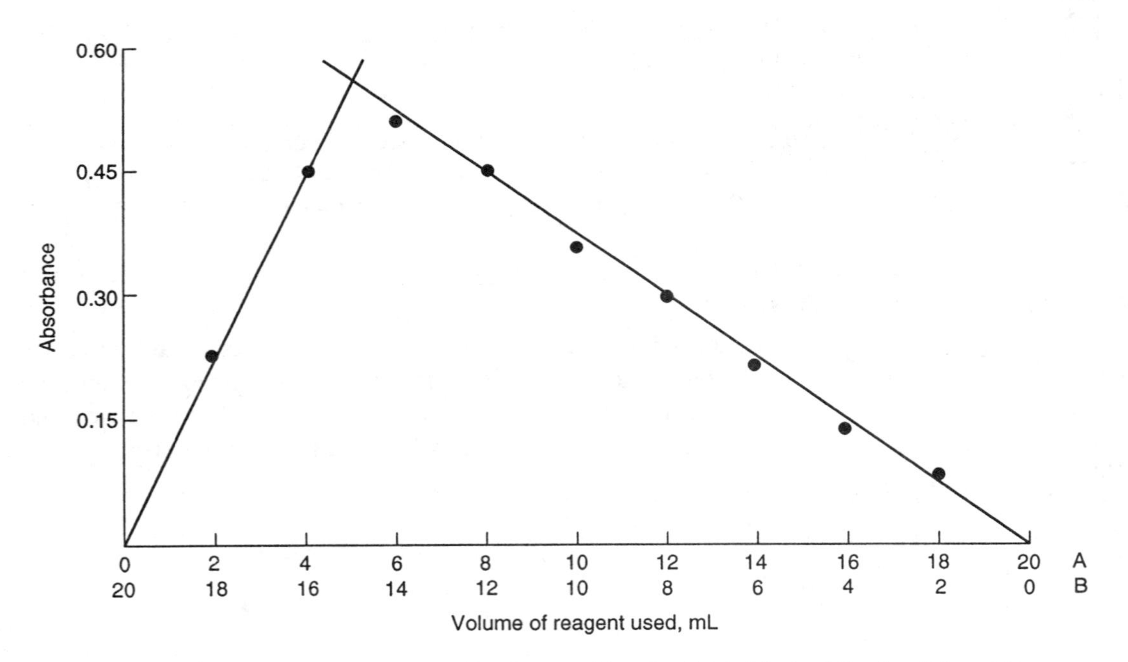

FIGURE 9.1 Graph of Absorbance versus Composition for Job's Method Study of A and B.

PROCEDURE

The sodium hydroxide provided for this experiment should be close to 1 M. Standardize this solution by titrating it against 6 g samples of KHP (which have been weighed to the nearest 0.1 mg on the analytical balance). Use phenolphthalein as the indicator. Review Experiment 8, if necessary.

Once the molarity of the sodium hydroxide solution has been determined, prepare 1.00 L of a solution of citric acid of exactly the same concentration. (The formula of citric acid monohydrate is $C_6H_8O_7{\cdot}H_2O$.) Be sure to mix the solution well. Once the solution of citric acid has been prepared, place the volumetric flask containing the solution and the bottle containing the sodium hydroxide solution in a water bath containing room-temperature water. Allow the solutions to equilibrate in the water bath for 10 to 15 minutes.

Prepare a series of coffee cup calorimeters by nesting sets of two cups together and placing the nested cups in empty beakers to enhance the stability of the calorimeter. Label the calorimeters with the volumes listed in Table 9.1. Use 100-mL graduated cylinders to measure the volumes of sodium hydroxide and of citric acid specified. Begin each run by placing the acid to be used in that run in a coffee cup calorimeter. Monitor the temperature of the acid. When it has reached a constant value, quickly add the amount of sodium hydroxide specified for the run. Stir this reaction mixture and monitor its temperature at 30-second intervals until the temperature begins to fall. Repeat this procedure until all mixtures listed in Table 9.1 have been studied. Take the difference between the maximum temperature observed for each mixture and its starting temperature as Δt. Plot Δt versus composition, as shown in Figure 9.1. ***Do this before you dispose of your solutions.*** If the data collected permit you to do so, determine the stoichiometric ratio for the neutralization of citric acid with sodium hydroxide. If you feel that you need to study additional solutions before you can make the determination of the stoichiometric ratio, prepare the mixtures you wish to study, measure Δt for each of the additional mixtures, and plot these Δt values along with the other ones on your graph.

TABLE 8.1 Volumes of citric acid and sodium hydroxide solution to be used in determining the stoichiometric ratio of their neutralization reaction.

Mixture	Volume of citric acid (aq), mL	Volume of NaOH (aq), mL
1	0.0	100.0
2	10.0	90.0
3	20.0	80.0
4	30.0	70.0
5	40.0	60.0
6	50.0	50.0
7	60.0	40.0
8	70.0	30.0
9	80.0	20.0
10	90.0	10.0
11	100.0	0.0

Date ______ Name ______________________ Section ______ Desk Number ______

PRE-LAB EXERCISES FOR EXPERIMENT 9

These exercises are to be completed after you have read the experiment but before you come to the laboratory to perform it.

Suppose you used Job's method to find the stoichiometry of the reaction:

$$a\ H_2\ (g) + b\ O_2\ (g) \longrightarrow \text{Product}\ (\ell)$$

The results of the study of seven mixtures are shown below.

Mixture	Initially Present Moles H_2	Initially Present Moles O_2	Moles of Product	Moles of Excess reactant
1	1.0	7.0	1.0	6.5
2	2.0	6.0	2.0	5.0
3	3.0	5.0	3.0	3.5
4	4.0	4.0	4.0	2.0
5	5.0	3.0	5.0	0.5
6	6.0	2.0	4.0	2.0
7	7.0	1.0	2.0	5.0

1. Use the form shown on page 98 to plot the yield of product (in moles) versus the composition for each of the seven mixtures. Draw two straight lines through the experimental points and, from their intersection, deduce the optimum ratio of H_2 (g) to O_2 (g). Find the formula of the product and write a balanced chemical equation for the reaction.

2. Repeat the operation performed in answer to Pre-lab Exercise 1, using the number of moles of unreacted gas (rather than the moles of product) to label the *y*-axis.

3. Do your answers for Pre-lab Exercises 1 and 2 agree? Explain your response, briefly.

Date ______ Name ________________________________ Section ______ Desk Number ______

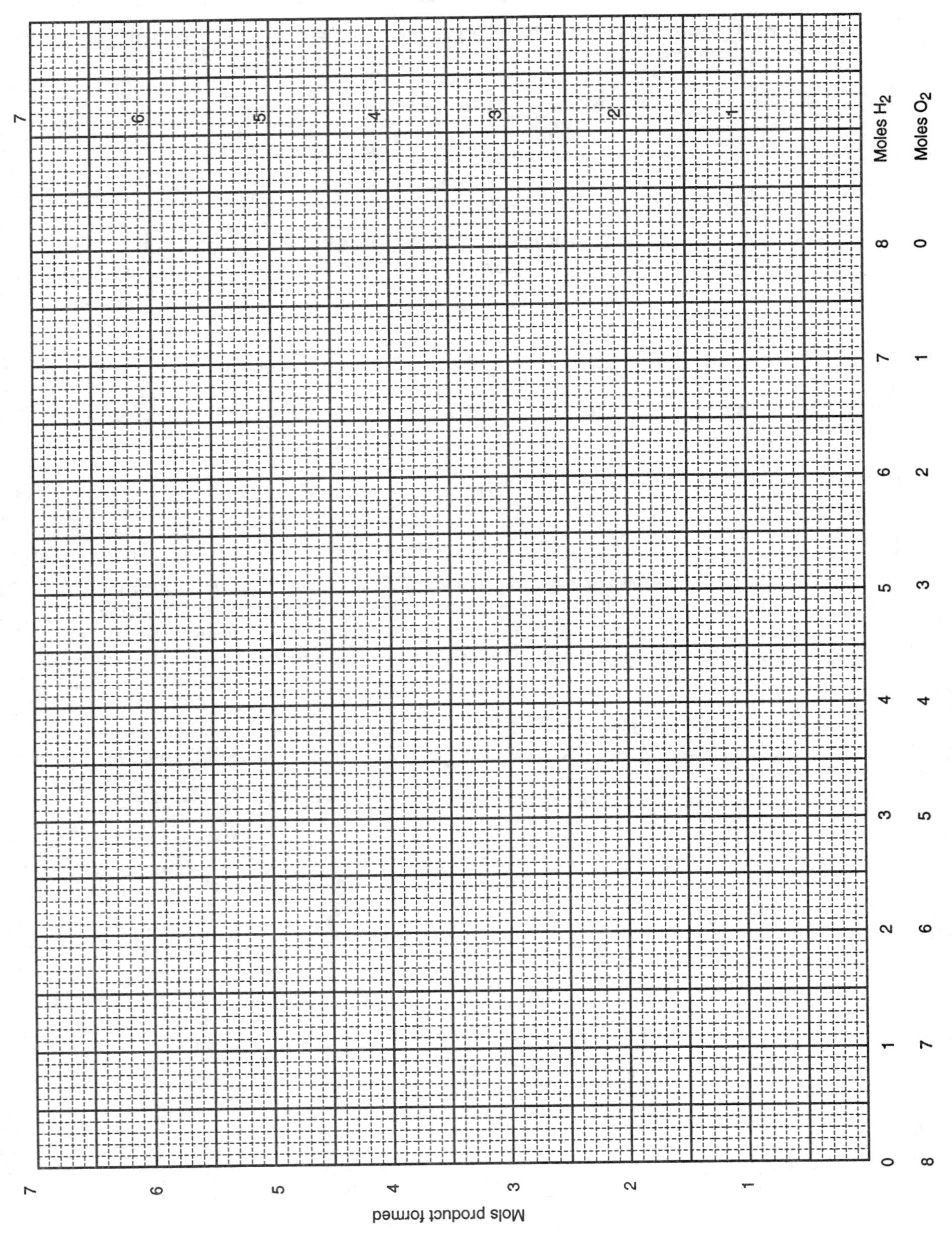

FIGURE 9.2 Grid for Plotting Data from Pre-Lab Exercises

Date ______ Name ______________________________ Section ______ Desk Number ______

SUMMARY REPORT ON EXPERIMENT 9

Standardization of Sodium Hydroxide

Mass of container plus KHP	______
Mass of container	______
Mass of KHP	______
Final buret reading	______
Initial buret reading	______
Volume of NaOH used	______
Molarity of NaOH	______

Preparation of citric acid solution

Mass of citric acid required to prepare 1.00 L of solution	______
Mass of container plus citric acid	______
Mass of container	______
Mass of citric acid	______

Determining the value of Δt

Mixture	______	Initial temperature	______
Volume NaOH	______	Volume citric acid	______
Temperature	______	Temperature	______
Temperature	______	Temperature	______
Temperature	______	Temperature	______
Temperature	______	Temperature	______
Temperature	______	Temperature	______
Temperature	______	Temperature	______

$\Delta t =$ ______

Mixture	______	Initial temperature	______
Volume NaOH	______	Volume citric acid	______
Temperature	______	Temperature	______
Temperature	______	Temperature	______
Temperature	______	Temperature	______

Temperature ____________ Temperature ____________

Temperature ____________ Temperature ____________

Temperature ____________ Temperature ____________

Δt = ____________

Mixture ____________ Initial temperature ____________

Volume NaOH ____________ Volume citric acid ____________

Temperature ____________ Temperature ____________

Temperature ____________ Temperature ____________

Temperature ____________ Temperature ____________

Temperature ____________ Temperature ____________

Temperature ____________ Temperature ____________

Temperature ____________ Temperature ____________

Δt = ____________

Mixture ____________ Initial temperature ____________

Volume NaOH ____________ Volume citric acid ____________

Temperature ____________ Temperature ____________

Temperature ____________ Temperature ____________

Temperature ____________ Temperature ____________

Temperature ____________ Temperature ____________

Temperature ____________ Temperature ____________

Temperature ____________ Temperature ____________

Δt = ____________

Mixture ____________ Initial temperature ____________

Volume NaOH ____________ Volume citric acid ____________

Temperature ____________ Temperature ____________

Temperature ____________ Temperature ____________

Date ______ Name ______________________________ Section ______ Desk Number ______

Temperature	____________	Temperature	____________
Temperature	____________	Temperature	____________
Temperature	____________	Temperature	____________
Temperature	____________	Temperature	____________

Δt = ____________

Mixture	____________	Initial temperature	____________
Volume NaOH	____________	Volume citric acid	____________
Temperature	____________	Temperature	____________
Temperature	____________	Temperature	____________
Temperature	____________	Temperature	____________
Temperature	____________	Temperature	____________
Temperature	____________	Temperature	____________
Temperature	____________	Temperature	____________

Δt = ____________

Mixture	____________	Initial temperature	____________
Volume NaOH	____________	Volume citric acid	____________
Temperature	____________	Temperature	____________
Temperature	____________	Temperature	____________
Temperature	____________	Temperature	____________
Temperature	____________	Temperature	____________
Temperature	____________	Temperature	____________
Temperature	____________	Temperature	____________

Δt = ____________

Mixture	____________	Initial temperature	____________
Volume NaOH	____________	Volume citric acid	____________
Temperature	____________	Temperature	____________

Temperature	________	Temperature	________
Temperature	________	Temperature	________
Temperature	________	Temperature	________
Temperature	________	Temperature	________
Temperature	________	Temperature	________

$\Delta t =$ ________

Mixture	________	Initial temperature	________
Volume NaOH	________	Volume citric acid	________
Temperature	________	Temperature	________
Temperature	________	Temperature	________
Temperature	________	Temperature	________
Temperature	________	Temperature	________
Temperature	________	Temperature	________
Temperature	________	Temperature	________

$\Delta t =$ ________

Mixture	________	Initial temperature	________
Volume NaOH	________	Volume citric acid	________
Temperature	________	Temperature	________
Temperature	________	Temperature	________
Temperature	________	Temperature	________
Temperature	________	Temperature	________
Temperature	________	Temperature	________
Temperature	________	Temperature	________

$\Delta t =$ ________

Mixture	________	Initial temperature	________
Volume NaOH	________	Volume citric acid	________

Temperature	______	Temperature	______
Temperature	______	Temperature	______
Temperature	______	Temperature	______
Temperature	______	Temperature	______
Temperature	______	Temperature	______
Temperature	______	Temperature	______

$\Delta t =$ ______

Plot the data you have acquired thus far. If you do not feel that these data will provide you with an unambiguous value for the stoichiometric ratio of sodium hydroxide to citric acid, decide on the composition of additional mixtures which you should prepare. Use the portion of the Summary Report Sheet given below to record this additional data.

Mixture	______	Initial temperature	______
Volume NaOH	______	Volume citric acid	______
Temperature	______	Temperature	______
Temperature	______	Temperature	______
Temperature	______	Temperature	______
Temperature	______	Temperature	______
Temperature	______	Temperature	______
Temperature	______	Temperature	______

$\Delta t =$ ______

Mixture	______	Initial temperature	______
Volume NaOH	______	Volume citric acid	______
Temperature	______	Temperature	______
Temperature	______	Temperature	______
Temperature	______	Temperature	______
Temperature	______	Temperature	______
Temperature	______	Temperature	______

Temperature	______	Temperature	______

Δt = ______

Mixture	______	Initial temperature	______
Volume NaOH	______	Volume citric acid	______
Temperature	______	Temperature	______
Temperature	______	Temperature	______
Temperature	______	Temperature	______
Temperature	______	Temperature	______
Temperature	______	Temperature	______
Temperature	______	Temperature	______

Δt = ______

Mixture	______	Initial temperature	______
Volume NaOH	______	Volume citric acid	______
Temperature	______	Temperature	______
Temperature	______	Temperature	______
Temperature	______	Temperature	______
Temperature	______	Temperature	______
Temperature	______	Temperature	______
Temperature	______	Temperature	______

Δt = ______

Mixture	______	Initial temperature	______
Volume NaOH	______	Volume citric acid	______
Temperature	______	Temperature	______
Temperature	______	Temperature	______
Temperature	______	Temperature	______
Temperature	______	Temperature	______

Temperature ____________ Temperature ____________

Temperature ____________ Temperature ____________

Δt = ____________

Mixture ____________ Initial temperature ____________

Volume NaOH ____________ Volume citric acid ____________

Temperature ____________ Temperature ____________

Temperature ____________ Temperature ____________

Temperature ____________ Temperature ____________

Temperature ____________ Temperature ____________

Temperature ____________ Temperature ____________

Temperature ____________ Temperature ____________

Δt = ____________

Stoichiometric ratio of NaOH to citric acid ____________

Attach your plot of Δt versus composition of the sodium hydroxide/citric acid mixtures.

EXPERIMENT 10

The Synthesis of Cobalt Oxalate Hydrate

Laboratory Time Required

One hour. May conveniently be combined with Experiment 11.

Special Equipment and Supplies

Analytical balance	Oxalic acid dihydrate
Centrifuge	Cobalt chloride hexahydrate
Litmus paper	Aqueous ammonia, concentrated

Safety

All of the chemicals used in this experiment are hazardous! Avoid getting the solutions on the skin, in the eyes, or in the mouth. Use the pipet bulb. Oxalic acid is a **poison.** Aqueous ammonia is **caustic.** Its vapor causes eye irritation. Cobalt chloride is **toxic.**

First Aid

If you get oxalic acid on your skin, flush the area with water. Then flush the skin with aqueous sodium bicarbonate.

If you ingest oxalic acid, drink a large quantity of water, followed by milk or milk of magnesia. See a doctor.

If you have contact with aqueous ammonia, flush the area with water. Use the eyewash fountain promptly to remove ammonia from the eyes. See a doctor if ammonia has gotten into your eyes.

In this experiment you are to synthesize a substance that will be given the tentative formula $Co_a(C_2O_4)_b{\cdot}cH_2O$. You may subsequently analyze the compound in Experiment 12, or your instructor may give you the values of *a*, *b*, and *c*.

PRINCIPLES

Hydrates

Compounds that contain water may have it present either in a variable or in a definite mass percentage. If the water content is variable, the substance may be merely wet; or the water may occupy channels, as in a zeolite; or a mixture of hydrates may be present. If the substance has a definite, nonvariable water content, it is called a **hydrate**, and the water molecules are found to occupy definite sites in the crystal. A few common hydrates and their formulas are: borax ($Na_2B_4O_7 \cdot 10H_2O$), plaster of Paris ($CaSO_4 \cdot {}^1/_2\,H_2O$), epsom salt ($MgSO_4 \cdot 7H_2O$), and blue copper sulfate pentahydrate ($CuSO_4 \cdot 5H_2O$).

Under appropriate conditions, most hydrated salts can be dehydrated to form either a lower hydrate or the anhydrous salt. For example, when the mineral gypsum is heated no hotter than about 160°C, it dehydrates to form plaster of Paris as shown in Equation 10.1.

$$CaSO_4 \cdot 2H_2O \rightleftharpoons CaSO_4 \cdot {}^1/_2\,H_2O + {}^3/_2\,H_2O \qquad (10.1)$$

This reaction is reversible, and the setting of plaster of Paris may be attributed to its rehydration to form crystalline gypsum. The equilibrium between the hydrated and the anhydrous forms of the salt is influenced both by the temperature of the system and by the relative humidity. Salts that can be reversibly hydrated are frequently used to control the amount of moisture in the air. Anhydrous salts that are commonly used as desiccants (drying agents) include $CaSO_4$, $CaCl_2$, and $Mg(ClO_4)_2$.

Calculation of Percent Yield

If stoichiometric quantities of chemicals are mixed together when a compound is being synthesized, then all reactants will be used up at the same time. No excess unreacted starting material will remain at the completion of the reaction. On the other hand, if the reactants are mixed in a nonstoichiometric ratio, then one reactant will be used up before the others. Because no more product can be formed when the supply of one of the reactants has been exhausted, the first reactant to be consumed limits the amount of product that can be formed and is designated as the **limiting reactant** or **limiting reagent**.

The amount of product formed when the limiting reagent has been completely consumed is called the **theoretical yield of the reaction**. In practice, the actual yield of product is frequently lower than this theoretical maximum (because side reactions occur, or product is lost in collection, or because the product is unstable). The percent yield of the reaction is a measure of success of the reaction. It is defined in Equation 10.2.

$$\text{Percent yield} = \frac{\text{actual yield}}{\text{theoretical yield}} \times 100\% \qquad (10.2)$$

If you will not be performing Experiment 12, your instructor will supply you with the values of a, b, and c in the formula $Co_a(C_2O_4)_b \cdot cH_2O$. You should then be able to calculate your percent yield of cobalt oxalate hydrate based on the amounts of the starting materials you used in this synthesis. If you will be doing Experiment 12, you will be able to calculate the percent yield after you have performed your analyses and found the empirical formula of the compound.

PROCEDURE

Synthesis of Cobalt Oxalate Hydrate

Place 100 mL of distilled water in a 250-mL (or 400-mL) beaker. Add 1.26 g of oxalic acid dihydrate ($H_2C_2O_4{\cdot}2H_2O$) and 1 mL of concentrated ammonia. Stir the mixture until the solid has dissolved completely.

Dissolve 2.34 g of cobalt chloride hexahydrate ($CoCl_2{\cdot}6H_2O$) in 100 mL of water in an Erlenmeyer flask. While stirring the oxalic acid solution constantly, add the cobalt chloride solution drop by drop. Let the mixture cool in an ice bath. A precipitate will form slowly.

After the precipitate has had a chance to settle, collect it by gravity filtration. Wash the collected solid sparingly with cold water. Allow the water to drain from the collected solid. Then transfer the filter paper and precipitate to a paper towel so that excess moisture can be absorbed. Before you leave the lab, place your precipitate (still in its filter paper cone) in a beaker and leave it in your desk to air dry for at least three days.

After your precipitate has dried thoroughly, weigh it on the analytical balance. Report your yield of product. If your instructor has given you the values of *a*, *b*, and *c*, determine the limiting reagent in your synthesis and report your percent yield. If you will be performing Experiment 12, save your product for analysis.

Disposal of Reagents

The filtrate from the synthesis of cobalt oxalate should not be poured down the drain before it has been tested for the complete removal of Co^{2+} ions. About 1 millimole or 60 mg of Co^{2+} typically remains in solution. To test, add 3 M NaOH by drops until the solution tests basic to litmus paper. If no precipitate forms, dilute and discard the solution. If a precipitate forms, allow it to settle. Then decant the clear solution into another beaker for dilution and disposal. Pour the $Co(OH)_2$ slurry into the designated collection bottle. At a later time, this may be filtered, ignited to Co_3O_4, and bottled for disposal in a hazardous waste landfill.

Leftover oxalic acid may be decomposed by acidification with H_2SO_4, followed by oxidation of the oxalate to CO_2 with $KMnO_4$. This may be accomplished by following the procedure given in Experiment 12 for the determination of oxalate. The resulting solution should be made just basic to precipitate $Mn(OH)_2$, which should then be placed in a collection bottle labeled $Mn(OH)_2$. This product can be subsequently ignited to produce Mn_3O_4, which may be placed in a hazardous waste landfill site.

Excess NH_3 solutions should be neutralized, diluted, and poured down the drain.

The cobalt oxalate hydrate product should be saved for analysis in Experiment 12 or, if Experiment 12 is not to be performed, placed in a separate collection bottle. The compound can be ignited to Co_3O_4 when convenient and then deposited in a hazardous waste landfill site.

Date ______ Name ______________________ Section ______ Desk Number ______

PRE-LAB EXERCISES FOR EXPERIMENT 10

These exercises are to be done after you have read the experiment but before you come to the laboratory to perform it.

The equation given below refers to the process of producing potassium chromate from the reaction of chromite ore ($FeCr_2O_4$) with potassium carbonate and oxygen gas at high temperatures.

$$4FeCr_2O_4(s) + 8K_2CO_3(s) + 7O_2(g) \longrightarrow 8K_2CrO_4(s) + 2Fe_2O_3(s) + 8CO_2(g)$$

In a particular experiment, 175.0 kg of chromite ore, 350.0 kg of potassium carbonate, and 83.0 kg of oxygen gas reacted to produce 190.0 kg of potassium chromate.

1. Find the limiting reagent in this situation.

2. Calculate the theoretical yield of potassium chromate.

3. Calculate the percent yield of potassium chromate.

Date ______ Name ________________________________ Section ______ Desk Number ______

SUMMARY REPORT ON EXPERIMENT 10

Mass of container plus oxalic acid dihydrate	______________
Mass of container	______________
Mass of oxalic acid dihydrate	______________
Mass of container plus cobalt chloride hexahydrate	______________
Mass of container	______________
Mass of cobalt chloride hexahydrate	______________
Mass of product	______________
Values of *a*, *b*, *c* (if supplied)	______________
Limiting reagent	______________
Percent yield	______________

EXPERIMENT 11

The Synthesis of a Nitrite Complex

Laboratory Time Required

One and one-half hours. May be combined with Experiment 10, provided that all flames are extinguished before acetone is used in the nitrite synthesis.

Special Equipment and Supplies

Analytical balance
Desiccator
Buchner funnel and vacuum flask
Ice bath

Transition metal chlorides
$CuCl_2 \cdot 2H_2O$
$CoCl_2 \cdot 6H_2O$
$NiCl_2 \cdot 6H_2O$
Alkaline earth metal chlorides
$CaCl_2 \cdot 2H_2O$
$SrCl_2 \cdot 6H_2O$
$BaCl_2 \cdot 2H_2O$
Potassium nitrite, KNO_2
Acetone

Safety

The KNO_2 is both **toxic** (when ingested) and a strong oxidizing agent. Take care to avoid contacting KNO_2 with your skin, mouth, or eyes. You should also dispose of excess KNO_2 in a way such that it does not come in contact with reducing agents.

The salts of Co^{2+}, Ni^{2+}, and Sr^{2+} are **toxic**.

Acetone is **flammable**. There should be no flames in the laboratory when acetone is in use.

First Aid

First aid for exposure to any of these chemicals involves thorough flushing of the skin or eyes with water. Chemicals may be removed from the stomach by drinking large amounts of water and inducing vomiting.

In this experiment, you will synthesize one of a group of compounds with the general formula $K_2MM'(NO_2)_6$, where M is an alkaline earth metal ion and M′ is a divalent transition metal ion. These compounds are interesting because they illustrate the general properties of complex ions or coordination compounds and also because a large number of compounds of the same general formula can be prepared by varying the identities of M and M′.

PRINCIPLES

Complex Ions

The structure of compound salts, such as $K_2BaCu(NO_2)_6$, was a question of great interest to inorganic chemists during the latter part of the nineteenth century. It was difficult to understand why compounds such as KNO_2, $Ba(NO_2)_2$, and $Cu(NO_2)_2$, which were perfectly stable in themselves, should combine further to form compound salts. The work of Alfred Werner and the concept of coordinate covalent bonding went a long way in solving the puzzle.

Werner postulated that compound salts contain complex ions, which are formed by the association of a metal ion with one or more molecular or ionic species, known as **ligands** or **complexing agents**. A familiar example is the $Ag(S_2O_3)_2^{3-}$ complex ion formed in the fixing process in photography by the reaction of hypo ($Na_2S_2O_3$ solution) with unexposed silver halide. Another familiar example is the deep blue $Cu(NH_3)_4^{2+}$ ion formed by the addition of ammonia to aqueous solutions containing Cu^{2+} ions. In countless other examples, complex ions play an important role in a technical process or an analytical procedure. In addition, many compounds of biochemical importance, such as chlorophyll and hemoglobin, contain complex ions.

In the compounds to be synthesized in this experiment, the complex ion is $M'(NO_2)_6^{4-}$, where M′ is a transition metal. Combined with certain nontransition metals and potassium ions, these complexes are incorporated into compounds that tend to have low solubilities in water. Hence they may be isolated from aqueous solution.

The bonding in complex ions varies from predominantly ionic to predominantly covalent, depending on the nature of the metal ion and ligand involved. The transition metal ions, in particular, show a strong tendency to form complexes with both negative ions and molecules serving as ligands. This results from the relatively small size and large charge of these metal ions as well as the presence of a partially filled set of d orbitals, which may participate in covalent bonding. Owing to their larger size and lower charge, the representative metal ions of group I and group II in general form less-stable complexes than do the transition metal ions and do so with a more restricted group of ligands.

The number of bonding positions around the central metal ion that are occupied by nearest ligand atoms is known as the **coordination number** of the metal ion. The coordination number of M′ in $M'(NO_2)_6^{4-}$ is six, with the $[O—N—O]^-$ ions arranged so that the six nitrogen atoms surround the metal ion in an octahedral arrangement. This coordination number is a common one for transition metal ions, although coordination numbers of two through twelve are known to occur. Examples of complexes illustrating coordination numbers two, four, six, and eight are shown in Figure 11.1.

Note that the characteristic geometry for coordination number six is octahedral, whereas, for coordination number four, both tetrahedral and square planar geometries are observed in different complexes.

Ligands such as Cl^-, NH_3, and H_2O can occupy only one coordination position around the central metal ion and are therefore known as **unidentate ligands**.

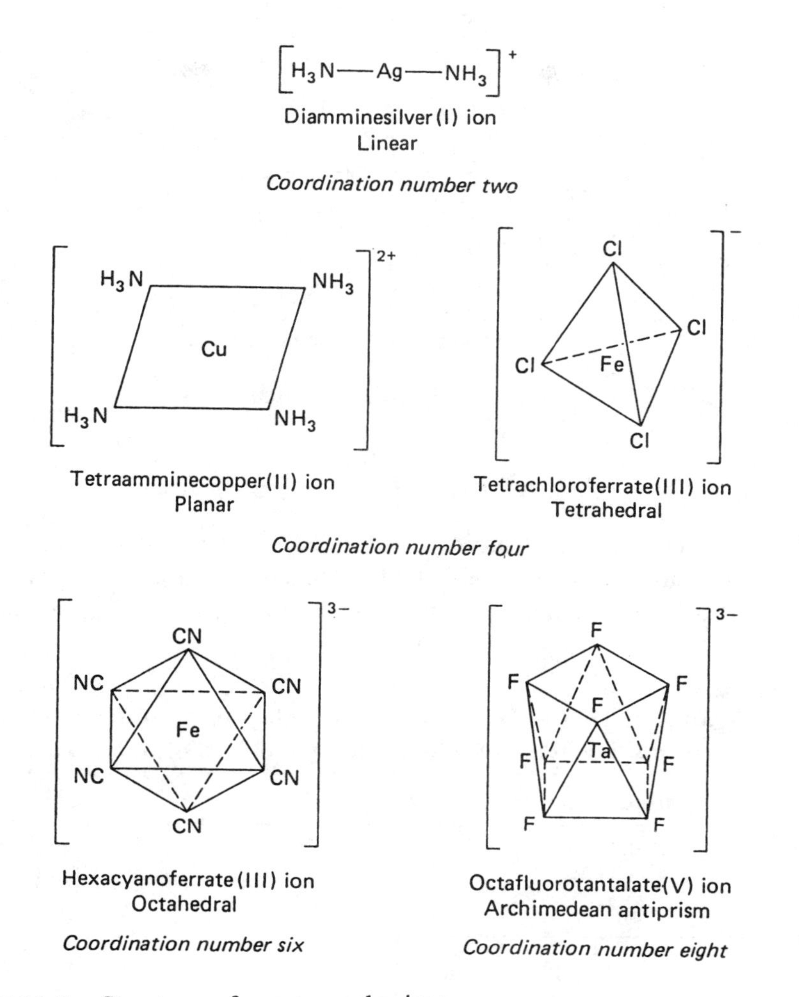

FIGURE 11.1 Structures of some complex ions.

Multidentate ligands contain two or more atoms, such as N, O, or S, which have unshared electron pairs that may be used to bond to a metal ion. These atoms are incorporated into a chain structure so that the ligand has several points of attachment to the metal ion. Ethylenediaminetetraacetic acid (EDTA, illustrated in Figure 11.2) is an example of a multidentate ligand. It is **hexadentate**, meaning that it can bind to a metal ion in six places. EDTA forms very stable complexes with a large number of metal ions and is a very useful reagent for analytical chemistry.

Preparation of $K_2MM'(NO_2)_6$ Compounds

The hexanitrite complexes of the divalent transition metal ions (such as Fe^{2+}, Co^{2+}, Ni^{2+}, or Cu^{2+}) are easily prepared by adding potassium nitrite to a solution of the desired transition metal ion in water. If the solution also contains a divalent nontransition metal ion (such as Ca^{2+}, Sr^{2+}, Ba^{2+}, or Pb^{2+}), the compound $K_2MM'(NO_2)_6$ precipitates as microcrystalline particles. The solid can be filtered and dried; the dried compound is stable in air. The solubility in water, however,

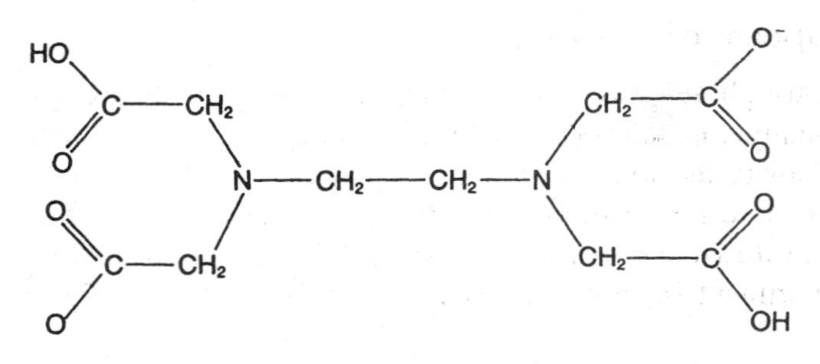

Dianion of ethylenediaminetetraacetic acid (H_2Y^{2-})

FIGURE 11.2 Ion of ethylenediaminetetraacetic acid (EDTA), which occurs in the salt $Na_2H_2Y{\cdot}2H_2O$.

depends on the identities of M and M′. Therefore the general procedure given here affords a reasonable yield of the solid product only for certain combinations of these ions. The recommended combinations are given in Table 11.1.

TABLE 11.1 Recommended Combinations of Transition Metal Ions and Nontransition Metal Ions to Prepare $K_2MM'(NO_2)_6$ Compounds

Transition Metal Ion, M′	Nontransition Metal Ions, M
Ni^{2+}	Ca^{2+}, Sr^{2+}, or Ba^{2+}
Co^{2+}	Sr^{2+} or Ba^{2+}
Cu^{2+}	Sr^{2+}

PROCEDURE

Obtain an assignment of transition metal and nontransition metal from your instructor. Weigh out 0.005 mole of the designated nontransition metal chloride and 0.005 mole of the designated transition metal chloride. Dissolve the two in a minimum volume (about 5 mL total) of distilled water, with low heat only if needed to dissolve the salts. Weigh out 0.05 mole of KNO_2, dissolve it in 5 mL of water, and then slowly add this solution, while stirring, to the solution of metal chlorides. Cool the mixture in an ice bath. Allow the solid to settle, then decant the supernatant liquid into a filter funnel, leaving most of the solid in the original beaker where it can be washed more efficiently. Add a 3-mL portion of cold water containing a small amount of KNO_2 to the solid, stir well, and pour the mixture onto the filter to collect the product. When the water has all drained out of the filter, pour 5 mL of acetone over the solid to help dry it. Pour a second 5 mL portion of acetone over the solid after the first has drained through the filter. Complete the drying of your compound by drawing a gentle stream of air through the filter using an aspirator. If no aspirator is available, open the filter paper and spread out the solid to dry in the air. The product is dry when it is powdery and has no odor of acetone.

After drying the solid, transfer it to a weighed bottle and reweigh. If your product loses weight during weighing, give the acetone more time to evaporate, then weigh the product again.

Disposal of Reagents

Although several of the chemicals used are hazardous, the small amounts remaining in solutions resulting from the synthetic procedure make it impractical to isolate the ions as insoluble salts. Consequently, the recommended disposal procedure for all solutions in this experiment is to dilute them tenfold and flush them down the drain with water. If you are to analyze your compound in Experiment 14, store it in a desiccator, over calcium chloride.

Questions

1. Determine which of the reactants used in your synthesis was the limiting reagent.
2. Calculate your percent yield of product.

Date ______ Name ________________________________ Section ______ Desk Number ______

PRE-LAB EXERCISES FOR EXPERIMENT 11

These exercises are to be completed after you have read the experiment but before you come to the laboratory to perform it.

1. Calculate the mass of potassium nitrite to be used in this experiment.

2. Robin Student calculated the mass of $CaC1_2 \cdot 6H_2O$ to be used in the synthesis of $K_2NiCa(NO_2)_6$ and subsequently weighed out that mass of $CaCl_2 \cdot 2H_2O$. Did Robin use too little or too much of the calcium salt, or did Robin's error have no effect on the mole ratio of K^+, Ni^{2+}, and Ca^{2+} in the reaction mixture? Explain your answer briefly.

Date ______ Name ______________________________ Section ______ Desk Number ______

SUMMARY REPORT ON EXPERIMENT 11

Assigned transition metal ion (M′) ____________________

Assigned nontransition metal ion (M) ____________________

Formula of reagent containing M′ ____________________________

Mass of container + reagent ____________________

Mass of container ____________________

Mass of reagent ____________________

Formula of reagent containing M ____________________________

Mass of container + reagent ____________________

Mass of container ____________________

Mass of reagent ____________________

Mass of product and bottle ____________________

Mass of bottle ____________________

Mass of product ____________________

Theoretical yield ____________________

Percent yield ____________________

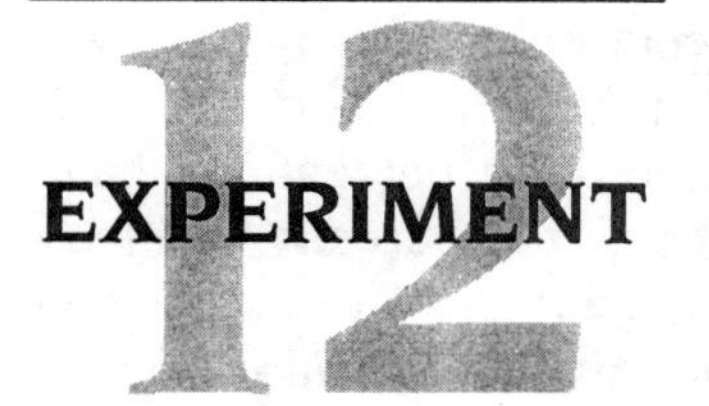

Determination of the Composition of Cobalt Oxalate Hydrate

Laboratory Time Required

Three hours

Special Equipment and Supplies

Analytical balance
Buret
Buret clamp
Litmus paper
0.02 M Potassium permanganate solution
0.1 M Sodium oxalate standard solution
6 M Sulfuric acid

Safety

All of the chemicals used in this experiment are hazardous! Avoid getting the solutions on the skin, in the eyes, or in the mouth. Use the pipet bulb.

Sodium oxalate is a **poison**.

Sulfuric acid is **corrosive** and may cause chemical burns.

Potassium permanganate is a strong oxidizing agent. It is **toxic.**

First Aid

If sodium oxalate gets on your skin, flush the area with water. Then flush with aqueous sodium bicarbonate.

If you ingest sodium oxalate, drink a large quantity of water, followed by milk or milk of magnesia. Then see a doctor.

If sulfuric acid gets on your skin, flush the area with water. Use the eyewash fountain promptly and see a doctor if sulfuric acid gets into your eyes.

Aqueous potassium permanganate may leave brown stains on the skin but normally causes no serious health problems. Rinsing with water should be sufficient treatment for potassium permanganate-stained skin. This may be followed by treatment with aqueous sodium hydrogen sulfite ($NaHSO_3$), which will reduce the $KMnO_4$ or MnO_2 to the colorless Mn^{2+}.

In this experiment, you will analyze a substance that has been given the tentative formula $Co_a(C_2O_4)_b{\cdot}cH_2O$. You may have synthesized the compound yourself in Experiment 10, or your instructor may provide you with a sample for analysis.

You will determine the percent cobalt (by mass) by a gravimetric method, and the percent oxalate (by mass) by an oxidation-reduction titration. When these percentages have been calculated, you will determine the percent water by difference (by subtracting the percentages of oxalate and cobalt from 100%). You will then use the composition by mass to calculate the actual formula of the compound. The experiment will be a test both of your proficiency and the applicability of the Law of Constant Composition. If the latter applies, the cobalt ions and oxalate ions combine in a definite ratio by mass. Because each ion has a characteristic mass, a definite number of cobalt ions combine with a definite number of oxalate ions. Hence *a* and *b* in the tentative formula should be integers.

PRINCIPLES

Thermal Decomposition of Cobalt Oxalate Hydrate

Because hydrates commonly lose all or part of their water when heated sufficiently, it would appear that the resulting mass loss could be used to determine the water content of hydrates that dehydrate easily. Unfortunately, this method cannot be used when thermal decomposition results in the formation of volatile products other than water. Metal oxalates, for example, generally decompose on being heated to yield gaseous CO or CO_2, as well as the metal or metal oxides. This is the case with cobalt oxalate hydrate, which decomposes to Co_3O_4 when heated in a crucible over a burner. Because you know the formula of the oxide, you can determine the mass of cobalt contained in a mass of Co_3O_4. You may then use the mass of cobalt and the mass of the original sample to calculate the percent Co in the cobalt oxalate hydrate.

Determination of Oxalate

The reaction used in the volumetric determination of oxalate is shown in Equation 12.1.

$$5C_2O_4^{2-} + 2MnO_4^- + 16H^+ \longrightarrow 10CO_2 + 2Mn^{2+} + 8H_2O \qquad (12.1)$$

No indicator is needed because the presence of excess permanganate, which has an intense violet color, is easily seen. Thus, at the end point, a color change of the solution from colorless to pink or violet is normally observed. However, the presence of Co^{2+} ions, which have an orange-pink color in solution, makes the detection of the end point slightly more difficult. In such cases, the end point will be indicated by a color change from orange-pink to rose-pink or violet. This color change is more easily detected visually than it would appear from its description.

The reaction between permanganate ions and oxalate ions tends to be very slow, particularly at first and, for this reason, the oxalate solution is heated to about 60°C and kept at this temperature during the titration. The solution must not be boiled because the oxalate might decompose at higher temperatures. The first addition of permanganate will give the solution a violet color that will persist for some time because the reaction with oxalate ions is slow even at 60°C. BE PATIENT. WAIT FOR EACH SMALL INCREMENT OF PERMANGANATE TO LOSE ITS COLOR BEFORE ADDING THE NEXT INCREMENT, OR YOU RISK PRECIPITATING MANGANESE DIOXIDE. As the titration proceeds, the permanganate additions will be decolorized more and more rapidly. This happens because the Mn^{2+} ions that form during the reaction serve as a catalyst for the reaction. At the end point of the titration, the addition of one drop of permanganate solution will give the titration mixture a permanent color change from orange-pink to violet because there will be no oxalate left to reduce the permanganate ion.

Determination of Water

As noted above, heating cobalt oxalate hydrate does more than simply drive off the water of hydration. However, the percent of water in the compound can be calculated by difference; that is, by subtracting the percentages of cobalt and oxalate from 100%.

PROCEDURE

Determination of Cobalt

Powder the solid cobalt oxalate hydrate (either your dried precipitate or the sample provided by your instructor) and use the analytical balance to weigh (to four significant figures) a 0.3 g sample of the compound. Transfer this sample completely to a preweighed crucible. Heat the crucible and sample to red heat until the sample has decomposed to Co_3O_4, which is a stable, black solid. Allow the crucible to cool and determine the mass of the crucible plus residue. Calculate the mass percent cobalt. Repeat the determination. Report both answers and their average. To conserve time, you should begin the oxalate analysis while waiting for the crucible to cool.

Determination of Oxalate

Prepare 300 mL of 0.5 M H_2SO_4 by **carefully** adding, while stirring, 25 mL of 6 M H_2SO_4 to a 400-mL beaker containing 275 mL of distilled water. Note the safety and first-aid procedures given at the beginning of this experiment.

Rinse a clean buret with two small portions of the approximately 0.02 M $KMnO_4$ solution and then fill the buret with the same solution. Accurately measure 10 to 15 mL of standard 0.1 M sodium oxalate solution into a 250-mL Erlenmeyer flask, add 100 mL of 0.5 M H_2SO_4 solution, and heat to about 60°C. Titrate with the potassium permanganate solution until a faint pink color, lasting about 30 seconds, is obtained. Calculate the molarity of the $KMnO_4$ solution. Repeat the standardization and calculate the average.

Weigh about 0.3 g of the cobalt oxalate hydrate to the nearest 0.1 mg and place it in a 250-mL Erlenmeyer flask. Add 100 mL of 0.5 M H_2SO_4 solution, stir to dissolve the solid, and heat to about 60°C. Titrate with the potassium permanganate solution until the color change signifying the end point is obtained. Calculate the mass percent oxalate in the compound. Repeat the titration and calculate the average.

Disposal of Reagents

Add 3 M NaOH, by drops, to the titration solution containing Co^{2+} and Mn^{2+} until the solution tests basic to litmus. If no precipitate forms, dilute and discard the solution. If a precipitate forms, allow it to settle. Then decant the clear solution into another beaker for dilution and disposal. Pour the $Co(OH)_2$, $Mn(OH)_2$, slurry into the designated collection bottle. At a later time, this may be filtered, ignited, and bottled for disposal in a hazardous waste landfill site. The ignited product should be labeled Co_3O_4, Mn_3O_4.

Excess $KMnO_4$ solution should be reduced to Mn^{2+} by adding solid $NaHSO_3$ in small portions until the violet color disappears. The Mn^{2+} may then be precipitated as the hydroxide from a solution that is just basic and placed in a second collection bottle. The ignited product should be labeled Mn_3O_4.

Leftover sodium oxalate may be decomposed by acidification with H_2SO_4 and subsequent oxidation of oxalate to CO_2 with $KMnO_4$, following the procedure given in the experiment for the determination of oxalate. The resulting solution should be made just basic to precipitate $Mn(OH)_2$, which should then be placed in the appropriate collection bottle. The solution may be flushed down the drain. Note that you should coordinate your disposal of excess oxalate and permanganate so that one is used in the treatment of the other.

Any unused cobalt oxalate hydrate should be placed in a third collection bottle. This may be ignited when convenient and added to the Co_3O_4 bottle for disposal in a hazardous waste landfill.

Excess H_2SO_4 should be neutralized, diluted, and poured down the drain.

Questions

1. Using the experimental values for the percent cobalt, oxalate, and water in your compound, calculate the formula of the compound.
2. Write an equation for the thermal decomposition of cobalt oxalate hydrate.

Date ______ Name ______________________________ Section ______ Desk Number ______

PRE-LAB EXERCISES FOR EXPERIMENT 12

These exercises are to be done after you have read the experiment but before you come to the laboratory to perform it.

1. Why is it unwise to boil the oxalate solution in preparation for its titration with permanganate?

2. A coordination compound is found by analysis to contain 23.4% Co, 27.1% NH_3, and 42.3% Cl. Assume the remainder, if any, is H_2O. Find the empirical formula for this coordination complex.

3. A 1.56 g sample of a hydrocarbon (a compound of carbon and hydrogen) was burned in air. The sample was completely converted to 5.28 g of carbon dioxide and 1.08 g of water. Find the empirical formula of the compound.

Date ______ Name ________________________ Section ______ Desk Number ______

SUMMARY REPORT ON EXPERIMENT 12

Determination of Cobalt

	Trial 1	*Trial 2*
Mass of crucible and cobalt oxalate	______	______
Mass of crucible	______	______
Mass of cobalt oxalate	______	______
Mass of crucible and residue	______	______
Mass of crucible	______	______
Mass of residue	______	______
% Co in cobalt oxalate	______	______
Average % Co	______	

Standardization of $KMnO_4$ Solution

	Trial 1	*Trial 2*
Concentration of standard $Na_2C_2O_4$ solution	______	
Final reading of $Na_2C_2O_4$ buret	______	______
Initial buret reading	______	______
Volume of $Na_2C_2O_4$ solution dispensed	______	______
Final reading of $KMnO_4$ buret	______	______
Initial buret reading	______	______
Volume of $KMnO_4$ solution required	______	______
Molarity of $KMnO_4$ solution	______	______
Average molarity of $KMnO_4$ solution	______	

Determination of Oxalate

	Trial 1	*Trial 2*
Mass of container and cobalt oxalate	______	______
Mass of container	______	______

Mass of cobalt oxalate

Final reading of $KMnO_4$ buret

Initial reading of KMnO4 buret

Volume of $KMnO_4$ solution use

% oxalate in cobalt oxalate

Average % oxalate

Determination of Water

Calculated% H_2O

Empirical formula of cobalt oxal:

58.93

Co_3O_4: 240.79

→ 73.42

42.18 %

EXPERIMENT 13

The Burning of a Candle

Laboratory Time Required

Two hours

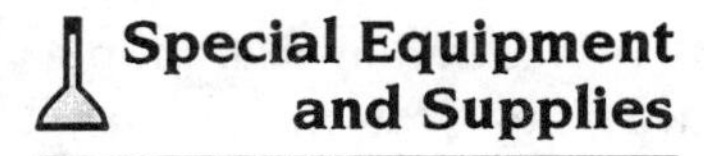

Special Equipment and Supplies

Balance
Glass plates
Shallow pan
Matches
250-mL Erlenmeyer flasks
Watch glass
Timer
Candle, 0.8 cm × 10 cm
Bromothymol blue indicator
Limewater
Ice

Safety

Exercise due caution in working with flames Do not leave flames unattended. Restrain long sleeves and hair near flames.

Limewater is a moderately strong base. Wear safety glasses when working with limewater. Rinse hands thoroughly with water after contact with limewater.

First Aid

Minor burns may be treated by immersing the area in cool water. More extensive burns will require a doctor's care.

If limewater gets in the eyes, flush them with water for at least 20 minutes and then seek medical attention. Limewater splashed on the skin should be washed off promptly.

The study of combustion reactions was of enormous importance in the development of modern chemistry. This experiment on the burning of a candle is a deceptively simple one that will allow you to develop your powers of observation and challenge you to analyze your data logically.

PRINCIPLES

Although electric lights and central heating have decreased the number of uses for fire in our homes, we still use combustion every day of our lives. We drive in cars powered by internal combustion engines; we may burn gas to cook our food or heat our homes; and we may light candles for festive occasions, religious ceremonies, or romantic evenings. If fire is important to us, it was all the more so to the ancients. The Greeks believed that fire was stolen from the gods by Prometheus, who was made to bear a terrible punishment for his crime.

Despite the fact that people have used fire for thousands of years, the true nature of combustion reactions was shrouded in mystery until the latter part of the eighteenth century. Lavoisier's discovery that combustion consists of the uniting of oxygen with other substances ranks with Dalton's atomic theory as the foundation of modern chemistry.

In this experiment, you will investigate the burning of a candle, collecting evidence to support or refute Lavoisier's theory of combustion. Qualitative observations will test the hypothesis that, during combustion, the carbon and hydrogen present in candle wax combine with oxygen of the air to form water, carbon dioxide, and possibly some elemental carbon as well. The physical properties of these reaction products will be used for identification. You might wish to review Experiment 7, if necessary, to recall the reactions of aqueous CO_2, with acid-base indicators and with limewater. You will use bromothymol blue indicator, which is blue in basic solutions and yellow in acidic solutions.

You will also study the reaction quantitatively to determine the chemical amount (in moles) of candle wax (assumed to be $C_{21}H_{44}$) and of oxygen consumed when the candle is burned in a limited amount of air. The experimental mole ratio will be compared with that obtained from a balanced equation for the reaction, which you will be asked to write.

The quantity of wax consumed will be determined by weighing the candle before and after the reaction. It will be slightly more difficult to find the quantity of oxygen consumed, but you can calculate this quantity using the Ideal Gas Law and Dalton's Law of Partial Pressures, shown in Equations 13.1 and 13.2, respectively.

$$PV = n\text{R}T \qquad (13.1)$$

$$P_{\text{tot}} = P_{\text{A}} + P_{\text{B}} \qquad (13.2)$$

You will begin the reaction by placing a flask of known volume over a burning candle that is standing in a dish of water. The number of moles of air in the flask (n) can be calculated using Equation 13.1, by substituting the volume of the flask in liters (V), the air pressure in atmosphere (P), the Kelvin temperature (T), and the value of the universal gas constant (R = 0.08206 L atm/mol K). The initial air pressure in the flask will be equal to atmospheric pressure in the laboratory, which is measured using a barometer.

As the candle burns, consuming oxygen, the water level in the flask will rise (Figure 13.1). This indicates that the pressure exerted by the gases in the flask has decreased to a value below atmospheric pressure. The water will rise until the pressure in the flask (given in Equation 13.3) equals atmospheric pressure.

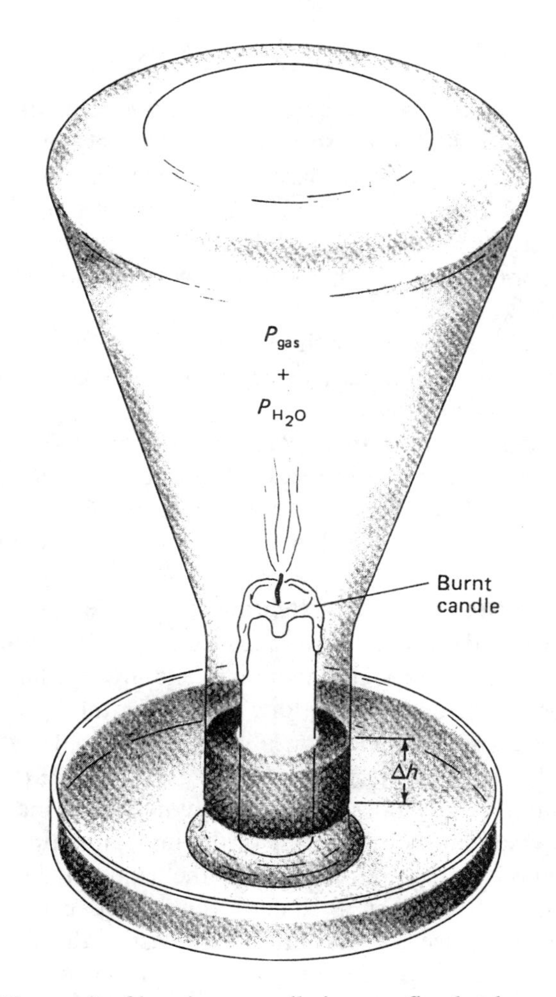

FIGURE 13.1 The result of burning a candle in a confined volume.

$$P_{atm} = P_{in\ flask} = P_{air} + P_{H_2O} + \frac{\Delta h}{13.6} \qquad (13.3)$$

In Equation 13.3, all pressures are given in units of torr, or mm Hg (760 torr = 1 atm). The symbol, P_{H_2O}, represents the partial pressure of water in the gas or vapor phase, resulting from the evaporation of liquid water. Fortunately, P_{H_2O} is constant at a given temperature. You may therefore obtain the value of P_{H_2O} from the table in Appendix B.

The final term in Equation 13.3, Δh /13.6, represents the pressure caused by the internal water level being Δh mm higher than the water level outside the flask. The 13.6 divisor is needed to convert from mm of H_2O to mm of Hg.

The number of moles of air in the flask after the candle is burned can be calculated using Equation 13.1, substituting the appropriate values for the air pressure, volume, and temperature. The air pressure can be calculated using Equation 13.3, as previously discussed. You can determine the final air volume by marking the water level on the flask after burning the candle and then measuring the volume of water needed to fill the flask to that mark. As a first approximation, you may assume that the amount of oxygen consumed by the burning candle corresponds to the difference between the initial and final amounts of air in the flask.

PROCEDURE

Obtain a candle and matches from your instructor. Hold the candle over a lit match for a few seconds so that the bottom of the candle begins to soften. Press the candle firmly onto a glass plate while the candle's bottom is still soft. The candle should be able to stand upright without support.

Light the candle. Hold a beaker in the flame for a few moments. Then allow the beaker to cool and wipe it with a white tissue. Note any evidence of the deposition of carbon on the beaker.

Fill a watch glass with ice and hold it for a few moments about 2 inches above the flame. Note any evidence for the condensation of water on the underside of the watch glass.

Extinguish the flame. Weigh the candle and plate on a balance. Light the candle once again and immediately invert a 250-mL Erlenmeyer flask carefully over the candle. The candle should continue to burn for a few seconds and then should go out. After the flame has burned out, weigh the candle and plate once again.

Repeat the process of lighting the candle and inverting a flask over it; use the timer to determine how long it takes for the flame to go out. Put another candle on the plate, light them both, and invert a flask over them. Use the timer, as before, to determine how long it takes for the flame to be extinguished.

Place a single candle on a glass plate and put the assemblage into a metal pan. Add water to the pan until the glass plate is just barely submerged about one-quater inch. Add some bromothymol blue indicator to the water and stir carefully so that the blue color of the indicator spreads throughout the water but the candle is not dislodged. Relight the candle and once again invert the 250-mL Erlenmeyer flask over it. Record your observations. After the candle has been extinguished, use a meter stick to measure the difference between the water levels inside and outside of the flask. Record this value of Δh. Use a glass-marking pen to place a line on the flask at the height of the water level inside the flask. Carefully lift the flask, with the glass plate remaining on its mouth, and invert and shake the flask gently. Record your observations.

Obtain another candle and repeat the procedure of mounting it on a glass plate and placing the plate in a pan. However, this time fill the pan with limewater until the plate is submerged about one-quarter inch. Light the candle and invert a dry, 250 mL Erlenmeyer flask over it. Record your observations. After the flame is extinguished, repeat the procedure of inverting and shaking the flask (with the glass plate over its mouth). Record your observations.

Finally, fill your first 250-mL Erlenmeyer flask to the pen mark with water poured from a large graduated cylinder. Record the volume of water required as the "volume of air after burning the candle." Then finish filling the flask using water from the graduated cylinder. Record the total water volume as the ''volume of air before burning candle." Also record the values of the barometric (atmospheric) pressure and ambient (room) temperature. Answer the questions on the Summary Report Sheet.

Disposal of Reagents

Be sure all matches are extinguished before they are discarded in the wastebaskets. Save the candles for reuse. Dispose of the limewater by flushing it down the drain with large amounts of water. Water containing bromothymol blue may also be poured down the drain.

Questions

1. This experiment is based on a phenomenon that is very often demonstrated in elementary school science classes. The fact that the water rises is attributed to its "replacing the 20% of air which is oxygen." Would that explanation agree with your results? Explain why or why not.
2. Is the water produced in the burning of a candle in its gaseous or liquid phase? What evidence do you have for your answer?
3. **a.** Consider the case of the candle burning while its base is submerged in water. Suppose the carbon dioxide dissolves in the water as soon as it is produced. Further suppose that only liquid water is produced as the candle burns. How high should the column of water rise under these circumstances?
   b. Suppose that the carbon dioxide does not dissolve, but remains in the gas phase. How high should the water column rise, if the water produced is actually in the liquid phase?
   c. Suppose that both the carbon dioxide and the water produced in the combustion of the candle are in the gas phase. How high should the water column rise, under these circumstances?
4. List at least five sources of error in this experiment that would affect your knowledge of the quantitative relationship between the amounts of wax and oxygen consumed in the combustion of the candle and the amounts of carbon dioxide and water produced.

Date ______ Name ______________________________ Section ______ Desk Number ______

PRE-LAB EXERCISES FOR EXPERIMENT 13

These exercises are to be completed after you have read the experiment but before you come to the laboratory to perform it.

1. Balance the equations shown below. Each shows the combustion of a hydrocarbon or carbohydrate.

$$CH_4 + O_2 \longrightarrow CO_2 + H_2O$$

$$C_8H_{18} + O_2 \longrightarrow CO_2 + H_2O$$

$$C_6H_{12}O_6 + O_2 \longrightarrow CO_2 + H_2O$$

2. Describe an experiment that would distinguish between a compound that is a hydrocarbon (a compound of C and H only) and a compound that is a carbohydrate (a compound of C, H, and O only).

3. A sample of nitrogen was collected over water on a day when the atmospheric pressure was 753.6 torr and the temperature was 24.0°C. A volume of 37.5 mL of gas was collected. The water level inside the collection vessel was 313 mm above the water level outside the vessel. Find the amount of N_2 (in moles) that was collected (1 atm = 760 torr).

Date ______ Name ______________________________ Section ______ Desk Number ______

SUMMARY REPORT ON EXPERIMENT 13

What did you observe when the beaker that had been held in the flame was wiped with a tissue? ______________________________

What can you infer from this? ______________________________

What did you observe when the ice-filled watch glass was held in the flame?

____

What can you infer from this? ______________________________

Mass of candle/glass plate after burning ______________

Mass of candle/glass plate before burning ______________

Change in mass ______________

What did you observe when the 250-mL Erlenmeyer flask was inverted over one candle in the absence of water? ______________________________

____

What can you infer from this? ______________________________

What did you observe when the 250-mL Erlenmeyer flask was inverted over two candles in the absence of water?

____

What can you infer from this? ______________________________

What did you observe when the 250-mL Erlenmeyer flask was inverted over the candle in the presence of water containing bromothymol blue indicator?

____

Value of Δh ______________

What did you observe when the flask containing water and bromothymol blue was shaken ? ______________________________

What can you infer from this? ______________________________

What did you observe when the flask was inverted over the candle in the presence of limewater? ______________________________

What did you observe when the flask containing the limewater was shaken?

____

What can you infer from this? ______________________________

Volume of air before candle was burnt ______________

Volume of air after candle was burnt ______________

Atmospheric pressure ______________

Room temperature ______________

Date ______ Name ______________________________ Section ______ Desk Number ______

Write a balanced equation for the combustion of candle wax, $C_{21}H_{44}$.

How many moles of candle wax burned? ________________

From the equation, how many moles of oxygen were consumed in combustion of the candle wax? ________________

Using the Ideal Gas Law, calculate the number of moles of air originally present in the 250-mL flask. ________________

Calculate the number of moles of air present in the flask after the candle was burnt. ________________

Calculate the number of moles of oxygen consumed based on the Ideal Gas Law calculations. ________________

EXPERIMENT 14

The Gasimetric Analysis of a Nitrite Complex

Laboratory Time Required

One and one-half hours.

Special Equipment and Supplies

Analytical balance
Aspirator
Small glass vials
Eudiometer tube
Glass beads
Sulfamic acid
$K_2MM'(NO_2)_6$

Safety

Sulfamic acid is a skin and eye irritant. The salts of Co^{2+}, Ni^{2+}, and Sr^{2+} are **toxic.**

First Aid

After exposure to any of these chemicals, thoroughly flush your skin or eyes with water. Remove chemicals from the stomach by drinking large amounts of water and inducing vomiting.

In this experiment, you will perform a partial analysis of one of a group of compounds with the general formula $K_2MM'(NO_2)_6$, where M is an alkaline earth metal ion and M′ is a divalent transition metal ion. You may have synthesized the compound you will analyze or your instructor may provide you with a sample for analysis. You will analyze the sample by decomposing it in a reaction with sulfamic acid and collecting the nitrogen gas evolved. You will use your data to calculate the percent nitrite in your sample and compare your experimental value to the value calculated from the formula of your compound. This will permit you to determine whether your compound's formula agrees with the general formula cited above.

PRINCIPLES

Nitrite Analysis

In the presence of acids, the nitrite complexes $K_2MM'(NO_2)_6$ readily release their nitrite ligands to form HNO_2 as shown in Equation 14.1.

$$M(NO_2)_6^{4-} + 6H^+ \longrightarrow 6HNO_2 + M^{2+} \tag{14.1}$$

The HNO_2 thus formed reacts quantitatively with sulfamic acid according to the reaction shown in Equation 14.2.

$$HNO_2 + NH_2SO_3^- \longrightarrow N_2 + HSO_4^- + H_2O \tag{14.2}$$

After you have collected the nitrogen gas and measured its volume, you can use the gas laws to calculate the number of moles of nitrite present in the sample. The apparatus needed is shown in Figure 14.1.

As shown in Figure 14.1, the nitrogen is collected by being bubbled through water into an eudiometer tube. As a result, the gas collected is wet and its total pressure is the sum of the pressure of the dry gas and the water vapor pressure. The vapor pressure of water depends, in turn, on the temperature. A table of vapor pressures of water at various temperatures is given in Appendix B.

In most cases, the volume of gas collected will not be sufficient to lower the water level inside the eudiometer tube to that of the water in the beaker in which the tube is mounted. Thus, the barometric pressure will not be completely balanced by the pressure of the wet gas and will still be able to support a column of water. The height of this column is designated as Δh in Figure 14.1.

The relation between the pressure of the dry gas, barometric pressure, water vapor pressure, and the height of the water column is given in Equation 14.3. The factor 13.6 is introduced to convert the height of the water column to the equivalent height of a mercury column. This is necessary because the other pressures are measured in units of torr (1 torr = 1 mm Hg).

$$P_{N_2} = P_{atm} - P_{water} - \frac{\Delta h}{13.6} \tag{14.3}$$

PROCEDURE

If you did not perform Experiment 11, obtain a sample of a compound salt from your instructor, who will give you its formula. Calculate the mass of compound that should be used in order to produce 30 to 40 mL of nitrogen gas and accurately weigh that amount into each of two glass vials. The vials should be small enough so that they can be upset easily after being inserted upright into a 50-mL Erlenmeyer flask. Dissolve 0.5 g of sulfamic acid in a 50-mL Erlenmeyer flask, using 20 mL of water. Carefully place one vial upright in the flask without mixing the solid and solution, and stopper the flask with a one-hole rubber stopper containing a short length of glass tubing. (Distilled water or glass beads may be added to the vial to reduce its tendency to tip over prematurely.) Attach one end of a rubber tube to the glass tube in the one-hole rubber stopper and insert the other end into a water-filled eudiometer tube. Mix the solid and solution by tilting the Erlenmeyer flask and then shaking the mixture gently until the solid has all reacted. Wait approximately 12 minutes and then read and record the

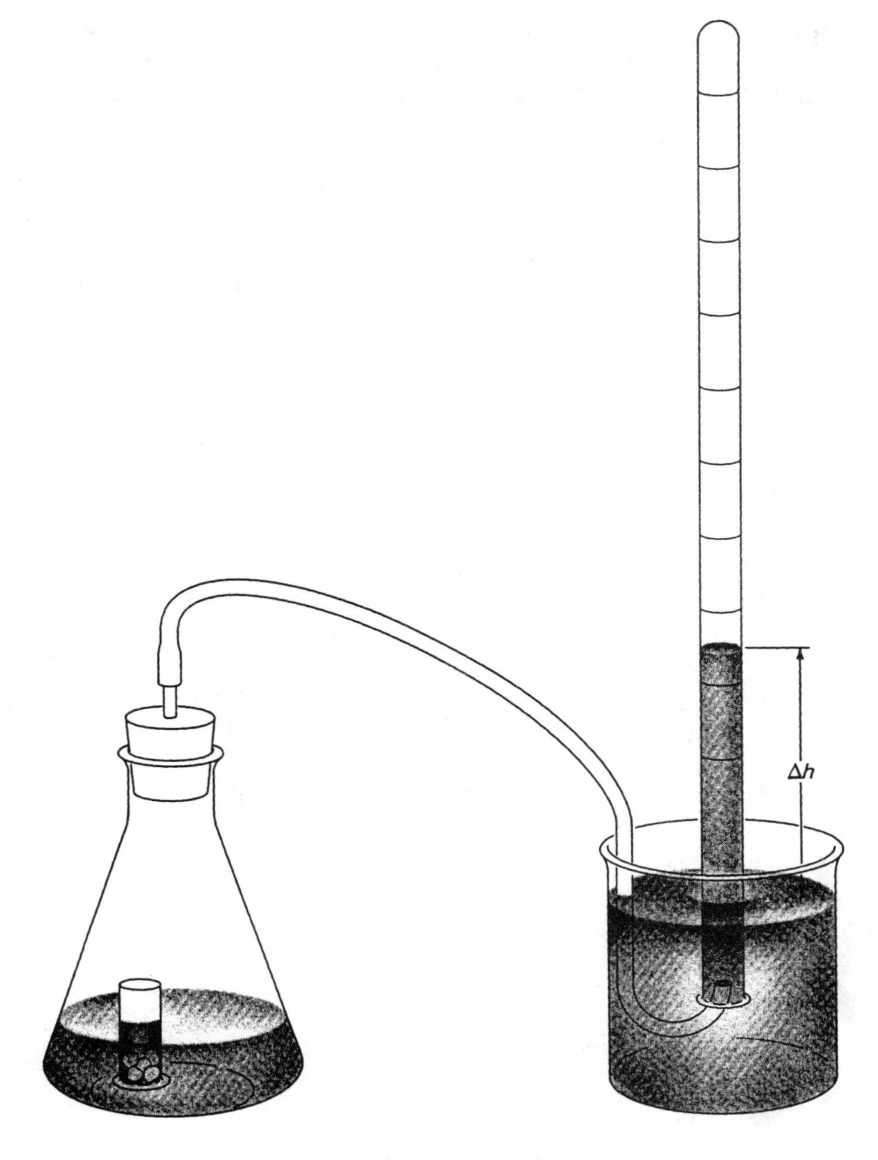

FIGURE 14.1 Reaction flask with eudiometer tube.

volume of gas collected in the eudiometer tube and also the difference (in millimeters) between the water level inside the eudiometer tube and the water level outside the tube. Use a meter stick to make the latter measurement. Repeat the analysis using the solid contained in the second vial.

Compute the mass percent nitrogen for each sample and average the results. Show your calculations clearly in your report, including corrections for the presence of water vapor in the gas collected and for hydrostatic pressure caused by the difference in water levels. Calculate the theoretical percent nitrogen and your percent error and include these in your report.

Disposal of Reagents

The small amounts of hazardous chemicals remaining in solution make it impractical to isolate the ions as insoluble salts. Consequently, the recommended disposal procedure for all solutions used in this experiment is to dilute them greatly and flush them down the drain with water. Excess nitrite salts can be stored in desiccators for analysis by other groups of students.

Date ______ Name ______________________________ Section ______ Desk Number ______

PRE-LAB EXERCISES FOR EXPERIMENT 14

These exercises are to be completed after you have read the experiment but before you come to the laboratory to perform it.

As a result of the reaction between a nitrite compound and sulfamic acid, a 47.6 mL sample of nitrogen was collected over water at 24.8°C. The height of the water column inside the eudiometer tube was 56.3 mm higher than the surface of the water in the surrounding beaker. Barometric pressure was 742.7 torr.

1. Show your work in finding the partial pressure of the nitrogen gas, P_{N_2}.

2. Show your work in finding the mass of the nitrogen gas collected.

3. The original sample, which decomposed to produce the nitrogen gas, weighed 0.1150 g. Find the percent nitrogen in the sample.

Date ______ Name ______________________________ Section ______ Desk Number ______

SUMMARY REPORT ON EXPERIMENT 14

	Trial 1	*Trial 2*	*Trial 3*
Mass of bottle (after removal of sample)	______	______	______
Mass of bottle (before removal of sample)	______	______	______
Mass of sample in vial	______	______	______
Volume of gas collected	______	______	______
Δh	______	______	______
Barometric pressure	______	______	______
Room temperature	______	______	______
Vapor pressure of water	______	______	______
Moles of nitrogen collected	______	______	______
Mass of nitrogen in sample	______	______	______
Percent nitrogen in sample	______	______	______
Average percent nitrogen		______	
Theoretical percent nitrogen		______	
Percent error		______	

EXPERIMENT 15

The Vapor Pressure of Water

Laboratory Time Required

Two hours. May be combined with Experiment 16.

Special Equipment and Supplies

Thermometer
Hot plate or burner
Graduated cylinder, 10-mL
Tall-form beaker, 1000-mL
Ice

Safety

This experiment involves moving a beaker full of hot water. Always remain alert and be cautious when handling hot water. Never leave a burner flame unattended.

Burns or electrical shock may be caused by electric heating devices that are poorly maintained or carelessly used. Avoid electric shock by using care when plugging in the power cord and by not using instruments that have frayed cords or that are wet from spilled liquids.

First Aid

You may soothe burnt fingers by immersing them in cool water. Seek medical attention for serious burns.

In case of electric shock, separate the victim from the source of electricity by turning off the power at the switch box, removing the power cord with a nonconducting tool, or by other means. If the victim is not breathing or has no heartbeat, CPR should be administered quickly by a qualified person. Get professional help *immediately*.

Although phase changes are not chemical changes, the examination of phenomena such as vaporization is an important part of the study of chemistry. A liquid's volatility, enthalpy of vaporization, and normal boiling point are characteristics that reflect the intermolecular forces present in the liquid. This experiment employs simple apparatus in the study of a one component system, water. Experiment 16 employs the same techniques, in scaled-down apparatus, to study vaporization of a two-component system.

PRINCIPLES

The atoms and molecules of any liquid are in constant motion, continually changing their molecular speeds and kinetic energies as a result of collisions. At any given temperature, a number of molecules may have sufficient kinetic energy to escape from the liquid at the surface, evaporating into the space above the liquid. Consequently, the particles remaining in the liquid have lower kinetic energy, and the temperature of the liquid decreases, unless the liquid absorbs energy from its surroundings. If the liquid is in an open container, allowed to absorb heat from the surroundings to maintain a constant temperature, evaporation will continue until no more liquid remains. If, however, the liquid evaporates in a closed container, an equilibrium is established in which the rate of escape from the liquid is balanced by the rate at which gas phase particles lose energy and return to the liquid phase. The pressure exerted on the walls of the container when equilibrium has been established is called the **equilibrium vapor pressure** of the liquid.

The value of the equilibrium vapor pressure increases with temperature for all liquids. When the vapor pressure reaches the value of the external pressure, the liquid boils. The temperature at which the vapor pressure equals 760 torr (one standard atmosphere) is called the **normal boiling point** of the liquid.

In this experiment, you will study the relationship between the vapor pressure of water and temperature by monitoring the volume of an air bubble that is surrounded by a water bath. At temperatures above 5°C, water has an appreciable vapor pressure and Dalton's Law of Partial Pressures is used to relate the partial pressure of air, the vapor pressure of water, and atmospheric pressure (see Equation 15.1).

$$P_{atm} = P_{air} + P_{H_2O} \qquad (15.1)$$

At temperatures below 5°C, the vapor pressure of water is negligibly small. Therefore, at low temperature, the bubble may be considered to contain only air. The Ideal Gas Law can be used to relate the amount of air (n_{air}) to the volume (V) of the bubble, the bath temperature (T), and the atmospheric pressure (P), as shown in Equation 15.2.

$$n_{air} = \frac{PV}{RT} \qquad T < 278 \text{ K} \qquad (15.2)$$

At temperatures above 5°C, the bubble becomes saturated with water vapor. However, the amount of air contained in the bubble is constant. The Ideal Gas Law can once again be used to obtain the partial pressure of air (P_{air}) from the number of moles (n_{air}) of air in the bubble, the volume (V) of the bubble, and the bath temperature (T), as shown in Equation 15.3.

$$P_{air} = \frac{n_{air}RT}{V} \qquad T > 278 \text{ K} \qquad (15.3)$$

The value of P_{H_2O}, the vapor pressure of water, is then obtained from Equation 15.1. Once the values for the vapor pressure at different temperatures have been obtained, they can be used to find two characteristic properties of water—its normal boiling point and its enthalpy of vaporization ΔH_{vap}. The enthalpy of vaporization is the heat that must be supplied to evaporate a mole of water at constant pressure. The relationship of ΔH_{vap} to the vapor pressure at different temperatures is given in Equation 15.4, where P_1 and P_2 represent the vapor pressure of water at temperature T_1 and T_2, respectively. The symbol "ln" denotes the natural logarithm. The constant, R, is the ideal gas constant with the value of 8.314 joule/K mole rather than the value 0.08206 L atm/K mole, which would be used in Equations 15.2 and 15.3.

$$\ln P_2 - \ln P_1 = \frac{-\Delta H_{vap}}{R}\left(\frac{1}{T_2} - \frac{1}{T_1}\right) \qquad (15.4)$$

The value of ΔH_{vap} is obtained by plotting $\ln P$ versus $1/T$. Such a plot should be a straight line, with slope equal to $-\Delta H_{vap}/R$. Once ΔH_{vap} has been obtained, one may solve the equation to find the value of T at which P would equal 760 torr.

PROCEDURE

Obtain a 10-mL graduated cylinder and a beaker large enough for the cylinder to be submerged in it. Fill the beaker half full with distilled water. Put enough distilled water in the 10-mL graduated cylinder to fill the cylinder to 90% capacity (ignoring graduations). Place your finger over the mouth of the graduated cylinder and invert the cylinder in the beaker. An air bubble, 4 to 5 mL in volume, should remain in the cylinder. Add distilled water until the graduated cylinder is covered completely, as shown in Figure 15.1. Heat the watei in the beaker to 75° or 80°C. The air sample should be allowed to extend beyond the calibrated portion of the cylinder without escaping. Remove the beaker from the heat when the desired temperature has been reached. Start recording the volume of the bubble and the water temperature when the air sample is contained completely within the calibrated portion of the cylinder. Take readings every 3°C, until the water temperature has cooled to 50°C. Then add ice to the beaker to lower the temperature below 5°C. Record the volume of the air bubble at low temperature. Also record the value of the barometric pressure.

Use the data obtained at low temperature to find the number of moles of air in your bubble. Then calculate the partial pressure of air at each of those temperatures. Prepare a table with columns for t(°C), T(K), $1/T$, P_{H_2O} and $\ln P$. Use your tabulated results to prepare a plot of $\ln P$ versus $1/T$. Use your plot to find the value of ΔH_{vap} for water. Predict the normal boiling point of water.

Disposal of Reagents

The water and ice in this experiment can be discarded in the sink.

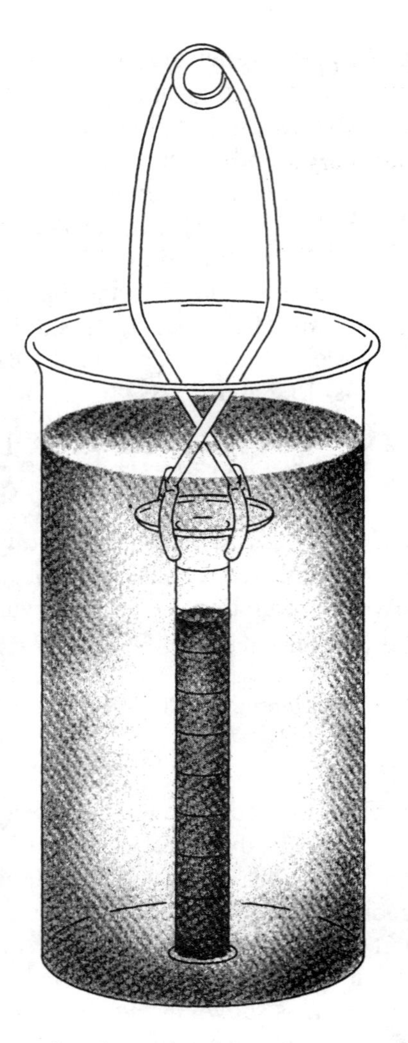

FIGURE 15.1 Apparatus used to determine the vapor pressure of water.

Questions

1. What is the uncertainty associated with each of your volume observations? How does this affect your value of P_{H_2O} at 50°C?
2. You could obtain ΔH_{vap} by inserting the data from two P_{H_2O} measurements into Equation 15.4, or from a plot of data from 8–10 measurements, as in this experiment. Which procedure is better? Why?

Date ______ Name ______________________ Section ______ Desk Number ______

PRE-LAB EXERCISES FOR EXPERIMENT 15

These exercises are to be completed after you have read the experiment but before you come to the laboratory to perform it.

1. The vapor pressure of acetone at 39.5°C is 400 torr. At 7.7°C, the vapor pressure of acetone is 100 torr. Use these data and Equation 15.4 to find the value of ΔH_{vap} for acetone.

2. Kim Pupil performed this experiment and prepared a plot of ln P versus $1/T$ with a temperature scale ranging from 1/273 K to 1/353 K. Fran Teacher penalized Kim 10 points. What was wrong with Kim's graph?

Date ______ Name ________________________ Section ______ Desk Number ______

SUMMARY REPORT ON EXPERIMENT 15

Observations

Barometric Pressure ______________ torr

t (°C)	V (mL)

Tabulated Results

n_{air} ______________

t(°C)	T(K)	$1/T$(K^{-1})	P_{air} (torr)	P_{H_2O} (torr)	ln$P$

ΔH_{vap} ____________ J/mol

Predicted normal boiling point of water ____________°C

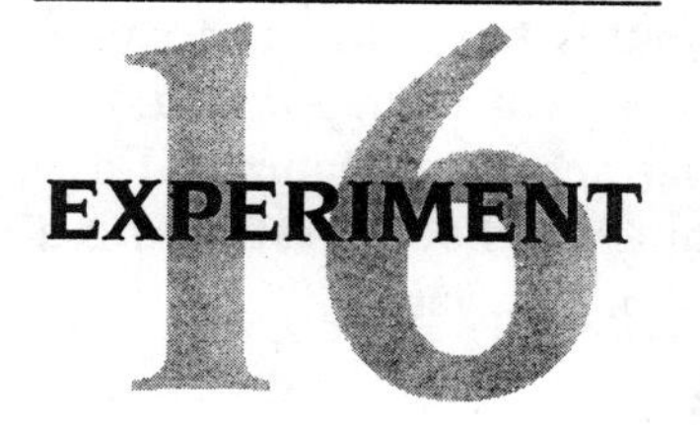

The Vapor Pressure of an Azeotropic Mixture of Isopropyl Alcohol and Water

Laboratory Time Required

One hour. May be combined with Experiment 15.

Special Equipment and Supplies

Graduated cylinder, 100-mL
Cylindrical container[1]
Shortened, graduated cylinder[2], 10-mL
Thermometer
Hot plate
Isopropyl alcohol

Safety

Burns or electrical shock may be caused by electrical heating devices that are poorly maintained or carelessly used. Avoid electrical shock by using care when plugging in the power cord and by not using instruments that have frayed cords or that are wet from spilled liquids

First Aid

In case of electrical shock, separate the victim from the source of electricity by turning off the power at the switch box, removing the power cord with a nonconducting tool, or by other means. If the victim is not breathing or has no heartbeat, CPR should be administered quickly by a qualified person. Get professional help *immediately*.

The vapor pressure of a liquid is a measure of the strength of the forces between molecules in the condensed phase. In **ideal solutions** of volatile liquids, the forces of attractions between solute and solvent molecules are comparable in strength to the forces of attraction between solute and solute, and between solvent and solvent. The vapor pressure of such

[1]The cylindrical container should have a volume of ~150 mL and can be made by sealing off a 6-inch-length of 40-mm tubing.
[2]The shortened, 10-mL graduated cylinder must fit completely inside the cylindrical container.

ideal solutions is predicted by Raoult's Law. Most mixtures of volatile liquids, however, are nonideal. They are said to exhibit negative or positive deviations from Raoult's Law. This experiment describes a simple method for studying a nonideal mixture of isopropyl alcohol and water.

PRINCIPLES

When a solution of a nonvolatile substance is heated, only solvent molecules gain enough energy to escape into the vapor phase, leaving behind an increasingly concentrated solution. In contrast, when a solution consists of a mixture of two volatile liquids, molecules of both components can escape into the gas phase, but not necessarily in the same proportions as in the original solution. The composition of the remaining liquid may, therefore, change.

Let us consider a mixture of 7.0 moles of toluene and 3.0 moles of benzene. Such a mixture is said to be an ideal solution because the interactions between the benzene and the toluene molecules are comparable in strength to those between one benzene molecule and another, and those between one toluene molecule and another. The partial pressure of any component in the gas phase above an ideal mixture of volatile liquids is given by Raoult's Law (see Equation 16.1).

$$P_i = X_i P_i^0 \qquad (16.1)$$

In Equation 16.1, P_i is the partial pressure of component i; X_i is the mole fraction of component i in the liquid phase; and P_i^0 is the vapor pressure of the pure substance i. Both P_i and P_i^0 are dependent on the phase temperature.

Let us assume that, at the temperature of our benzene and toluene mixture, the vapor pressure of pure benzene is 73 torr, while that of toluene is 27 torr. We may use Raoult's Law to find the partial pressure of both benzene and toluene above the liquid solution (see Equation 16.2).

$$P_T = \frac{7.0}{10.0}\ (27\ \text{torr}) = 19\ \text{torr} \qquad (16.2)$$

$$P_B = \frac{3.0}{10.0}\ (73\ \text{torr}) = 22\ \text{torr}$$

(Note that in Equation 16.2, the subscript T refers to toluene and the subscript B refers to benzene.)

How does the composition of the vapor phase compare to that of the liquid phase? We may use Dalton's Law of Partial Pressures to find a relation between the partial pressures of the gases and the composition of the vapor phase (see Equations 16.3 through 16.6).

$$P_{\text{tot}} = P_i + P_j + \ldots \qquad (16.3)$$

$$P_{\text{tot}} = \frac{n_{\text{tot}} RT}{V} \qquad (16.4)$$

$$P_i = \frac{n_i RT}{V} \tag{16.5}$$

$$X_i^V = \frac{n_i}{n_{tot}} = \frac{P_i}{P_{tot}} \tag{16.6}$$

(Note that the subscript tot refers to the total pressure or the total number of moles. The subscripts i and j refer to the partial pressure or number of moles of components i and j, respectively. The symbol X_i^V stands for the mole fraction of component i in the vapor phase.)

Substituting the partial pressures of benzene and toluene obtained in Equation 16.2 into Equation 16.6 yields the desired information concerning the composition of the vapor phase (see Equation 16.7).

$$X_B^V = \frac{22 \text{ torr}}{41 \text{ torr}} = 0.54 \tag{16.7}$$

$$X_T^V = \frac{19 \text{ torr}}{41 \text{ torr}} = 0.46$$

Because X_B^V is higher than X_B, we say that the vapor is enriched in benzene. This is an example of a general rule concerning ideal solutions: the vapor phase is enriched in the more volatile component, while the liquid is depleted of that component. The process of fractional distillation makes use of this property. Vapor fractions are repeatedly condensed and reevaporated, until they are essentially pure samples of the more volatile component.

There are many liquids that mix together in nonideal fashion. Such nonideal solutions are said to exhibit deviations from Raoult's Law. Such deviations may be positive (i.e., the vapor pressure of the solution is higher than that predicted by Raoult's Law) or negative. Mixtures of chloroform and acetone exhibit negative deviations from Raoult's Law, whereas mixtures of methanol and water exhibit positive deviations.

When a solution deviates strongly from Raoult's Law, it may form an **azeotrope:** a mixture that evaporates without any change in composition. If the deviation of the mixture from Raoult's Law is positive, the boiling point of the azeotrope will be lower than that of either of the pure components. If the deviation of the mixture from Raoult's Law is negative, the boiling point will be higher than that of either of the pure components. (By contrast, mixtures that are ideal solutions boil at temperatures that are intermediate between the boiling points of the pure components.)

In this experiment, you will be working with an azeotropic mixture of isopropyl alcohol, $(CH_3)_2CHOH$, and water. Using the method outlined in Experiment 15, you will determine the vapor pressure of the azeotrope at various temperatures and predict the mixture's boiling point. Because the procedure and data analysis used in this experiment are very similar to those of Experiment 15, it would be a good idea to review the equations used in that experiment before you do the calculations for this experiment.

PROCEDURE

Prepare a mixture of 135 mL of isopropyl alcohol and 15 mL of distilled water in a clean, dry beaker. Stir the mixture to ensure homogeneity. Pour approximately 100 mL of the mixture into the cylindrical container and place approximately 9 mL of the mixture into the shortened, 10-mL graduated cylinder. Cover the top of the graduated cylinder with your finger and invert the graduated cylinder in the cylindrical container. Add enough of the mixture to cover the graduated cylinder completely. Use the hot plate to heat the solution in the cylindrical container to roughly 65°C.

CAUTION THE ISOPROPYL ALCOHOL IS FLAMMABLE. DO NOT SUBSTITUTE A GAS BURNER FOR THE HOT PLATE.

The air sample should extend beyond the calibrated portion of the graduated cylinder but should not be allowed to escape. After the desired temperature has been reached, remove the cylindrical container from the hot plate. When the air sample is once again contained completely within the calibrated portion of the graduated cylinder, record the volume every 2°C until the temperature reaches 50°C. Then add ice to the cylindrical container to lower the temperature of the mixture to 5°C or lower and make a final volume measurement. Record the barometric pressure.

Use the data obtained at low temperature to find the number of moles of air in your sample. Then find the partial pressure of air at each of the higher temperatures for which you recorded volume data. Also find the vapor pressure of the azeotrope at each of those temperatures. Prepare a table with columns for t(°C), T(K), V_{air} (mL), $P_{\text{azeotrope}}$ and ln$P$. Use your tabulated results to prepare a plot of ln$P$ versus $1/T$. Use your plot to find the normal boiling point of the azeotrope. (The normal boiling point is defined as the temperature at which the vapor pressure of a liquid equals 760 torr.)

Use the data in Table 16.1 and equation 15.4 to calculate the vapor pressure of pure isopropyl alcohol and the vapor pressure of pure water at one of the temperatures for which you collected volume data. Combine these vapor pressures of the pure components with the appropriate mole fractions to obtain the Raoult's Law prediction for the vapor pressure of the azeotropic mixture. Compare the predicted vapor pressure with your experimental value for the vapor pressure at the same temperature. Is the direction of the deviation what you would expect, based on a comparison of the normal boiling points of isopropyl alcohol, water, and the azeotrope?

TABLE 16.1 Vapor Pressures and Densities of Isopropyl Alcohol and Water at Selected Temperatures

Vapor Pressure		Density at 20°C	
Isopropyl Alcohol	*Water*	*Isopropyl Alcohol*	*Water*
P = 400 torr at 67.8°C	P = 400 torr at 83.0°C	d = 0.785 g/mL	d = 0.998 g/mL
P = 760 torr at 82.3°C	P = 760 torr at 100.0°C		

Disposal of Reagents

All liquids used in this experiment may safely be flushed down the drain.

Date ______ Name ________________________________ Section ______ Desk Number ______

PRE-LAB EXERCISES FOR EXPERIMENT 16

These exercises are to be completed after you have read the experiment but before you come to the laboratory to perform it.

1. When solutions exhibit negative deviations from Raoult's Law, this is an indication that it is harder for the molecules to escape from the solution than from either of the pure components. What molecular interactions would account for the negative deviations by mixtures of chloroform, $HCCl_3$, and acetone, $O{=}C(CH_3)_2$?

2. When solutions exhibit positive deviations from Raoult's Law, this is an indication that it is easier for the molecules to escape from the solution than from the pure components. What molecular interactions would account for the positive deviations exhibited by mixtures of methanol, CH_3OH, and water?

Date ______ Name ______________________________ Section ______ Desk Number ______

SUMMARY REPORT ON EXPERIMENT 16

t(°C)	T(K)	$1/T$(K^{-1})	V_{air} (mL)	$P_{azeotrope}$ (torr)	ln$P$
______	______	______	______	______	______
______	______	______	______	______	______
______	______	______	______	______	______
______	______	______	______	______	______
______	______	______	______	______	______
______	______	______	______	______	______
______	______	______	______	______	______
______	______	______	______	______	______
______	______	______	______	______	______
______	______	______	______	______	______
______	______	______	______	______	______

Barometric pressure ______ torr

T_b for azeotrope ______ °C

Raoult's Law prediction of $P_{azeotrope}$ at ______ °C ______ torr

Experimental value of $P_{azeotrope}$ at ______ °C ______ torr

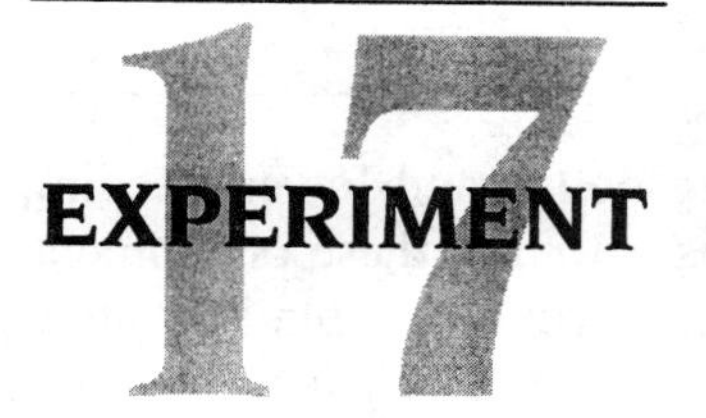

Determination of Molecular Mass by Freezing-Point Depression

Laboratory Time Required

Two hours

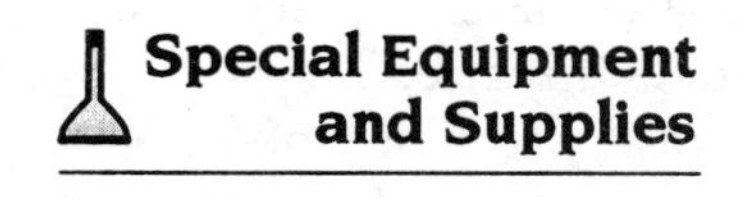

Special Equipment and Supplies

- Buret
- Buret clamp
- Graduated cylinder, 25-mL
- Freezing-point depression apparatus:
  - Large test tube
  - Two-hole stopper
  - Wire stirrer
  - Thermometer
  - Tall-form beaker, 1000-mL
- Timer
- Isopropyl alcohol
- Ice
- Rock salt

Safety

Although isopropyl alcohol is a household chemical, it is meant for external use only. Follow standard laboratory procedures: wear safety glasses and do not eat or drink in the lab.

Be careful not to break the thermometer or other glass apparatus. Clean up broken glass carefully to avoid being cut. Consult your instructor about cleaning up spilled mercury.

First Aid

If alcohol spills on your skin or enters your eyes, wash out the alcohol with copious amounts of water. Seek medical attention if eye irritation results.

Determining the molecular mass of an unknown substance by freezing-point depression is a classical, but still useful, technique. In this experiment, you will determine the molecular mass of isopropyl alcohol.

PRINCIPLES

Solutions of nonvolatile solutes have higher boiling points and lower freezing points than the solvents used to prepare the solutions. Both of these phenomena result from the fact that the vapor pressure of a solution (of a nonvolatile solute) is lower than the vapor pressure of the pure solvent.

It is relatively easy to see why the boiling point of a solution will be higher than that of the pure solvent. The vapor pressure of a liquid is determined by the ability of particles at the liquid's surface to escape into the vapor phase. In solutions of the type we are considering, some of the spaces at the surface are occupied by the nonvolatile solute particles. As a result, fewer solvent particles are in positions from which they can enter the vapor phase, and the solution's vapor pressure is lower than that of the pure solvent at all temperatures. This means that the solution will have to be heated to a higher temperature before its vapor pressure reaches that of the atmosphere. Hence, solutions of nonvolatile solutes have higher boiling points than do the pure solvents.

Although it may not seem as readily apparent, the same phenomenon (vapor pressure lowering) that causes the boiling-point elevation described above also leads to freezing-point depression. Figure 17.1 illustrates this effect.

The freezing point of a solution is the temperature at which the solvent in a solution and the pure solid solvent have the same vapor pressure. Vapor-pressure lowering results in freezing-point depression provided that the solution does not freeze as a solid solution. The solid that forms must be pure solvent.

Quantitatively, the magnitude of freezing-point depression is proportional to the amount of solute present in a given mass of solvent. The concentration unit employed is **molality** symbolized by a lowercase m and defined as the number of moles of solute per kilogram of solvent (see Equation 17.1).

$$m = \frac{\text{moles solute}}{\text{kg solvent}} \tag{17.1}$$

The difference between the freezing point of the solution and that of the solvent is called the freezing-point depression and is given the symbol Δt_f. If Δt_f

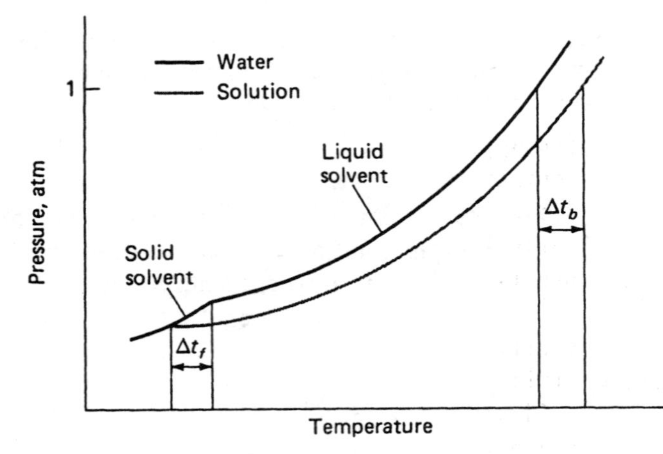

FIGURE 17.1 Vapor pressure diagram for water and aqueous solutions.

is calculated by subtracting the freezing temperature of the solvent from that of the solution, it will have a negative value (see Equation 17.2).

$$\Delta t_f = t_{f\,\text{solution}} - t_{f\,\text{solvent}} \tag{17.2}$$

The extent to which the freezing point is depressed for a solution of given molality is determined by the nature of the solvent. A 1.00 m aqueous solution would, ideally, have a freezing point of –1.86°C because the freezing point of pure water is 0.00°C. A 1.00 m solution in which camphor is the solvent would have a freezing point of 139°C, which represents a depression by 40°C (the freezing point of pure camphor is 179°C). Thus, the freezing-point depression constant, k_f, for water is –1.86°C/m, while k_f for camphor has a value of –40°C/m. The unit, °C/m, is identical to the unit, °C kg/mol. The relation between Δt_f, k_f, and m is shown in Equation 17.3.

$$\Delta t_f = k_f m \tag{17.3}$$

Equation 7.3 reveals that by measuring the freezing-point depression of a solution you can determine the solution's molality. Knowing this experimentally determined molality and knowing how the solution was prepared (i.e., what mass of solute was combined with what mass of solvent), you can obtain the molecular mass of the solute. A computational scheme for obtaining the molecular mass is shown in Equations 17.4 through 17.6.

$$m_{\text{expt}\ell} = \Delta t_f / k_f \tag{17.4}$$

$$(m_{\text{expt}\ell})(\text{kg solvent}) = \text{moles solute} \tag{17.5}$$

$$\frac{\text{mass solute}}{\text{moles solute}} = \text{molecular mass of solute} \tag{17.6}$$

In this experiment, you will determine the molecular mass of isopropyl alcohol by measuring the freezing-point depression observed in an aqueous solution of the alcohol. This approach may be used very successfully, even though isopropyl alcohol, like most alcohols, is more volatile than water. The experiment works because, at low temperatures, the vapor pressure of isopropyl alcohol is small enough (8 torr at 0°C) to be considered negligible.

PROCEDURE

Obtain a freezing-point depression apparatus like the one shown in Figure 17.2. If the thermometer has not been inserted into the stopper, insert it now. Most likely one side of the stopper will have been cut. The best way to insert the thermometer is to pull open the cut side of the stopper and place the thermometer inside the opening at an appropriate distance from the bottom of the thermometer.

Prepare an ice/water/rock salt bath by placing equal amounts of ice and rock salt and a small amount of tap water in the 1000-mL tall-form beaker and stirring the mixture. Check the temperature of the bath and be sure it is at least as low as –10°C. If it is not, pour off some of the water, add more ice and rock salt, and

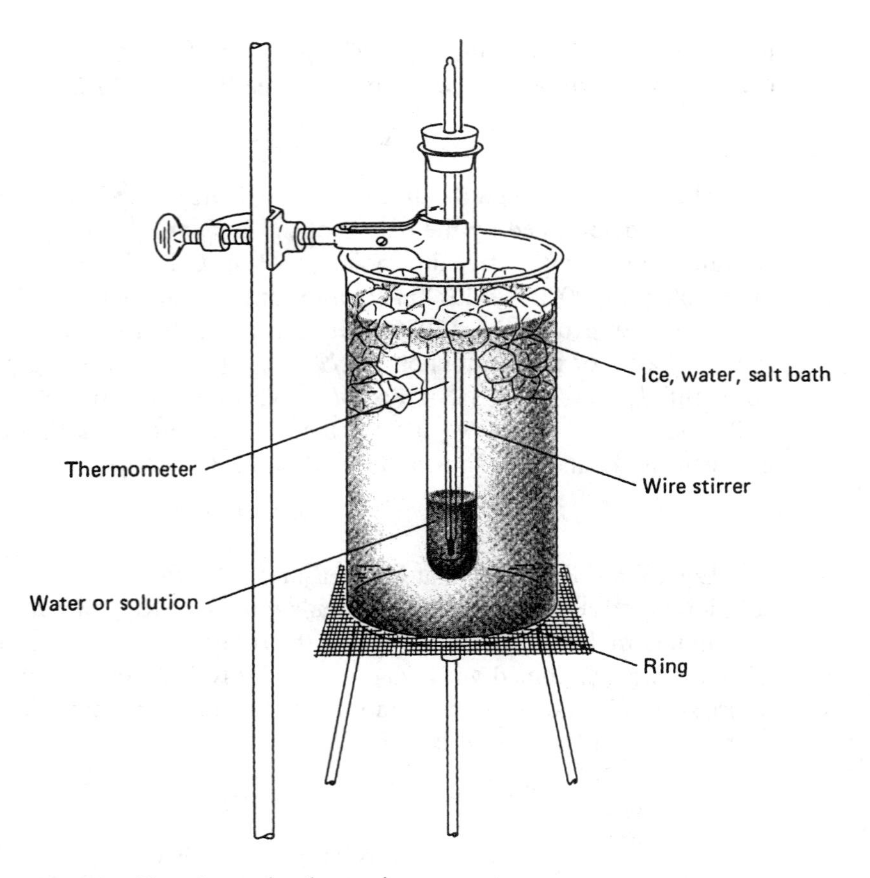

FIGURE 17.2 Freezing-point depression apparatus.

stir the mixture again. Continue this process until the temperature of the bath is –10°C. Measure 25.0 mL of distilled water with the graduated cylinder and transfer it completely into the freezing-point depression test tube. Stopper the test tube with the stopper holding the wire stirrer and thermometer. Place the test tube in the ice/water/rock salt bath. Move the stirrer up and down constantly to agitate the water, pausing regularly, at 30-second intervals, to record the time and the temperature. Continue until the temperature readings have been constant for five consecutive times or until the water has frozen sufficiently that stirring has become difficult. Allow the frozen water to thaw and return to room temperature. Then repeat the procedure given above.

After the water has thawed and returned to room temperature for a second time, use the buret to dispense 6 to 7 mL of isopropyl alcohol into the test tube. Record the exact volume taken. Stopper the test tube and place it in the bath. Repeat the process of agitation, recording the temperature every 30 seconds. Continue until five consecutive readings are the same or the mixture has frozen sufficiently that agitation becomes difficult. Allow the frozen mixture to thaw and return to room temperature. Then repeat the procedure.

If the water or alcohol/water mixture does not freeze at a constant temperature, plot the time/temperature data. The points should fall on two straight lines (see Figure 17.3) with the intersection of the lines marking the freezing point of the material under study. The shaded portions of the plots in Figure 17.3 indicate

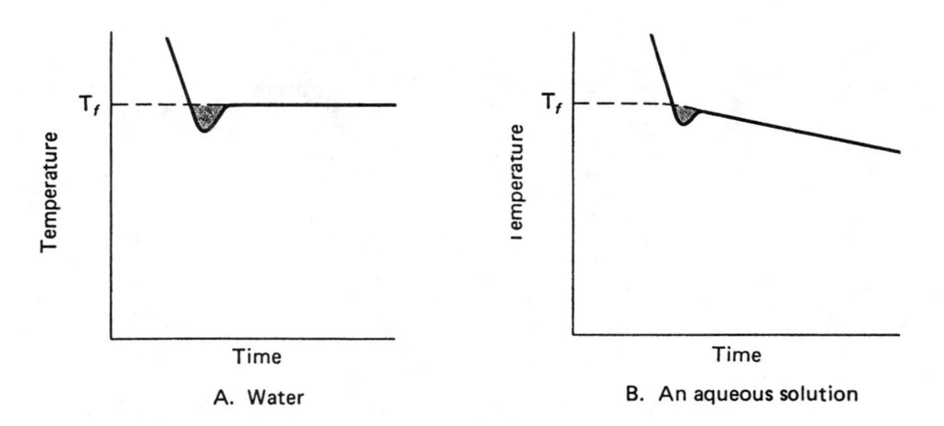

FIGURE 17.3 Cooling curves for (A) water and (B) for an aqueous solution.

supercooling, a phenomenon in which the liquid does not solidify until it is cooled to a temperature that is slightly below its freezing point.

Disposal of Reagents

Pour off the water from the bath and, if any solid rock salt remains, place the ice/rock salt mixture in styrofoam tubs for reuse by later classes.

The mixture of isopropyl alcohol and water may be diluted and poured down the drain.

Date ______ Name ______________________________ Section ______ Desk Number ______

PRE-LAB EXERCISES FOR EXPERIMENT 17

These exercises are to be completed after you have read the experiment but before you come to the laboratory to perform it.

1. A solution is prepared by mixing 2.17 g of an unknown nonelectrolyte with 225.0 g of chloroform is –63.5°C. The value of k_f for chloroform is –4.68°C/*m*. Find the molecular mass of the unknown.

2. Why is the freezing point of pure water determined experimentally rather than just assigned a value of 0°C?

Date ______ Name ______________________________ Section ______ Desk Number ______

SUMMARY REPORT ON EXPERIMENT 17

Time/Temperature Readings for Pure Water

TRIAL 1		TRIAL 2	
Time	*Temperature*	*Time*	*Temperature*

Date ______ Name ______________________________ Section ______ Desk Number ______

Final buret reading ____________________

Initial buret reading ____________________

Volume used, isopropyl alcohol ____________________

Time/Temperature Readings for Mixture

TRIAL 1		TRIAL 2	
Time	*Temperature*	*Time*	*Temperature*

Date ______ Name ________________________________ Section ________ Desk Number ________

	TRIAL 1	TRIAL 2
Density of isopropyl alcohol	0.785 g/mL	
Mass of isopropyl alcohol used		
t_f, water		
t_f, mixture		
Δt_f		
m_{exptl}		
Moles alcohol in sample		
Molecular mass of isopropyl alcohol		

EXPERIMENT 18

Acid/Base Titration of Ascorbic Acid

Laboratory Time Required

Three hours

Special Equipment and Supplies

Teflon-stoppered buret
Buret clamp
Powdered ascorbic acid
Pineapple or grapefruit juice
0.1 M sodium hydroxide
Phenolphthalein indicator

Safety

Remember, bases such as sodium hydroxide can burn your skin and are especially hazardous to the eyes.

Although it may be tempting, **do not** drink any unused portion of your juice. Drinking is not permitted in the lab. Any unused juice must be discarded.

First Aid

Following skin contact with sodium hydroxide, wash the area thoroughly with water.

If sodium hydroxide solution gets in your eyes, rinse them thoroughly with water (at least 20 minutes of flushing with water is recommended). Then seek medical attention.

Ascorbic acid is a water-soluble vitamin (vitamin C) that must be consumed regularly to ensure proper body function. Lack of vitamin C may result in scurvy, a disease with symptoms that include diarrhea, ulcerated and bleeding gums, and hemorrhage. Sailors on long sea voyages used to be very susceptible to this disease (have you ever seen old films in which a sea captain refers to his crew as a "scurvy lot"?). Scurvy was eliminated from British ships with the introduction of "limes" (which we call "lemons" today) into the sailors' daily rations, leading to the nickname "Limey."

Citrus fruits other than limes and lemons, tomatoes, cantaloupes, and fresh vegetables are also good sources of vitamin C. Although most people can obtain their minimum daily requirement (MDR), equal to 60 mg of vitamin C per day, from these sources quite readily, many people prefer to consume larger quantities of this vitamin. The practice of consuming megadoses (500 mg or more) of vitamin C was popularized in the 1960s by Linus Pauling, Nobel Laureate in Chemistry, who claimed that consumption of larger doses of ascorbic acid would help cure the common cold and prevent certain forms of cancer. Recent demographic studies have suggested that men who took daily megadoses of vitamin C had a significantly longer life expectancy than men who did not. The questions, however, remain: is it necessary to consume vitamin C in pill form, or can one obtain the desired dosages via ingestion of foods naturally rich in vitamin C? How do we determine the amount of vitamin C contained in a given portion of food? To answer these questions, you will perform an acid/base titration of pineapple or grapefruit juice. You may also combine your study of the vitamin C content of the juice with a redox titration, as described in Experiment 27.

PRINCIPLES

The formula for ascorbic acid is $C_6H_8O_6$. Only one of ascorbic acid's hydrogens is capable of reacting with a base in a titration. To emphasize this point, the formula of ascorbic acid is often written as $HC_6H_7O_6$. The salt produced in the titration of ascorbic acid with sodium hydroxide is sodium ascorbate. The balanced chemical equation corresponding to this reaction is given in Equation 18.1.

$$HC_6H_7O_6 + NaOH \longrightarrow H_2O + NaC_6H_7O_6 \tag{18.1}$$

This equation suggests that we can determine the amount of vitamin C in a sample of juice via a titration with a standardized solution of sodium hydroxide. Because the mole ratio of ascorbic acid to sodium hydroxide is 1:1, the number of moles of acid present in the sample is equal to the number of moles of base that must be added to reach the equivalence point of the titration (see Equation 18.2).

$$M \times V \text{ (in liters)} = \text{moles (base)} = \text{moles (acid)} \tag{18.2}$$

One problem in performing the titration of acids with stock solutions of sodium hydroxide is that the concentration of sodium hydroxide solutions tends to change with time. This results from the reaction between aqueous sodium hydroxide and atmospheric carbon dioxide, which produces sodium hydrogen carbonate (sodium bicarbonate). For this reason, you will standardize your NaOH solution by a procedure similar to the one described in Experiment 8.

PROCEDURE

A. Standardization of Sodium Hydroxide

Place an accurately weighed 0.5 g sample of powdered ascorbic acid in a clean,

250-mL Erlenmeyer flask. Dissolve the ascorbic acid in approximately 50 mL of distilled water. Add 3 drops of phenolphthalein indicator.

Prepare a Teflon-stoppered buret for titration by rinsing it with two small portions of distilled water, followed by two 10-mL portions of the sodium hydroxide solution. Fill the buret and follow the usual procedures for eliminating air bubbles and setting the initial level (see Experiment 8). Record the initial buret reading.

If the approximate concentration of the NaOH is known, estimate the approximate volume of this solution that will be required to react completely with the ascorbic acid. Run somewhat less than this amount into the flask containing the ascorbic acid, while stirring by swirling the flask. Use your wash bottle to clean the flask walls of drops of base that may have splattered out of the titration mixture. Continue adding the NaOH solution. As you approach the volume estimated to be needed for complete reaction, add the NaOH more slowly, while continuing to swirl the flask and wash down the walls. Stop the titration when the addition of a single drop of base changes the color of the solution to a light pink, indicating you have reached the end point. Record the final buret reading.

Repeat the titration. Report the molarity of the NaOH from each trial and the average molarity.

B. Titration of a Sample of Juice

Use a graduated cylinder to measure a 20-mL sample of pineapple or grapefruit juice. Transfer the juice to a clean, 250-mL Erlenmeyer flask. Add roughly 75 mL of distilled water to the juice.

Add 3 drops of phenolphthalein to the flask containing the diluted sample. Titrate, as above, with sodium hydroxide, stopping when the titration mixture has acquired a permanent, pinkish-orange color.

Repeat the titration. Calculate the amount of acid in each sample of juice. If all the acid is ascorbic acid, how many mg of vitamin C are there in a 6-oz. serving of juice?

Disposal of Reagents

Excess ascorbic acid should be placed in the containers used for solid waste. Solutions should be neutralized and diluted. They may then be flushed down the drain.

Questions

1. Could the titration described in this experiment be performed with orange juice or grape juice? Explain your answer, briefly.
2. How did citrus fruits get their name? Does this have any bearing on the results you have obtained in performing this experiment?
3. List the differences you observed between the titration of powdered ascorbic acid and the titration of juice. How might the characteristics of the juice sample affect the results of your titration?

Date ______ Name ______________________ Section ______ Desk Number ______

PRE-LAB EXERCISES FOR EXPERIMENT 18

These exercises are to be completed after you have read the experiment, but before you come to the laboratory to perform it.

1. A 0.5013 g sample of ascorbic acid required 22.65 mL of sodium hydroxide before the phenolphthalein end point was reached. Another 0.4985 g sample of the same material required 22.54 of the same base.

   a. Calculate the molarity of the sodium hydroxide based on the results of these trials. Do the two trials agree with acceptable precision?

   b. What is the uncertainty associated with the average molarity of the sodium hydroxide as obtained from these trials? Assume an uncertainty of ±0.0005 g for each mass and ±0.06 mL for each volume.

2. A certain juice claims that a 6-oz. serving provides the MDR of ascorbic acid. Calculate the molarity of ascorbic acid in the juice.

Date ______ Name ______________________ Section ______ Desk Number ______

SUMMARY REPORT ON EXPERIMENT 18

A. Standardization of the Sodium Hydroxide

	Trial 1	*Trial 2*	*Trial 3**
Mass of vial (before sample removal)	______	______	______
Mass of vial (after sample removal)	______	______	______
Mass of sample	______	______	______

	Trial 1	*Trial 2*	*Trial 3**
Final buret reading, NaOH	______	______	______
Initial buret reading, NaOH	______	______	______
Volume used, NaOH	______	______	______
Molarity of NaOH	______	______	______
Average molarity of NaOH		______	

B. Titration of a Sample of Juice

	Trial 1	*Trial 2*	*Trial 3**
Juice used	______________________		
Volume of juice	______	______	______
Final buret reading, NaOH	______	______	______
Initial buret reading, NaOH	______	______	______
Volume of NaOH used	______	______	______
Amount of acid	______	______	______
mg Ascorbic acid in a 6 oz. serving of juice‡	______	______	______

*Optional.
‡ Calculate on the assumption that the only acid in the juice is ascorbic acid.

EXPERIMENT 19

Absorption Spectroscopy and Beer's Law

Laboratory Time Required Three hours

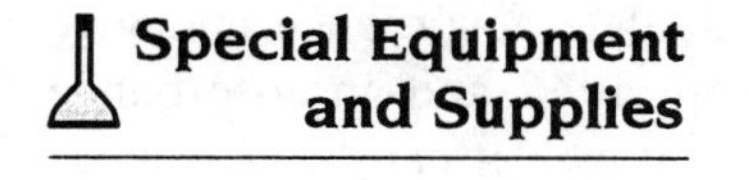

Special Equipment and Supplies

- Spectrophotometer
- Buret or Mohr pipet
- Volumetric flask, 100-mL
- Pipet bulb
- Spectrophotometer cells
- $CoCl_2{\cdot}6H_2O$
- $CuSO_4{\cdot}5H_2O$
- $Ni(NO_3)_2{\cdot}6H_2O$

Safety The $CoCl_2{\cdot}6H_2O$ and $Ni(NO_3)_2{\cdot}6H_2O$ are classified as **poisons** and, consequently, should not be ingested or discarded in the sanitary sewer. Clean up spills promptly and avoid exposing your skin to the solid reagents or their solutions.

First Aid **Rinse the exposed area thoroughly with water.**

Light brings us news of the universe. Our perception of light is mostly of the visible range of radiant energies, the slot between roughly 400 and 800 nm within the electromagnetic spectrum. But because most light is invisible, it must be detected by other means. Through the transmission, absorption, and scattering of light, we can gain information about the hidden world beyond what usual observations reveal.

In this experiment, you will study the Beer-Lambert Law, one of the most fundamental and widely applied spectroscopic laws. In the first part of the experiment, you will determine where in the visible range of the electromagnetic spectrum a solution of a transition metal salt absorbs light. In the second part, you will determine the nature of the relationship between absorbance and the concentration of the solution. In the third part of the experiment, you will study a mixture of the solutions and determine how the appearance of the mixture relates to its spectrum.

PRINCIPLES

Spectroscopy is the study of the interaction of electromagnetic radiation with matter. In spectroscopy, two terms are inescapable: transmittance and absorbance. **Transmittance** (T) is defined as the ratio of the intensity of light after it passes through the medium being studied (I) to the intensity of light before it encountered the medium (I_o), as shown in Equation 19.1.

$$T = I/I_o \qquad (19.1)$$

Spectroscopists more commonly refer to percent transmittance (%T), which is simply: $I/I_o \times 100\%$. Often the same spectroscopic information that is reported as the percent transmittance is more conveniently expressed as absorbance (A):

$$A = -\log(I/I_o) \qquad (19.2)$$

Note that

$$A = 2 - \log(\%T) \qquad (19.3)$$

If one knows the percent transmittance, one can calculate absorbance and vice versa. Some non-digital spectrophotometers have both a %T and an absorbance scale displayed on a meter. Because the %T scale is linear, it can be read with good precision over the entire range of transmittances. However, the absorbance scale is a logarithmic scale and cannot be read with precision at high absorbance values. Therefore, if the absorbance is larger than 0.7, it is preferable to calculate the absorbance, using the %T, rather than to read the absorbance directly. Modern digital spectrophotometers may be programmed to display either absorbance or percent transmittance by simply touching a mode selection button.

An operational statement of the Beer-Lambert Law can be represented as

$$A = \varepsilon c \ell \qquad (19.4)$$

where c is the concentration of some absorbing substance in solution, ℓ is the optical path length, and ε is the molar absorptivity. The molar absorptivity is a constant that depends on the nature of the absorbing system (the solute–solvent combination) and the wavelength of the light passing through it. A plot that shows the dependence of A (or ε) on wavelength is called a **spectrum**. In one part of this experiment, you will determine the visible spectrum of an aqueous solution of cobalt chloride, copper sulfate, or nickel nitrate.

When absorbance measurements are made at a fixed wavelength in a cell of constant path length, ε and ℓ are constant. Therefore, the absorbance, A, should be directly proportional to c, the concentration of the solute. A solution that shows such a linear relation between A and c is said to obey the Beer-Lambert Law, which is the optimum situation for a spectrophotometric method of analysis. In the second part of this experiment, you will attempt to verify whether some or all the aqueous solutions studied obey the Beer-Lambert Law.

It is also possible, although less desirable, to perform spectrophotometric analyses in systems that deviate from the Beer-Lambert Law, as the Cu^{2+}/NH_3 system is shown to do in Figure 19.1.

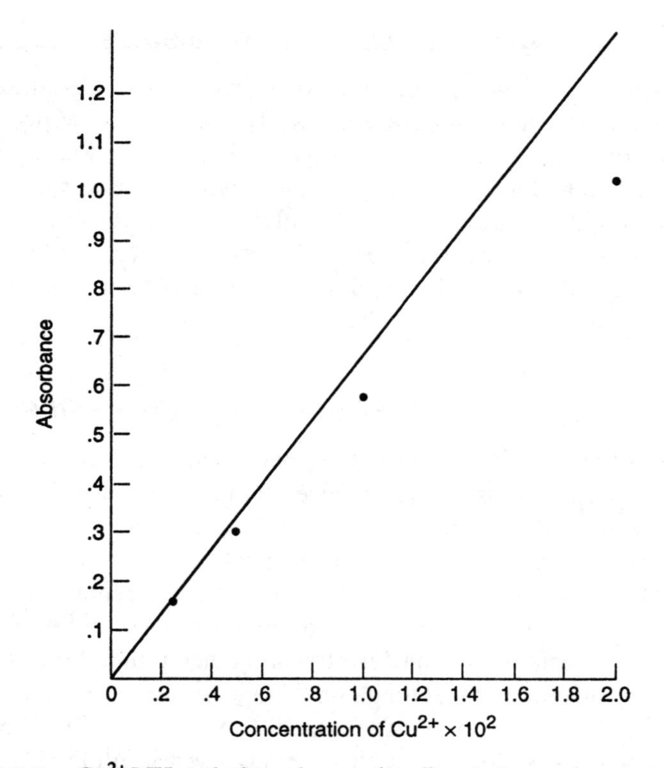

FIGURE 19.1 Cu^{2+}/NH_3 solutions do not obey Beer's Law. A proper calibration curve would be a smooth line drawn through the data points.

The deviation results from the equilibrium distribution of Cu^{2+} between the absorbing species $Cu(NH_3)_4^{2+}$ and several nonabsorbing complexes of general formula $Cu(NH_3)_{4-i}(H_2O)_i^{2+}$. In such cases, the absorbance versus concentration measurements result in a calibration curve, rather than the linear plot predicted by the Beer-Lambert Law.

PROCEDURE

A. Operation of the Spectrophotometer

The large variety of spectrometers available in undergraduate teaching laboratories makes it impossible for us to give operating instructions for each type. Your instructor will demonstrate appropriate techniques for the instrument you will use in this experiment.

You will be asked to determine the wavelengths of monochromatic light which correspond to certain colors of the visible spectrum. In some instruments you may accomplish this task by placing a spectrophotometer cell containing a 3/4-inch piece of chalk in the sample compartment, and slowly changing the wavelength over the 400- to 700-nm range while looking into the test tube. You may need to raise, lower, or turn the test tube to obtain the best reflection from the chalk surface. In other spectrophotometers, you may be able to hold a lint-free tissue in the path of the light while changing the wavelength; the color of the light should be clearly visible on the tissue. Record your observations on the Summary Report Sheet.

B. Preparation of 0.200 M Aqueous Salt Solution

Obtain a salt assignment from your instructor. Calculate the mass of your assigned salt needed to prepare 100.00 mL of a 0.200 M solution of your salt and weigh the necessary mass on the analytical balance. Place your sample in a clean beaker and add about 50 mL of water. Swirl the beaker until the solid has totally dissolved; then carefully pour the solution into a 100-mL volumetric flask. Rinse the beaker with several small volumes of water and add each rinsing to the solution in the flask. Finally, dilute the solution in the volumetric flask to exactly 100 mL, and mix well.

C. Determination of the Spectrum of the Salt Solution

Obtain two matched spectrophotometer cells and fill one of them half full with distilled water to use as a reference. Set the wavelength of the spectrophotometer at 390 nm; then insert the cell in the cell compartment, taking care to align the reference marks. Close the compartment door and set the instrument to exactly 100%*T* or zero absorbance. Remove the reference cell and adjust the zero setting, if necessary. (Note: Some modern spectrophotometers do not require manual adjustment of the zero transmittance point.) Check the zero and 100%*T* (or zero absorbance) settings until they are reproducible. Next rinse the sample cell with a small portion of the 0.200 M solution and discard the rinsing. Fill the cell about two-thirds full with solution, insert it in the sample compartment, and align the reference marks. Close the sample door and read the %*T* (or absorbance) as accurately as possible.

Change the wavelength to 400 nm and again adjust the zero and 100% *T* (or zero absorbance) settings using the reference cell as described in the previous paragraph. Replace the reference cell with the sample cell and record the %*T* (or absorbance) of the solution. Repeat this procedure at 10-nm intervals over the 390 to 600 nm range.

Prepare a plot of absorbance versus wavelength, using Equation 19.3 if necessary, to convert from %*T* to absorbance. Determine the wavelength of maximum absorbance for your salt. Obtain the absorbance maxima for the other salts from your classmates who studied them.

D. Beer's Law

Place four clean, dry test tubes in a test tube block. Use a buret or Mohr pipet to dispense 6 mL, 4 mL, 2 mL, and 1 mL volumes of the your 0.200 M solution into the four test tubes. In the same way, add 2, 4, 6, and 7 mL volumes of distilled water, respectively, to the test tubes, and swirl or thump the tubes to mix the solutions. Each test tube should now contain 8 mL of solution. Calculate the concentration of your assigned salt in each test tube.

Set the spectrophotometer wavelength to the absorbance maximum and adjust the zero and 100% *T* (or zero absorbance) setting, using the reference cell as usual. Rinse and then fill the sample cell with one of the diluted cobalt chloride solutions and measure the %*T* (or absorbance) of the solution. Record your data on the Summary Report Sheet. Repeat this procedure for each of the your diluted solutions. Plot absorbance (obtained from %*T* via Equation 19.3 if necessary) versus the concentration of your salt for the five solutions (including the original 0.200 M solution). If the data display a linear relationship, draw the best straight line through the experimental points, including the origin.

Obtain absorbance (or %*T*) and concentration data from classmates who studied the other salts and prepare plots of these data as well. Fit the best straight line to each set of data to determine its linearity and comment on whether the relationship between absorbance and concentration is linear or not.

E. Determination of the Spectrum of a Mixture

Work with classmates who studied salts other than the one you studied to prepare a mixture specified by your instructor. Note the color of your mixture on the Summary Report Sheet and obtain its spectrum by the procedure specified in Procedure C.

Disposal of Reagents

Any solutions containing cobalt or nickel ions should be placed in labeled collection bottles.

Questions

1. If the solution obeys the Beer-Lambert Law, why is it better to plot the absorbance versus concentration data for several concentrations, rather than using only a single solution plus the origin to determine the straight line?
2. What does it mean if the experimental points obviously curve away from the expected straight line?
3. Calculate the molar absorptivity, at the absorbance maximum, for each of the salts using a path length of $^1/_2$ inch (1.27 cm) for test tube spectrophotometer cells or 1.00 cm for square spectrophotometer cells.
4. Which of the salts, if any, give spectra that do not obey Beer's Law?
5. Although Cu^{2+}/NH_3 solutions do not obey Beer's Law, they are usually used in preference to simple aqueous solutions of Cu^{2+} when copper ion concentration is being determined spectrophotometrically. Explain why.

Date ______ Name ______________________________ Section ______ Desk Number ______

PRE-LAB EXERCISES FOR EXPERIMENT 19

These exercises are to be completed after you have read the experiment but before you come to the laboratory to perform it.

1. Why is it necessary to recalibrate the spectrophotometer against a reference cell each time the wavelength is changed?

2. If a solution measures 52.3% *T*, what is the absorbance of the solution?

3. A solution of K_2CrO_4 has an absorbance of 0.512 in a cell of 1.00 cm pathlength at a wavelength of 370 nm. The value of ε (the molar absorptivity) is 4.84×10^3 M^{-1} cm^{-1} for K_2CrO_4 at that wavelength. What is the concentration of the solution?

Date ______ Name ______________________________ Section ______ Desk Number ______

SUMMARY REPORT ON EXPERIMENT 19

A. Operation of the Spectrophotometer

LIGHT TRANSMITTED	WAVELENGTH, nm
Red	________
Orange	________
Yellow	________
Green	________
Blue	________
Violet	________

B. Preparation of 0.200 M Aqueous Solution

Assigned salt ________

Mass of beaker plus salt ________

Mass of beaker ________

Mass of salt ________

C. Determination of the Spectrum of Your 0.200 M Solution

Wavelength	390	400	410	420	430	440	450
% Transmittance	____	____	____	____	____	____	____
Absorbance	____	____	____	____	____	____	____

Wavelength	460	470	480	490	500	510	520
% Transmittance	____	____	____	____	____	____	____
Absorbance	____	____	____	____	____	____	____

Wavelength	530	540	550	560	570	580	590	600
% Transmittance	____	____	____	____	____	____	____	____
Absorbance	____	____	____	____	____	____	____	____

What is the wavelength of the absorbance maximum? _____

Other salts

Name of salt ________________

Obtained from: ________________

Mass used in preparing solution ________________

Wavelength of maximum absorbance ________________

Name of salt ________________

Obtained from: ________________

Mass used in preparing solution ________________

Wavelength of maximum absorbance ________________

Date ______ Name ______________________________ Section ______ Desk Number ______

D. Beer's Law

Salt ______________________

Wavelength used ______________________

Concentration (M)	0.200	____	____	____	____
% Transmittance	____	____	____	____	____
Absorbance	____	____	____	____	____

Salt ______________________

Wavelength used ______________________

Concentration (M)	0.200	____	____	____	____
% Transmittance	____	____	____	____	____
Absorbance	____	____	____	____	____

Salt ______________________

Wavelength used ______________________

Concentration (M)	0.200	____	____	____	____
% Transmittance	____	____	____	____	____
Absorbance	____	____	____	____	____

E. Determination of the Spectrum of a Mixture

Specifications for the mixture ______________________________

____

Appearance of the mixture ______________________________

____

Wavelength	390	400	410	420	430	440	450
% Transmittance	____	____	____	____	____	____	____
Absorbance	____	____	____	____	____	____	____

Wavelength	460	470	480	490	500	510	520
% Transmittance	____	____	____	____	____	____	____
Absorbance	____	____	____	____	____	____	____

Wavelength	530	540	550	560	570	580	590	600
% Transmittance	____	____	____	____	____	____	____	____
Absorbance	____	____	____	____	____	____	____	____

Date ______ Name ________________________________ Section ______ Desk Number ______

Is the spectrum of the mixture what you would have expected based on its color? Explain your answer briefly.

Chemical Equilibria

Laboratory Time Required

Three hours

Special Equipment and Supplies

Buret
Buret clamp
Test tubes
Stoppers
Pipet
Pipet bulb

Glacial acetic acid
Ethyl acetate
Ethanol
18 M sulfuric acid
0.1 M sodium hydroxide
Potassium hydrogen phthalate (KHP)
Phenolphthalein

Safety

Both glacial acetic acid and sulfuric acid are **corrosive** and capable of causing burns to the skin or possible blindness. Sodium hydroxide presents a similar hazard. Take every precaution against skin or eye exposure to chemicals. Safety glasses with side shields and laboratory aprons are mandatory.

Ingestion of glacial acetic acid, sulfuric acid, or sodium hydroxide is **extremely** dangerous.

CAUTION MOUTH PIPETING IS FORBIDDEN

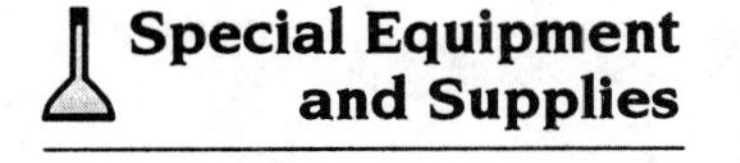

First Aid

Rinse the exposed skin or eyes thoroughly with water. See a doctor if chemicals have entered the eyes of if skin irritation is extensive.

In principle, all chemical reactions are reversible. It is frequently found that the conversion of reactants to products is incomplete because the product formed decomposes, once again producing reactant molecules. When the rates of product formation and product decomposition are the same, the system has reached a state of **dynamic equilibrium** in which there is no net change in the concentration of any of the materials (reactants or products) involved in the reaction.

In this experiment, the esterification reaction between acetic acid and ethanol to produce ethyl acetate and water is studied. The reverse of this reaction—the hydrolysis of ethyl acetate—is also studied, to illustrate the point that equilibrium may be approached from either side.

PRINCIPLES

Alcohols react with carboxylic acids to produce esters. Esters often have very pleasant odors and tastes and are found in many flowers and fruits. The reaction of ethanol (ethyl alcohol) with acetic acid produces ethyl acetate, which occurs naturally in bananas.

The process of esterification is acid-catalyzed. In addition to the ester, water is a product of the reaction of the alcohol and carboxylic acid. In the reverse reaction, called **hydrolysis**, the elements of water are added to the ester, resulting in the regeneration of the acid and the alcohol. Like esterification, hydrolysis is acid-catalyzed. The reversible reaction of ethanol with acetic acid is shown in Equation 20.1.

$$\underset{\text{acetic acid}}{CH_3CO_2H} + \underset{\text{ethanol}}{CH_3CH_2OH} \overset{H^+}{\rightleftharpoons} \underset{\text{ethyl acetate}}{CH_3CO_2CH_2CH_3} + \underset{\text{water}}{H_2O} \quad (20.1)$$

The equilibrium constant for this reaction is given in Equation 20.2. The square brackets indicate that concentrations are in units of moles/Liter.

$$K_c = \frac{[\text{ethyl acetate}][\text{water}]}{[\text{acetic acid}][\text{ethanol}]} \quad (20.2)$$

The concentrations of acetic acid in the esterification mixture and in the hydrolysis mixture are easily determined via titration with standardized sodium hydroxide. Concentrations of the other reactants and products may be determined from initial conditions and the stoichiometry of the reaction. Information given in Table 20.1 may prove useful in obtaining the necessary concentrations.

TABLE 20.1 Molar Masses and Densities of Reagents.

Reagent	Molar Mass	Density
Acetic acid	60.0	1.05 g/mL
Ethyl acetate	88.1	0.90 g/mL
Water	18.0	1.00 g/mL
Ethanol	46.0	0.80 g/mL

PROCEDURE

A. Standardization of Sodium Hydroxide Solutions

Because carbon dioxide from the air has some tendency to react with sodium hydroxide in dilute aqueous solution, it will be necessary to begin each week's work by standardizing the NaOH (aq). This is easily accomplished via titration of potassium hydrogen phthalate (commonly denoted as KHP). Consult Experiment 8 for the details of the standardization procedure. Note that the molar mass of KHP is 204.22 and that one mole of NaOH reacts with one mole of KHP. Only one buret is needed for the standardization and subsequent titration of the hydrolysis and esterification mixtures. Record the standardization data on the Summary Report Sheet.

B. Initial Conditions of Esterification Reaction

Weigh (**CAUTIOUSLY!**) 1.4–1.6 g of glacial acetic acid into a clean dry test tube. Add 10.0 mL of ethanol and 10.0 mL of deionized water. Stopper the test tube and invert it several times to ensure a homogeneous mixture. Use a 2-mL pipet to transfer a sample of the mixture to a clean flask. Add 10 mL of distilled water and 3 drops of phenolphthalein to the solution in the flask. Then, titrate to the phenolphthalein end point with the 0.1 M sodium hydroxide solution that you have standardized. Record the titration data on the Summary Report Sheet.

CAREFULLY add 5 drops of concentrated (18 M) sulfuric acid to the test tube containing the esterification mixture. Mix thoroughly once again. Then use the 2-mL pipet to transfer another sample of the esterification mixture to a new clean flask. As before, add 10 mL of distilled water and 3 drops of phenolphthalein indicator to the solution in the flask. Titrate this second sample with the standardized base and record the titration data on the Summary Report Sheet.

Obviously, the titration of the mixture containing the sulfuric acid catalyst will require more base than the titration of the esterification mixture before the catalyst was added. The portion of the acidity contributed by the sulfuric acid catalyst will remain constant throughout the course of the reaction because a catalyst is not consumed as the reaction proceeds.

Label and stopper the test tube containing the esterification mixture and place it in your drawer until your next laboratory class.

C. Initial Conditions of Hydrolysis Reaction

Weigh 2.1–2.3 g of ethyl acetate into a second test tube. Add 10.0 mL of ethanol and 10.0 mL of deionized water. Then, add 5 drops of concentrated H_2SO_4, stopper the tube, and mix its contents by inverting the tube several times. Pipet a 2-mL sample into a clean flask, add 10 mL of water, 3 drops of phenolphthalein, and titrate, as before, with standardized NaOH to determine the concentration of protons provided by the sulfuric acid. Record the titration data on the Summary Report Sheet.

Label and stopper the test tube containing the hydrolysis mixture and allow it to remain in your drawer until your next laboratory class.

D. Titration of Equilibrium Mixtures

At your next laboratory class, repeat the standardization of the sodium hydroxide solution. Record the standardization data on the Summary Report Sheet.

Pipet a 2-mL sample from each of your stored test tubes into a separate clean flask. Add 10 mL of distilled water and 3 drops of phenolphthalein to each flask. Titrate the contents of each flask with your standardized NaOH to find the equilibrium concentration of acetic acid in each. Remember that you must allow for the acidity that arises from the presence of the sulfuric acid catalyst.

Questions

Use your data to calculate K_c, as defined in Equation 20.2, for each mixture. Do your values agree? If they do not, suggest reasons for the discrepancy.

Date ______ Name ______________________________ Section ______ Desk Number ______

PRE-LAB EXERCISES FOR EXPERIMENT 20

These exercises are to be completed after you have read the experiment, but before you come to the laboratory to perform it.

Initial Conditions

1. When one studies equilibria of aqueous solutions, one usually omits $[H_2O]$ from the equilibrium expression. Why is $[H_2O]$ included in K_c for the esterification/hydrolysis studied in this experiment?

2. What should happen to the acidity of the esterification mixture as the system moves from initial conditions to equilibrium? To the acidity of the hydrolysis mixture?

3. Calculate [ethanol] and [water] in each initial mixture.

Equlibrium Mixture

Pat Student used 27.26 mL of 0.1025 M NaOH to titrate 2.00 mL of esterification mixture to the phenolphthalein end point. After sulfuric acid was added to the mixture, Pat used 34.61 mL of NaOH to titrate a second 2.00-mL sample. Pat left the mixture to equilibrate for one week and then found it took 30.03 mL of 0.1037 M NaOH to titrate a fresh 2.00-mL sample. Find the equlibrium concentration of acetic acid in Pat's esterification mixture.

Date ______ Name ______________________________ Section ______ Desk Number ______

SUMMARY REPORT ON EXPERIMENT 20

Determining the Initial Conditions

Standardization of the Sodium Hydroxide

	Trial 1	*Trial 2*	*Trial 3**
Mass of vial (before removal of KHP)			
Mass of vial (after removal of KHP)			
Mass of KHP			

	Trial 1	*Trial 2*	*Trial 3**
Final buret reading			
Initial buret reading			
Volume used			
Molarity of NaOH			
Average molarity of NaOH			

Preparation of the Esterification Mixture

Mass of container plus acetic acid ______________

Mass of container ______________

Mass of acetic acid ______________

Preparation of the Hydrolysis Mixture

Mass of container plus ethyl acetate ______________

Mass of container ______________

Mass of ethyl acetate ______________

Titration of Esterification Mixture before the addition of H_2SO_4

Final buret reading ______________

Initial buret reading ______________

Volume of NaOH used ______________

Titration of Esterification Mixture after the addition of H_2SO_4

Final buret reading ______________

Initial buret reading ______________

Volume of NaOH used ______________

Titration of the Hydrolysis Mixture

Final buret reading ______________

Initial buret reading ______________

Volume of NaOH used ______________

*Optional

Date ______ Name ______________________________ Section ______ Desk Number ______

Determining the Equilibrium Conditions

Standardization of the Sodium Hydroxide

	Trial 1	*Trial 2*	*Trial 3**
Mass of vial (before removal of KHP)	______	______	______
Mass of vial (after removal of KHP)	______	______	______
Mass of KHP	______	______	______

	Trial 1	*Trial 2*	*Trial 3**
Final buret reading	______	______	______
Initial buret reading	______	______	______
Volume used	______	______	______
Molarity of NaOH	______	______	______
Average molarity of NaOH		______	

Titration of the Esterification Mixture

Final buret reading ______________

Initial buret reading ______________

Volume of NaOH used ______________

Titration of the Hydrolysis Mixture

Final buret reading ______________

Initial buret reading ______________

Volume of NaOH used ______________

*Optional

Complete the table shown below. Attach a separate sheet showing your calculations.

	[acetic acid]	[ethanol]	[ethyl acetate]	[water]
Initial	______	______	______	______
Change	______	______	______	______
Final	______	______	______	______

From the esterification mixture K_c = ______________

From the hydrolysis mixture K_c = ______________

EXPERIMENT 21

Determination of the Dissociation Constant of an Acid/Base Indicator

Laboratory Time Required

Three hours

Special Equipment and Supplies

Spectrophotometer
Sample cells (cuvets)
Buret
Mohr pipets
Buffer solutions, varying in pH from 3.5 to 5.5
1 M HCl Solution
1 M NaOH Solution
Indicator solutions

Safety

Acids and bases are harmful to the skin and eyes. Avoid contact with acids and bases.

First Aid

Wash off any acids or bases spilled on the skin with copious amounts of water. Should acid or base get in your eyes, flush the chemicals out by washing your eyes with water for 20 minutes. Seek medical attention.

An acid/base indicator is a weak acid or weak base whose protonated (HIn) and deprotonated (In^-) forms have different colors in aqueous solution. This difference in colors is obviously necessary for the substance to be used as a visual indicator. Because the different colors result from the two forms of the indicator differing in their ability to absorb specific wavelengths of light, the absorption of light by an indicator solution may be used to study its ionization reaction. In this experiment, you will use absorption spectrophotometry to measure the dissociation constant for one of several indicators.

PRINCIPLES

The acid/base indicators to be studied in this experiment are weak acids, which undergo dissociation reactions of the type shown in Equation 21.1. The corresponding equation for K_a for these indicators is shown in Equation 21.2.

$$HIn + H_2O \rightleftharpoons H_3O^+ + In^- \quad (21.1)$$

and

$$K_a = \frac{[H_3O^+][In^-]}{[HIn]} \quad (21.2)$$

Because HIn and In^- have different absorptivities at a selected wavelength, absorption spectrophotometry may be used to measure the change in concentration with pH of each form of the indicator. From this information, we can calculate the dissociation constant of the indicator.

Assume that the Beer-Lambert Law is obeyed by both HIn and In^-. If that is so, the relation $A = \varepsilon cl$ may be employed to study solutions containing HIn or In^- or both. (As noted in Experiment 19, the Beer-Lambert Law relates the absorbance, A, of a solution to the absorbtivity, ε, of a species that is present in solution at some concentration, c, when light of a particular wavelength traverses a path of length, l, through the solution.)

In the course of this experiment, you will study several solutions. Each will contain the same amount of indicator, but will differ from the others in pH. For instance, one solution will contain a concentration, c, of indicator in the presence of strong acid. In this low pH mixture, virtually all of the indicator will be protonated (i.e., will be in the HIn form). The absorbance of this solution is given in Equation 21.3.

$$A_{\text{low pH}} = A_{\text{HIn}} = \varepsilon_{\text{HIn}} c \ell \quad (21.3)$$

A second solution will contain a concentration, c, of indicator in the presence of strong base. In this high pH mixture, virtually all of the indicator will be deprotonated (i.e., will be in the In^- form). The absorbance of this solution is given by Equation 21.4.

$$A_{\text{high pH}} = A_{\text{In}} = \varepsilon_{\text{In}} c \ell \quad (21.4)$$

In a mixture buffered at intermediate pH, substantial amounts of both HIn and In^- may be present. The concentration of HIn and In^- are then equal to $X_{\text{HIn}}c$ and $(1 - X_{\text{HIn}})c$, respectively, where X_{HIn} represents the mol fraction of indicator in the HIn form and $(1 - X_{\text{HIn}})$ represents the mol fraction of indicator in the In^- form. The absorbance of such a mixture will be the sum of the absorbances of HIn and In^-, as shown in Equation 21.5.

$$A_{\text{intermediate pH}} = A_{\text{HIn}} + A_{\text{In}} = \varepsilon_{\text{HIn}} X_{\text{HIn}} c \ell + \varepsilon_{\text{In}}(1 - X_{\text{HIn}}) c \ell \quad (21.5)$$

The object of this experiment is to evaluate X_{HIn} for a solution of known pH. Once a value of X_{HIn} has been obtained, it is easy to determine the concentrations of HIn and In^- and evaluate K_a. A simple rearrangement of Equation 21.5 shows the relationship between the absorbances of the various solutions and the value of X_{HIn} (see Equations 21.6 and 21.7).

$$A_{\text{intermediate pH}} = X_{\text{HIn}}[\varepsilon_{\text{HIn}}c\ell] + (1 - X_{\text{HIn}})\,[\varepsilon_{\text{In}}c\ell] \quad (21.6)$$

$$A_{\text{intermediate pH}} = X_{\text{HIn}}A_{\text{low pH}} + (1 - X_{\text{HIn}})\,A_{\text{high pH}} \quad (21.7)$$

Before making any of the absorbance measurements, you must choose an appropriate wavelength to use. Maximum sensitivity is obtained if the wavelength chosen has the greatest difference between A_{HIn} and A_{In}. This optimum wavelength is easily determined by comparing the spectrum of an indicator in acid solution with the corresponding spectrum of the indicator in basic solution. Figure 21.1 shows the spectra of three different indicators that are suitable for this experiment.

PROCEDURE

Obtain an indicator assignment from your instructor. Refer to Figure 21.1 and choose the optimum wavelength for your absorbance measurements. Clean five 10-mL test tubes, a buret, and a 2-mL pipet. Use the buret and pipet, respectively, to dispense 5.00 mL of your indicator solution and 2.00 mL of water into each of the five test tubes.

Add 1.00 mL of 1 M HCl to one test tube, 1.00 mL of 1 M NaOH to a second tube, and 1.00 mL of a different buffer solution to each of the remaining test tubes. The buffer solutions to be used will be labeled with nominal pH values. However, the actual pH of each indicator/buffer mixture will be slightly higher than that of pure buffer, owing to the dilution of the buffer with indicator and water. If the actual pH of each diluted buffer is not specified, you will need to measure it accurately using a pH meter.

If possible, familiarize yourself with the operation of the spectrophotometer by studying the instruction manual. Review the material on absorption spectrometry in Experiment 19, if necessary. Follow your instructor's directions concerning the use of the spectrophotometer.

Clean and rinse two matched test tubes or spectrophotometer cells. Fill each cell with water and dry the cell on the outside using a soft, absorbent tissue. Inspect each cell to ensure that no dirt, air bubbles, or scratches are in the light path. Insert each cell in turn into the spectrophotometer and observe the deflection of the meter pointer. The cell that has the lower absorbance should be noted and used as your reference. With that reference cell in the cell compartment, close the cover, and set the light control so that the meter reads 100% T or zero absorbance. Replace the reference with the sample cell, close the cover, and read and record its absorbance.

Rinse the sample cell with two successive small portions of one of the five indicator solutions prepared previously. Transfer the rest of the solution to the sample cell and insert it into the spectrophotometer. Read and record the absorbance. Repeat these operations for the four remaining indicator solutions. Before reading each sample, check the 0 and 100% T settings of the instrument, using your reference cell. Correct the absorbance of each indicator solution for the absorbance of the sample cell.

Using the absorbances of the HCl, the NaOH, and the buffered solutions, calculate the mol fractions of both HIn (X_{HIn}) and In^- (X_{In}) present in each buffered mixture. Report the values of K_a obtained and the average value of K_a. Examine your data carefully before you discard your solutions. If you have not

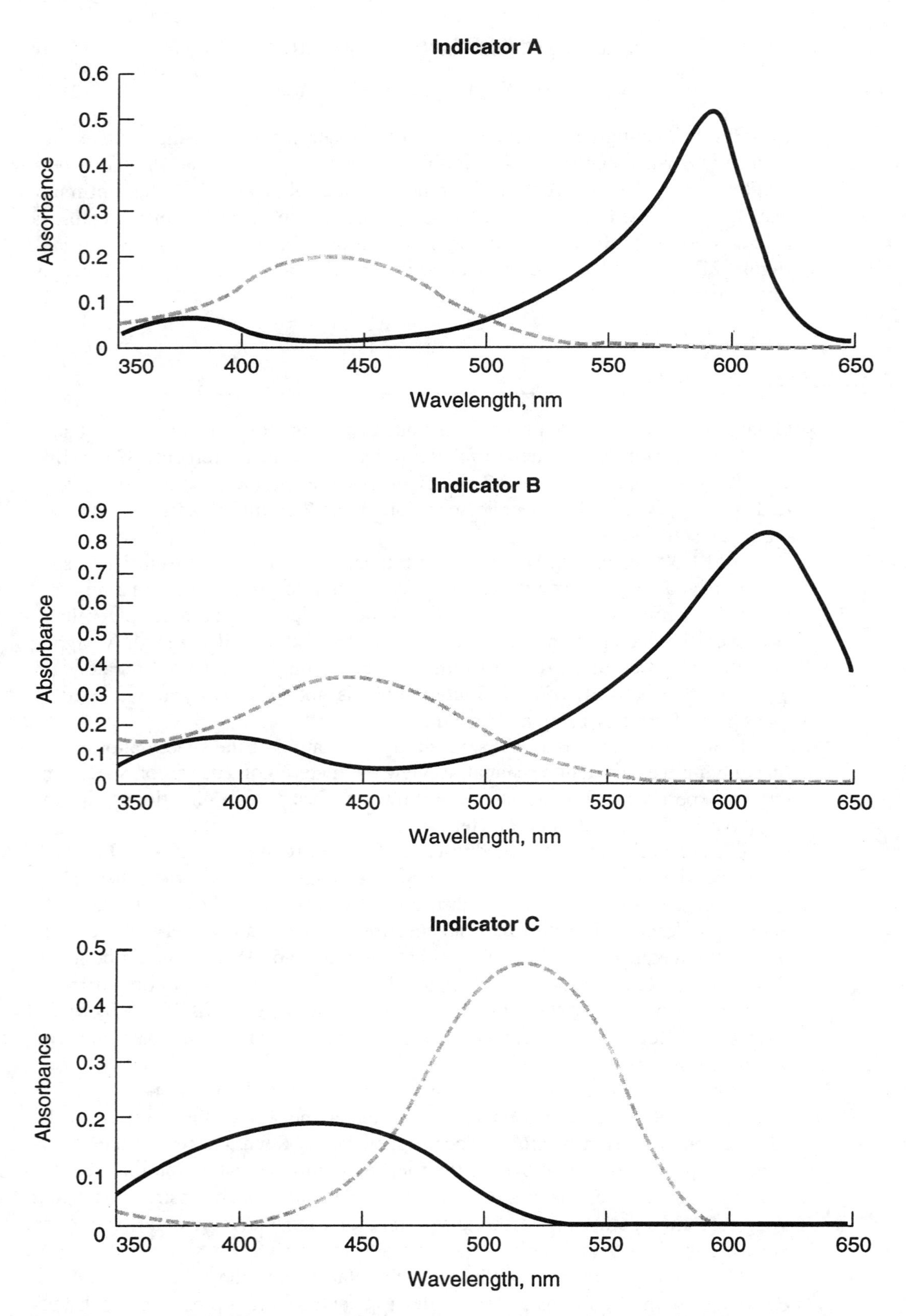

FIGURE 21.1 Absorbtion spectra of three acid-base indicators in 0.125 M HCl(---) and in 0.125 M NaOH (—).

obtained data that will give meaningful values of K_a, obtain or prepare buffers with the pH's you need to make mixtures containing substantial portions of both protonated and deprotonated indicators. Your instructor will give you directions for preparing buffers of intermediate pH from the buffers that have been supplied to you.

Disposal of Reagents

All reagents used in this experiment may be flushed down the drain after they have been neutralized.

Questions

1. Would the indicator that you studied in this experiment be suitable for indicating the equivalence point in the titration of:
   a. a strong acid with a strong base?
   b. a weak acid with a strong base?
   c. a weak base with a strong acid? Justify each answer, briefly.
2. If two or more light-absorbing materials are present together, the total absorbance is the sum of the individual absorbances. Using the relation between A and $\%T$, show that the total $\%T$ is not the sum of the individual $\%T$ values.
3. The most accurate results for this experiment are obtained by studying a mixture in which HIn and In^- are present in equal concentration. What is the relation between K_a and $[H^+]$ at this point? Specify the pH of the buffer that should be used to obtain a mixture where $[HIn] = [In^-]$.
4. Is the indicator whose spectrum is shown in Figure 21.2 more highly colored in acid or basic solution? Explain your answer briefly.

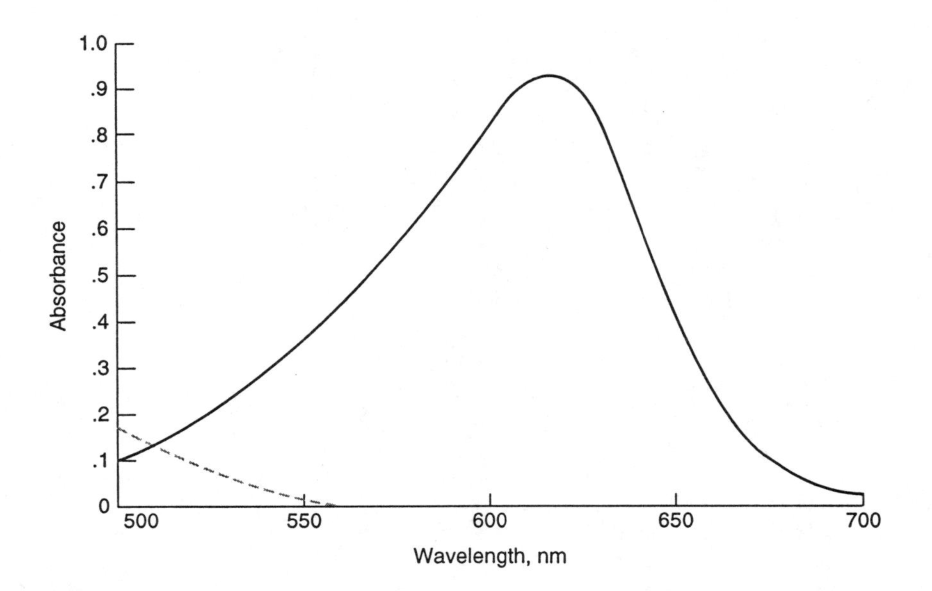

FIGURE 21.2 Absorption spectrum of an acid-base indicator in 0.1 M HCl (---) and in 0.1 M NaOH (—).

Date ______ Name ________________________________ Section ______ Desk Number ______

PRE-LAB EXERCISES FOR EXPERIMENT 21

These exercises are to be completed after you have read the experiment but before you come to the laboratory to perform it.

1. A certain indicator has an absorbance of 0.800 in 1 M NaOH and an absorbance of 0.200 in 1 M HCl. In a solution buffered to a pH of 3.98, the indicator has an absorbance of 0.552. Assume the path length and total concentration of indicator are the same for all mixtures studied and find the value of K_a for the indicator.

2. Jan Student found that the indicator had an absorbance of 0.214 when 5.00 mL of indicator had been mixed with 2.00 mL of water and 1.00 mL of 1 M HCl. Jan also found that the indicator had an absorbance of 0.218 when 5.00 mL of indicator had been mixed with 2.00 mL of water and 1.00 mL of pH 2 buffer (pH of diluted buffer, 2.13). Explain briefly why Jan's results cannot be used to find the value of K_a for the indicator. Give a possible reason for Jan's anomalous results.

Date ______ Name ______________________________ Section ______ Desk Number ______

SUMMARY REPORT ON EXPERIMENT 21

Indicator studied ______________

Wavelength used ______________

Absorbance of sample cell versus reference cell ______________

Solution Number	Actual pH	Absorbance	Corrected Absorbance
1			
2			
3			
4			
5			
6*			
7*			
8*			

*Optional—mixtures to be studied if the others do not give satisfactory results.

	$[H^+]$	X_{HIn}	X_{In}	K_a
3				
4				
5				
6				
7				
8				

$K_{a,av}$ ______________

Hydrolysis of Salts

Laboratory Time Required

Three hours

Special Equipment and Supplies

pH Meter
Electrodes
Volumetric flask, 100-mL
Balance
Burner or hot plate

$NaCl$
$NaC_2H_3O_2 \cdot 3H_2O$
NH_4Cl
Na_2CO_3
$NaHCO_3$
$NaH_2PO_4 \cdot H_2O$
$Na_2HPO_4 \cdot 7H_2O$
$Na_3PO_4 \cdot 12H_2O$
$Fe(NO_3)_3 \cdot 9H_2O$
$CuSO_4 \cdot 5H_2O$
$Zn(NO_3)_2 \cdot 6H_2O$
$KAl(SO_4)_2 \cdot 12H_2O$
Buffer solution, pH 6–7

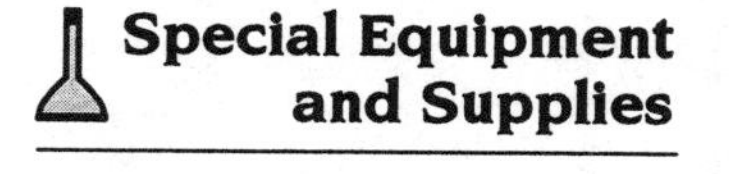

Safety

None of these reagents is very hazardous, but you should still exercise care to avoid getting chemicals on your skin, mouth, or eyes.

First Aid

Thoroughly flush the affected area with water.

The first acid/base concept to which students are usually introduced is the Arrhenius model, which views acids as proton donors and bases as hydroxide ion donors. In this classical model, acids and bases react to produce salts and water. Although such a view is helpful in considering the stoichiometry of acid/base interactions, it fails to account for the basic properties of materials, such as ammonia and the amines, which do not contain hydroxide ions. The model also does not explain why many salts exhibit acid or alkaline properties in aqueous solution and why some acids and bases are stronger than others. Later models, such as the Bronsted-Lowry concept and the Lewis acid/base theory, are better suited to explaining these phenomena. These models are used in the interpretation of the experimental results obtained in this experiment on the hydrolysis of salts.

PRINCIPLES

Salts are formed when acids react with bases. Because such reactions are referred to as "neutralizations," one might expect that salt solutions would be "neutral" with regard to pH. However, this is not always the case. It is true that solutions of sodium chloride in water are neutral; however, aqueous solutions of ammonium chloride are observed to be acidic whereas aqueous solutions of sodium acetate are alkaline (basic).

These observations are best understood by considering the Bronsted-Lowry theory of acid/base behavior. According to Bronsted and Lowry, an acid is a proton donor and a base is a proton acceptor. Each acid/base reaction involves two acid/base conjugate pairs, as illustrated by Equation 22.1.

$$\underset{\text{acid 1}}{HCl(g)} + \underset{\text{base 2}}{H_2O(\ell)} \longrightarrow \underset{\text{acid 2}}{H_3O^+} + \underset{\text{base 1}}{Cl^-(aq)} \quad (22.1)$$

The process described by Equation 22.1 is the formation of hydrochloric acid from the dissolution of hydrogen chloride gas in water. In this process, hydrogen chloride acts as a strong acid, donating virtually all of its protons to water, which acts as a base. The hydronium ion, H_3O^+, is the conjugate acid of water while the chloride ion is the conjugate base of HCl. Because HCl is a strong acid, its conjugate base is an extremely poor proton acceptor; thus, Cl^- is too weak a base to accept protons from the fairly strong acid, H_3O^+. Therefore, chloride ions are certainly not expected to react with water, which is a poorer proton donor than the hydronium ion (see Equation 22.2).

$$Cl^-(aq) + H_2O(\ell) \longrightarrow NR \quad (22.2)$$

Similarly, Na^+, as the (extremely weak) conjugate acid of the strong base, NaOH, does not react with water to produce free protons or hydronium ions (see Equation 22.3). Consequently, solutions of NaCl should be neutral.

$$Na^+(aq) + H_2O(\ell) \longrightarrow NR \quad (22.3)$$

On the other hand, the ammonium ion, NH_4^+, as the conjugate acid of a weak base, is strong enough to donate some protons to water. Aqueous solutions of ammonium chloride are acidic, because the ammonium ion functions as an acid, while the chloride ion does not assert itself as a base (see Equations 22.4–22.6).

$$NH_4Cl(s) \xrightarrow{H_2O} NH_4^+(aq) + Cl^-(aq) \quad (22.4)$$

$$NH_4^+(aq) + H_2O(\ell) \rightleftharpoons H_3O^+ + NH_3(aq) \quad (22.5)$$

$$Cl^-(aq) + H_2O(\ell) \longrightarrow NR \quad (22.6)$$

Likewise, the acetate ion, $C_2H_3O_2^-$, as the conjugate base of a weak acid, is strong enough to accept some protons from water. Aqueous solutions of sodium acetate are basic because the acetate ion functions as a base, whereas the sodium ion does not assert itself as an acid (see Equations 22.7–22.9).

$$NaC_2H_3O_2{\cdot}3H_2O(s) \xrightarrow{H_2O} Na^+(aq) + C_2H_3O_2^-(aq) \quad (22.7)$$

$$C_2H_3O_2^-(aq) + H_2O(\ell) \rightleftarrows HC_2H_3O_2(aq) + OH^-(aq) \quad (22.8)$$

$$Na^+(aq) + H_2O(\ell) \longrightarrow NR \quad (22.9)$$

The arguments cited above are frequently summarized in a few simple rules. Salts produced by the reaction of a strong acid with a strong base are neutral with respect to acid/base character. Salts produced by the reaction of a strong acid with a weak base will themselves behave as acids. Salts produced by the reaction of a weak acid with a strong base will themselves behave as bases. These rules are adequate for prediction of the acid/base character of many salts in aqueous solutions. There are, however, some salts that need further consideration.

The so-called acid salts are those, such as KH_2PO_4 and $NaHSO_3$, which are produced when some—but not all—of the protons from a polyprotic acid are reacted with a base. The hydrogen-containing anions act as both acids and bases, as illustrated in Equations 22.10 and 22.11.

$$\underset{\text{acid 1}}{HSO_3^-} + \underset{\text{base 2}}{H_2O} \rightleftarrows \underset{\text{acid 2}}{H_3O^+} + \underset{\text{base 1}}{SO_3^{2-}} \quad (22.10)$$

$$\underset{\text{base 1}}{HSO_3^-} + \underset{\text{acid 2}}{H_2O} \rightleftarrows \underset{\text{acid 1}}{H_2SO_3} + \underset{\text{base 2}}{OH^-} \quad (22.11)$$

Whether the resulting solution is acidic or basic depends on whether the reaction shown in Equation 22.10 or that shown in Equation 22.11 occurs to the greater extent. A comparison of K_a for Equation 22.10 with K_b for Equation 22.11 will provide the answer.

Still another case involves salts, such as $FeCl_3$ and $Cu(NO_3)_2$, which form $Fe^{3+}(aq)$, $Cl^-(aq)$, $Cu^{2+}(aq)$, and $NO_3^-(aq)$, in water solution. We have previously shown that ions such as $Cl^-(aq)$ and $NO_3^-(aq)$ are not significantly basic in water. However, there is the possibility that $Fe^{3+}(aq)$ and $Cu^{2+}(aq)$ will behave as acids. According to the Lewis definition, such cations are acids because they attract electrons from their waters of hydration and increase the tendency of those water molecules to ionize.

The $Cu(H_2O)_4^{2+}$ ion can ionize in two successive steps, as shown in Equations 22.12 and 22.13, respectively. This Lewis acid explanation would lead us to expect that small, highly charged cations, such as Al^{3+}, would be most acidic, while large, univalent cations, such as Na^+, would have little acidic character.

$$Cu(H_2O)_4^{2+} + H_2O \rightleftarrows Cu(OH)(H_2O)_3^+ + H_3O^+ \quad (22.12)$$

$$Cu(OH)(H_2O)_3^+ + H_2O \rightleftarrows Cu(OH)_2(H_2O)_2 + H_3O^+ \quad (22.13)$$

In this experiment, your class will be preparing solutions of several salts and measuring the pH's of those solutions. You will use the values of the appropriate equilibrium constants (such as K_a, K_b, K_w) to predict these pH's and you will be expected to comment on any discrepancies between the predicted and the observed values.

PROCEDURE

Boil 150 mL of distilled water for approximately 10 minutes. This should remove dissolved carbon dioxide that would otherwise give the water an acidic pH.

Obtain an assignment of a salt from your instructor and weigh out the mass needed for the preparation of 100 mL of a 0.1 M solution. (Be sure to check the reagent bottle to determine the exact formula of your salt; the number of waters of hydration is sometimes variable.) Record the actual mass of salt weighed out on the analytical balance and dissolve the sample completely in 50 mL (or less) of distilled water. Transfer this solution to a 100-mL volumetric flask, rinsing the beaker that held the solution originally and adding the rinsing to the volumetric flask. Swirl the flask carefully to promote mixing; then carefully add distilled water up to the mark on the volumetric flask. Cap the flask and invert it several times to ensure good mixing.

Set up your pH meter. Immerse the electrodes into a buffer solution. Allow two minutes for thermal equilibrium to be reached. Then measure the temperature of the buffer solution with a thermometer. Set the manual temperature compensator dial on the pH meter accordingly. Adjust the "calibrate" control until the meter reading matches the pH of the buffer exactly. Wait 5 seconds to be sure that the reading remains constant.

Turn the function switch to "standby." Rinse and dry the electrodes. Return the function switch to "pH." Measure the pH of your boiled and cooled distilled water and your salt solution. When you have finished with your pH measurements, return the meter to "standby" and unplug it.

A class data sheet will be posted in the laboratory so that you may provide your classmates with information concerning the salt solution you prepared. When all students have entered their data, copy the information to the data sheet provided in your lab manual.

In your report, write the best net ionic equation of the reaction for each salt (not just the one you were assigned) whose solution has a pH that indicates a reaction occurred between an ion of the salt and water (an ion that reacts with water is said to "hydrolyze" and such a reaction is referred to as "hydrolysis"). Use the appropriate equilibrium constant to rationalize the observed pH of the solution. Your textbook, the *CRC Handbook of Chemistry and Physics*, and similar reference texts, may be used to obtain the values of the equilibrium constants.

Disposal of Reagents

All solutions may be diluted and poured down the drain.

Date ______ Name ______________________ Section ______ Desk Number ______

PRE-LAB EXERCISES FOR EXPERIMENT 22

These exercises are to be completed after you have read the experiment but before you come to the laboratory to perform it.

1. In Equation 22.5, NH_4^+ is shown acting as an acid. Write the expression for K_a of NH_4^+ and show how the value of K_a is related to K_b for NH_3.

2. In Equation 22.8, $C_2H_3O_2^-$ is shown acting as a base. Write the expression for K_b of $C_2H_3O_2^-$ and show how the value of K_b is related to the value of K_a for $HC_2H_3O_2$.

3. In Equations 22.10 and 22.11, HSO_3^- ion is shown behaving as both an acid and base. Write the expressions for K_a and K_b for Equations 22.10 and 22.11, respectively. Evaluate these K's by relating them to K_{a1} and K_{a2} for H_2SO_3. Is HSO_3^- stronger as an acid or as a base?

4. Write the equilibrium expression for the reaction of $Cu(H_2O)_4^{2+}$ and water to produce the complex ion, $Cu(OH)(H_2O)_3^+$. Use the information given below to evaluate the equilibrium constant.

$$Cu(H_2O)_4^{2+} + OH^- \rightleftharpoons Cu(OH)(H_2O)_3^+ + H_2O \qquad K_{form} = 1 \times 10^7$$
$$2H_2O \rightleftharpoons H_3O^+ + OH^- \qquad K_w = 1 \times 10^{-14}$$

5. Use your answers to the preceding questions to predict the pH values of 0.1 M solutions of NH_4Cl, $NaC_2H_3O_2$, $NaHSO_3$, and $Cu(NO_3)_2$. In making your prediction concerning $NaHSO_3$ you may use the relation:

$$[H^+] = \sqrt{K_{a1} K_{a2}}$$

where K_{a1} and K_{a2} are the acid dissociation constants for H_2SO_3. Does your predicted value of $[H^+]$ agree with your answer to Exercise 3?

Date ______ Name ______________________________ Section ______ Desk Number ______

SUMMARY REPORT ON EXPERIMENT 22

Salt assigned ______________

Mass of salt + container ______________

Mass of container ______________

Mass of salt ______________

pH of boiled and cooled water ______________

pH of solution ______________

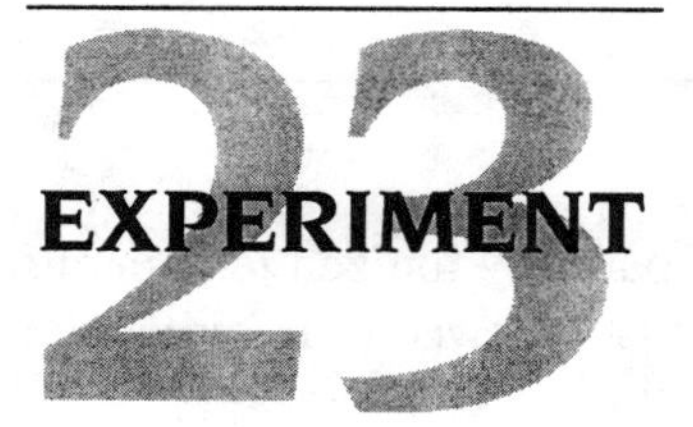

pH Titration of Cola Drinks

Laboratory Time Required

Three hours

Special Equipment and Supplies

pH meter
Glass electrode
Calomel electrode
Magnetic stirrer
Stirring bar
Buret
Buret clamp
Pipet
Pipet bulb

Cola drink
0.01 M Sodium hydroxide
Potassium hydrogen phthalate (KHP)
pH 4.0 buffer (optional)
pH 7.0 buffer

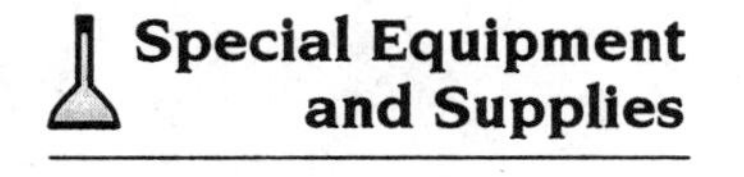

Safety

The sodium hydroxide solution is **caustic.** Avoid getting it on your skin or in your eyes. Basic solutions in general are hazardous to the eyes and can cause serious vision impairment or even blindness.

Although it may be tempting, do not drink any unused portion of your cola sample. Drinking is not permitted in the laboratory. Any excess cola must be discarded.

First Aid

Flush the skin or eyes thoroughly with water. If sodium hydroxide has gotten in your eyes, see a doctor.

Cola drinks commonly contain phosphoric acid, which imparts a pleasant tartness to their taste. Because phosphoric acid is an acid, it would seem that you could determine the concentration of phosphoric acid in such beverages by titration. However, an indicator titration cannot be performed because the dark color of the drink would mask any change in the color of the indicator. Therefore, a pH meter will monitor the acidity of the solution throughout the titration, a plot of pH versus volume of base added will be prepared, and the concentration of the phosphoric acid in the drink will be obtained from the analysis of this plot.

PRINCIPLES

Weak Acids

A weak acid differs from a strong acid in being only partially ionized in aqueous solution. In such solutions, a dynamic equilibrium exists between the molecular and ionized forms of the acid, as represented by Equation 23.1.

$$HA + H_2O \rightleftharpoons H_3O^+ + A^- \tag{23.1}$$

The equilibrium expression for this reaction is shown in Equation 23.2, where K_a is the acid dissociation constant for the weak acid.

$$K_a = \frac{[H_3O^+][A^-]}{[HA]} \tag{23.2}$$

When the weak acid, HA, is titrated with sodium hydroxide, the concentration of HA progressively decreases, whereas the concentration of A^- progressively increases, because the acid reacts with the base to form the strong electrolyte NaA (see Equation 23.3):

$$HA(aq) + Na^+(aq) + OH^-(aq) \longrightarrow H_2O + Na^+(aq) + A^-(aq) \tag{23.3}$$

In order for the titration to be successful, the reaction shown in Equation 23.3 must go to completion.

This condition will be met if the acid's K_a value is several orders of magnitude larger than K_W, the ion product for water (see Equation 23.4).

$$K_W = [H_3O^+]\,[OH^-] \tag{23.4}$$

To see how the values of K_a and K_W are involved in determining whether the titration reaction will go to completion, begin by assuming the reaction shown in Equation 23.3 results in an equilibrium situation (see Equation 23.5).

$$HA + OH^- \rightleftharpoons H_2O + A^- \tag{23.5}$$

Next, set up an equilibrium expression corresponding to Equation 23.5. Then multiply both the numerator and denominator by $[H_3O^+]$ (see Equation 23.6).

$$K = \frac{[A^-]}{[HA][OH^-]} \times \frac{[H_3O^+]}{[H_3O^+]} = \frac{K_a}{K_W} \tag{23.6}$$

The value of K_W is 1.0×10^{-14} at 25°C. For a typical weak acid ($K_a = 1 \times 10^{-5}$), K for the titration reaction will have a value on the order of 10^9, certainly high enough to ensure that the reaction does indeed go to completion.

Very often, a weak acid is essentially unionized at the start of the titration. At the halfway point in the titration, exactly half of the HA originally present will have been neutralized and, therefore, the concentrations of HA and A^- will be equal. A useful relation (Equation 23.7) is obtained by substituting this information into the equilibrium expression, Equation 23.2.

$$K_a = \frac{[H_3O^+]_{1/2}[A^-]_{1/2}}{[HA]_{1/2}} = [H_3O^+]_{1/2} \tag{23.7}$$

Thus the dissociation constant of the acid is numerically equal to the hydronium ion concentration at the halfway point in the titration of a weak acid. Because the pH at the halfway point in a pH titration is easily determined, and because pH equals $-\log[H_3O^+]$, pH titrations provide an accurate and convenient experimental technique for measuring dissociation constants of weak acids. The only caution against using Equation 23.7 is that its derivation is based on the assumption that the initial dissociation of the acid is negligible. The degree of dissociation will always increase as the initial concentration of acid is decreased. When the solution is sufficiently diluted that the weak acid is appreciably dissociated, the pH at the halfway point in the titration bears no relation to the value of K_a.

Examination of the pH curve shown in Figure 23.1 reveals that, near the halfway point, the pH of the solution changes only slightly on the addition of small increments of NaOH solution. This is because the concentrations of HA and A^- present are both relatively large compared to the amount of base added, and the ratio $[A^-]/[HA]$ therefore changes only slightly. Examination of the equilibrium expression reveals that, if the change in the $[A^-]/[HA]$ ratio is small, the change in $[H_3O^+]$ or in the pH of the solution must be small also. (Therefore, at this point in the titration, the reaction mixture constitutes a good buffer solution.)

As the equivalence point in the titration is approached, however, the concentration of unreacted HA becomes progressively smaller, so that successive increments of NaOH neutralize a greater fraction of the remaining HA. This produces a large change in the $[A^-]/[HA]$ ratio and, therefore, in the pH of the solution. At the equivalence point, there is a steep increase in pH, producing an inflection point in the pH curve.

Beyond the equivalence point, the pH is determined by the ion product for water (see Equation 23.4). The first small excess of NaOH greatly increases the

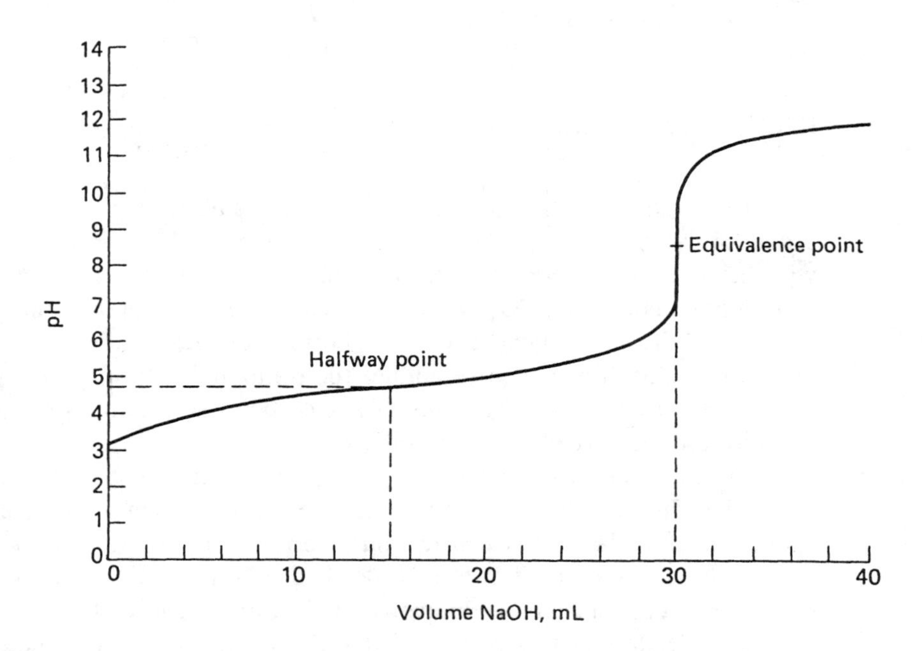

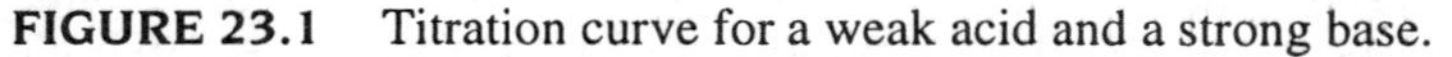

FIGURE 23.1 Titration curve for a weak acid and a strong base.

concentration of OH^- in the solution, with a correspondingly large decrease in the H_3O^+ concentration. A large pH change therefore occurs. Well past the equivalence point, however, the concentration of OH^- in solution is large, and further small additions of NaOH produce only slight changes in the OH^- concentration and consequently in the pH.

It is interesting to note that the pH at the equivalence point is greater than 7. This is so because A^-, the conjugate base of HA, reacts with water, liberating hydroxide ions (see Equation 23.8).

$$A^-(aq) + H_2O \rightleftharpoons HA(aq) + OH^-(aq) \quad (23.8)$$

The equilibrium expression for this reaction is shown in Equation 23.9.

$$K_b = K_w/K_a = ([HA][OH^-])/[A^-] \quad (23.9)$$

It is thus possible to calculate K_a using the pH of the solution at the equivalence point and the concentration of A^- present. However, the large slope of the curve makes it difficult to evaluate the pH accurately at the equivalence point, and so the procedure is less reliable than the method which uses the pH at the halfway point to calculate K_a.

Polyprotic Acids

The weak acid discussed above is a monoprotic acid, meaning that it has one ionizable proton. Phosphoric acid is a triprotic acid, meaning that it has three ionizable protons. The reaction showing the dissociation of each of these protons has its own value of K_a, as shown below (see Equations 23.10–23.12).

$$H_3PO_4 + H_2O \rightleftharpoons H_3O^+ + H_2PO_4^- \qquad K_{a1} = \frac{[H_3O^+][H_2PO_4^-]}{[H_3PO_4]} \quad (23.10)$$

$$H_2PO_4^- + H_2O \rightleftharpoons H_3O^+ + HPO_4^{2-} \qquad K_{a2} = \frac{[H_3O^+][HPO_4^{2-}]}{[H_2PO_4^-]} \quad (23.11)$$

$$HPO_4^{2-} + H_2O \rightleftharpoons H_3O^+ + PO_4^{3-} \qquad K_{a3} = \frac{[H_3O^+][PO_4^{3-}]}{[HPO_4^{2-}]} \quad (23.12)$$

If the values of K_{a1}, K_{a2}, and K_{a3} are separated from each other by several orders of magnitude, the dissociation may be regarded as taking place in a stepwise fashion (i.e., all H_3PO_4 molecules are transformed to $H_2PO_4^-$ ions before any substantial amount of HPO_4^{2-} is formed). This is, in fact, true for phosphoric acid. As a result, the titration curve for phosphoric acid, shown in Figure 23.2, consists of two parts, each of which resembles the titration curve of a monoprotic weak acid with a strong base. The value of K_{a3} is such that it is not possible to titrate the third proton of H_3PO_4 successfully.

The titration curve you obtain using your drink sample should look very much like the one shown in Figure 23.2. However, because colas contain substances, such as lactic acid, citric acid, and potassium hydrogen phosphate, that will also react with sodium hydroxide, you may find that there are significant differences between your experimental titration curve and the one shown in Figure 23.2. Nonetheless, you should still be able to use your data to find not only the concentration of phosphoric acid in your cola sample, but also the

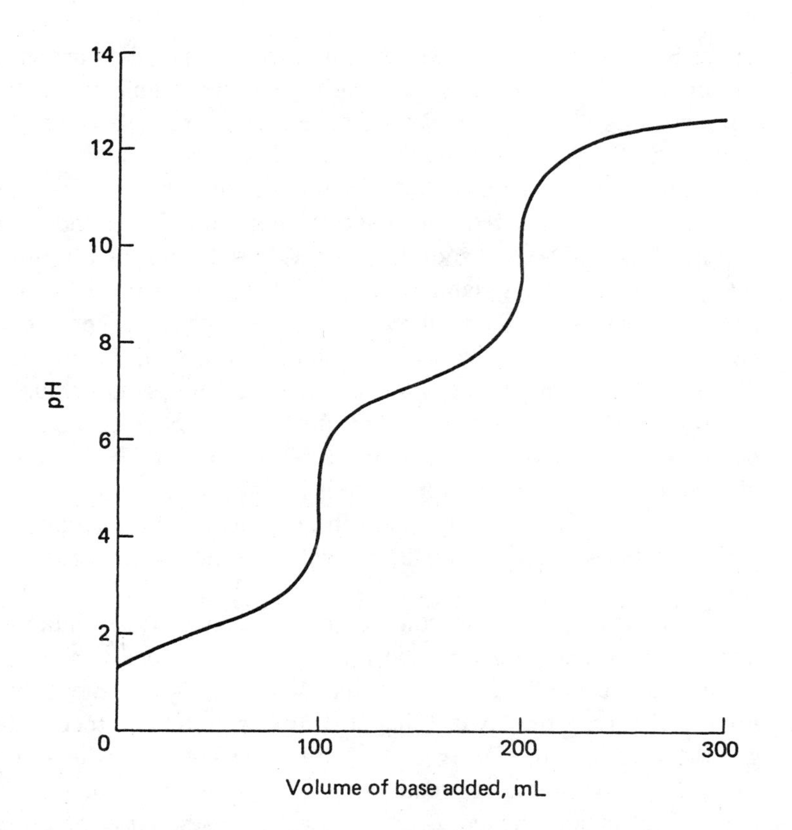

FIGURE 23.2 Titration curve for 100 mL of 0.4 M H_3PO_4, titrated with 0.4 M NaOH.

values of K_{a1} and K_{a2} for phosphoric acid. These quantities are often determined by pH titration to identify unknown weak acids or bases.

Sample Preparation

You can simplify the analysis of your titration curve if you select a drink that contains phosphoric acid as the only acid. (The contents should be listed on the label.) Regardless of the brand of drink you select, you will need to remove the dissolved carbon dioxide, which is the source of its effervescence. This is necessary because aqueous carbon dioxide reacts with sodium hydroxide; this would alter the appearance of the titration curve. In addition, the presence of bubbles in the cola sample would make it impossible to measure the volume of the sample accurately.

PROCEDURE

Prepare your pH meter for use as you did in Experiment 22. Connect the electrodes according to the instructions and leave them suspended with their tips immersed in water. Connect the line cord and turn on the instrument to allow a brief warm-up period.

Place approximately 100 mL of your cola drink in an Erlenmeyer flask. Mark the liquid level with a label and cover the mouth of the flask with a watch glass. Simmer the sample for 20 minutes to expel the carbon dioxide from the sample. After you have stopped heating and the sample has cooled down to room temperature, add distilled water, if necessary, to restore the liquid level to its preheating height.

Rinse a buret once with distilled water and three times with small amounts of the solution of NaOH. Then fill the buret with NaOH and mount it in the buret clamp. Standardize the sodium hydroxide solution by titrating it against accurately weighed 0.06 g samples of KHP (see Experiment 8 for details). Record the standardization data on the Summary Report Sheet. Refill the buret, as necessary, for the standardizations and the titration of the cola sample.

Rinse a 10-mL pipet once with distilled water and three times with small amounts of the decarbonated cola drink. Use the pipet twice to transfer 20.00 mL of the decarbonated cola drink into a 50-mL beaker. Save the remainder of your decarbonated cola sample for analysis in Experiment 24.

Calibrate the pH meter using the pH 7.0 buffer. (Increased accuracy is possible if a pH 4.0 buffer is used as well to confirm the linearity of the pH measurements.)

Place a stirring bar in your sample. Place the flask containing the sample on a magnetic stirrer. Immerse the tips of the electrodes in the decarbonated cola. Adjust the speed of the stirrer so that good mixing is achieved without endangering the electrode tips by a flying stirring bar. Read and record the initial pH. Begin the titration by adding 1- to 2-mL increments of the standardized NaOH. Record the pH and buret readings after each addition of base. As each equivalence point is approached, decrease the size of the base increments until, finally, single drops are being added. Continue the titration until well past the second equivalence point.

Prepare a plot of pH versus volume of base added. Compare this plot to the one in Figure 23.2 and discuss possible reasons for the differences in their appearance.

Use your plot to locate the first equivalence point. Combine the molarity of the base with the volume used to reach the equivalence point to find the concentration of phosphoric acid in the drink.

Find the pH at the halfway point for the first portion of the curve to find a value of K_{a1} for phosphoric acid. Compare this value with the literature value.

Now calculate an independent value of K_{a1} by combining the initial pH of the drink with the concentration of phosphoric acid in Equation 23.10. Compare this value of K_{a1} with the literature value. Find the percent ionization of the phosphoric acid in the drink.

Evaluate K_{a2} by finding the pH halfway between the first and second equivalence points. Compare this value with the literature value. Calculate the percent ionization of $H_2PO_4^-$ in your sample.

Calculate an independent value for the concentration of phosphoric acid in the drink, based on the volume of base used to reach the second end point.

Disposal of Reagents

All of the solutions used in this experiment can safely be poured down the drain after they have been neutralized and diluted in the usual manner.

Questions

1. If 0.100 M HCl were titrated with 0.100 M NaOH, what would be the pH of the solution:
   a. before the addition of NaOH?
   b. at the equivalence point?
   c. at the halfway point?
2. Makers of a popular carbonated beverage claim that none of their colas contain more than 30 mg of phosphorus per 6-oz. serving. How does this figure compare with your calculation of the percent of phosphoric acid in the drink you analyzed?
3. What is the equivalent mass of phosphoric acid (i.e., the mass, in grams, of phosphoric acid that reacts with one mole of hydroxide ions)?
4. Can Equation 23.7 be used to evaluate K_{a1} for phosphoric acid? Explain your answer briefly.

Date ______ Name ______________________________ Section ______ Desk Number ______

PRE-LAB EXERCISES FOR EXPERIMENT 23

These exercises are to be completed after you have read the experiment but before you come to the laboratory to perform it.

1. Consider the dissociation of the tricrotic acid, H_3Y.

$$H_3Y + H_2O \rightleftarrows H_3O^+ + H_2Y^- \qquad K_{a1} = 2.0 \times 10^{-4}$$

$$H_2Y^- + H_2O \rightleftarrows H_3O^+ + HY^{2-} \qquad K_{a2} = 8.0 \times 10^{-8}$$

$$HY^{2-} + H_2O \rightleftarrows H_3O^+ + Y^{3-} \qquad K_{a3} = 4.0 \times 10^{-13}$$

Use the values of K_{a1}, K_{a2}, K_{a3}, and K_w to evaluate the equilibrium constants associated with each of the reactions shown below:

$$H_3Y + OH^- \rightleftarrows H_2O + H_2Y^-$$

$$H_2Y^- + OH^- \rightleftarrows H_2O + HY^{2-}$$

$$HY^{2-} + OH^- \rightleftarrows H_2O + Y^{3-}$$

2. Use your answers to Question 1 to explain why only two equivalence points would be observed in the titration of H_3Y with 0.1 M sodium hydroxide.

3. A 20.00-mL sample of H_3Y was titrated with 0.1000 M NaOH. The first equivalence point was reached after the addition of 14.98 mL of base. The second equivalence point was reached after the addition of 30.03 mL. Find two values for the molarity of H_3Y in the original sample.

4. Find the initial pH of the solution of H_3Y considered in the previous questions.

Date ______ Name ______________________________ Section ______ Desk Number ______

SUMMARY REPORT ON EXPERIMENT 23

Standardization of the Sodium Hydroxide

	Trial 1	*Trial 2*	*Trial 3**
Mass of vial (before removal of KHP)	______	______	______
Mass of vial (after removal of KHP)	______	______	______
Mass of KHP	______	______	______

	Trial 1	*Trial 2*	*Trial 3**
Final buret reading	______	______	______
Initial buret reading	______	______	______
Volume used	______	______	______
Molarity of NaOH	______	______	______
Average molarity of NaOH		______	

Titration of the decarbonated cola

	Trial 1	*Trial 2*	*Trial 3**
Volume of cola sample	______	______	______
Concentration of standardized NaOH	______	______	______
Initial buret reading	______	______	______

*Optional

Buret reading, mL	*Volume of base added, mL*	*pH*
______	______	______
______	______	______
______	______	______
______	______	______
______	______	______
______	______	______
______	______	______
______	______	______
______	______	______
______	______	______
______	______	______
______	______	______
______	______	______
______	______	______
______	______	______
______	______	______
______	______	______

Date ______ Name ______________________________ Section ______ Desk Number ______

Date ______ Name ______________________________ Section ______ Desk Number ______

	Trial 1	Trial 2	Trial 3*
Volume of NaOH added at first equivalence point			
Molarity of H_3PO_4			
pH at point halfway to first equivalence point			
Experimental value of K_{a1}			
Literature value of K_{a1}			
Value of K_{a1} from initial pH			
Percent ionization of H_3PO_4			
Value of K_{a2} from titration curve			
Literature value of K_{a2}			
Percent ionization of $H_2PO_4^-$			
Volume of NaOH to reach second equivalence point			
Molarity of H_3PO_4			

*optional

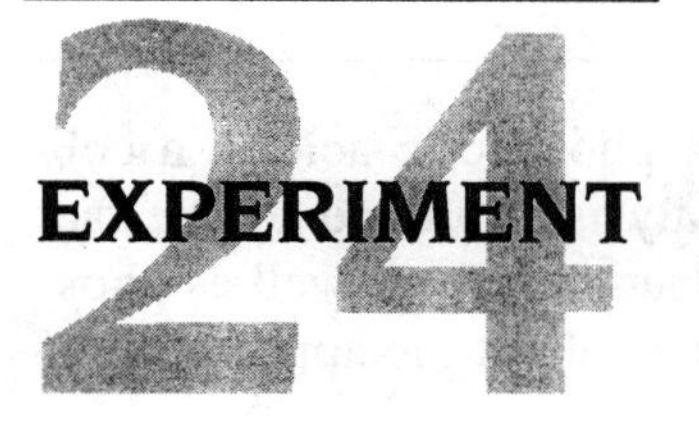
EXPERIMENT 24

Spectrophotometric Determination of the Percentage of Phosphoric Acid in Cola Drinks[1]

Laboratory Time Required

Three hours

Special Equipment and Supplies

Spectrophotometer
Spectrophotometer cells
Graduated cylinder, 100-mL
Volumetric flask, 100-mL
Commercially available cola
Standardized phosphate solution
Ammonium molybdate/ammonium metavanadate color-enhancing reagent

Safety

Vanadium compounds are **toxic.** Avoid exposure from skin contact, ingestion, or inhalation of fine powders. Although it may be tempting, do not drink your unused cola sample. Discard it down the drain.

First Aid

Thoroughly flush the affected area with water to remove chemicals from the skin.

If vanadium compounds have been taken orally, drink a large amount of water or milk and see a physician.

The exact recipe for preparing Coca Cola® is a closely guarded secret. However, cans of Coca Cola® do list some of the ingredients used in preparing the product. One of these ingredients, also found in other cola drinks, is phosphoric acid, H_3PO_4. This acid is used to impart a pleasant tartness to the taste. In this exercise, you will determine the concentration of phosphoric acid in a cola drink by a spectrophotometric method.

[1]Adapted from an article by J. Murphy, *J.Chem. Educ.*, **60** (1983) p. 420.

PRINCIPLES

In Experiment 23 you determined the percentage of phosphoric acid in a cola drink by pH titration. The utility of that form of analysis is limited by the fact that many colas contain acids (such as citric acid or lactic acid), as well as phosphoric acid. The presence of these acids will, of course, affect the appearance of the titration curve.

Spectrophotometric analysis provides a more accurate determination of phosphoric acid. Here, too, there are minor problems. Fortunately, they are easily remedied.

The first problem is that the molar absorptivity of phosphoric acid at all wavelengths is quite small. However, addition of an ammonium molybdate/ammonium metavanadate mixture to solutions containing phosphate species converts them to the intensely yellow heteropoly acid, $(NH_4)_3PO_4 \cdot NH_4VO_3 \cdot 16MoO_3$, which absorbs strongly at a wavelength of 400 nm.

The second problem is that most cola drinks are highly colored and contain light absorbers other than H_3PO_4. This problem is circumvented by diluting the cola fiftyfold and using the diluted drink as a **color blank**. This means that the spectrophotometer can be set to regard the diluted cola as a solution that does not absorb at all. Any absorbance of a diluted cola sample that has been treated with the color-enhancing reagent is then regarded as arising from the presence of the phosphoric acid alone.

PROCEDURE

The first step is to decarbonate your beverage (if you did this in Experiment 23, you may skip to the next paragraph). Place approximately 100 mL of cola in an Erlenmeyer flask. Use a label to mark the liquid level. Cover the mouth of the flask with a watch glass and simmer the cola for 20 minutes. This will expel the carbon dioxide dissolved in the drink so that the normal effervescence of the drink will not interfere with pipetting. Remove the flask from the heat and allow the decarbonated cola to cool to room temperature. If much liquid has evaporated during the simmering process, add distilled water to return the level to that shown by your label.

Preparation of the Calibration Curve

Clean and rinse two matched spectrophotometer cells. Fill each cell with water and dry the cell on the outside using a soft, absorbent tissue. Inspect each cell to ensure that no dirt, air bubbles, or scratches are in the light path. Insert each cell in turn into the spectrophotometer and measure the absorbance. The cell that has the lower absorbance should be noted and used as your reference. With that reference cell in the cell compartment, close the cover and set the light control so that the meter reads 100% T (or zero absorbance). Replace the reference with the sample cell, close the cover, and read and record its % T (or absorbance). The difference between the absorbance of the sample cell and the absorbance of the reference cell should be used as a correction to each absorbance measured in the sample cell.

Pipet 2.00 mL of the cooled, decarbonated cola into a 100-mL volumetric flask. Add 10 mL of the color-enhancing reagent and dilute to the mark with distilled water. Cap and invert the flask several times to ensure complete mixing. Transfer some of this solution to a clean, dry, 6-inch test tube. Set the tube aside to allow the color to develop. Be sure to label the test tube. Prepare a color blank by placing 2 mL of cooled, decarbonated cola in a 100-mL graduated cylinder. Dilute to the 100 mL mark with distilled water. Stir the mixture with a stirring rod and save some of this solution in another clean, dry, labeled, 6-inch test tube.

Preparation of the Phosphate Standards

Prepare a set of phosphate standards as described below.

1. Record the concentration of the standard phosphate solution provided.
2. Rinse your 2-mL pipet and 10-mL graduated cylinder once with distilled water and three times with small portions of the standard phosphate solution.
3. Rinse your 100-mL volumetric flask three times with distilled water.
4. Pipet 0.50 mL of standard phosphate solution into the 100-mL volumetric flask. Add 10.0 mL of the color-enhancing reagent to the flask. Then dilute to the mark with distilled water. Cap and invert the flask several times to ensure complete mixing.
5. Transfer some of the diluted standard to a clean, dry, labeled, 6-inch test tube. Set the solution aside to allow time for the color to develop.
6. Repeat Steps 3 through 5 (substituting first 1.00 mL, then 1.50 mL, and finally, 2.00 mL of undiluted standard in Step 4).

Determining of the Phosphate Concentration

When the color of the solutions has had a chance to develop fully, measure the absorbances with the spectrophotometer, following your instructor's directions. The instrument should be set to a wavelength of 400 nm. Use distilled water as a reference for the standards; use the color blank as a reference for the cola sample. Prepare a plot of absorbance versus concentration for the phosphate standards. Use your plot to determine the molarity of phosphoric acid in your diluted cola sample. Calculate the molarity of phosphoric acid in the original cola sample. Also determine the percentage of phosphoric acid in the drink (assume the density of the drink is 1.00 g/mL). If you performed Experiment 23, compare your results.

Disposal of Reagents

All solutions, including the color-enhancing reagent, may be diluted and flushed down the drain. The low concentration of ammonium metavanadate in the phosphate color standards makes isolation of this substance impractical.

Date ______ Name ______________________________ Section ______ Desk Number ______

PRE-LAB EXERCISES FOR EXPERIMENT 24

These exercises are to be completed after you have read the experiment but before you come to the laboratory to perform it.

1. Why must the cola sample be heated?

2. Why must the cola sample be diluted fiftyfold?

3. Why is the ammonium molybdate/ammonium metavanadate solution used?

4. What is a color blank? How does the color blank used in this experiment differ from the actual sample?

Date ______ Name ________________________________ Section ______ Desk Number ______

SUMMARY REPORT ON EXPERIMENT 24

Concentration of phosphate standard ____________________

Absorbance of sample cell versus reference cell ____________________

Volume of standard solution	$M_{H_3PO_4}$	A
0.50 mL	______________	______________
1.00 mL	______________	______________
1.50 mL	______________	______________
2.00 mL	______________	______________
Diluted cola sample	______________	______________

Molarity of H_3PO_4 in original cola drink ______________

Percentage of H_3PO_4 in original cola drink ______________

Molarity of H_3PO_4 in original cola drink (determined in Experiment 23) ______________

EXPERIMENT 25

Temperature Change and Equilibrium

Laboratory Time Required

Three and one-half hours, divided between two periods.

Special Equipment and Supplies

For preparing saturated solutions

Buret	0.200 M $CuSO_4$
	0.400 M KIO_3

For evaluating K_{sp}

Buret	0.200 M $CuSO_4$
Buret clamp	0.5 M Aqueous NH_3
Pipet, 5 mL	Ice
Pipet bulb	Rock salt
Thermometer	
Water baths	

Safety

None of the chemicals used in this experiment poses a serious health hazard. However, the KIO_3 is an oxidizer and aqueous NH_3 is a base. Avoid exposing your skin or eyes to contact with these or other chemicals.

First Aid

Rinse the exposed area thoroughly to flush away chemicals.

Our experience indicates that most physical and chemical changes are affected by temperature. More than a century ago, observations of temperature effects on everyday phenomena led to a formulation of the laws of thermodynamics, which are mathematical descriptions of these observations. The first law tells us about energy; the second, about reversibility and the direction of spontaneous change; and the third, about absolute zero.

The laws of classical thermodynamics apply specifically to the equilibrium condition of a system. When these laws are applied to chemical reactions, they relate the equilibrium condition and the temperature of a given reaction to the thermodynamic quantities, $\Delta H°$, $\Delta S°$, and $\Delta G°$, which specify respectively the change in enthalpy, entropy, and free energy for the system under standard conditions. By experimentally studying the effect of temperature on the equilibrium condition for a chemical reaction, we should obtain the fundamental information that we need to calculate these important thermodynamic functions.

In this experiment, you will investigate the effect of temperature on the solubility of a sparingly soluble salt in water. The solubility data will permit the calculation of the equilibrium constant for the reaction, K_{sp}. Then the experimental values of $\Delta G°$, $\Delta H°$, and $\Delta S°$ will be found through graphing.

PRINCIPLES

The Solubility Product

The solubility product constant, K_{sp} , is the equilibrium constant for a chemical equation that shows one mole of a salt as the reactant and the salt's separated ions as the products. Equilibrium is achieved when the solid is in contact with a saturated solution of the ions. For copper iodate, the K_{sp} expression is shown in Equation 25.1 and the corresponding chemical equation is given in Equation 25.2.

$$K_{sp} = [Cu^{2+}][IO_3^-]^2 \tag{25.1}$$

$$Cu(IO_3)_2(s) \rightleftharpoons Cu^{2+}(aq) + 2IO_3^-(aq) \tag{25.2}$$

Copper iodate is a good salt to use in studying the relationship between temperature and K_{sp}. Its solubility is low enough that the saturated solutions will be fairly dilute and thus likely to behave ideally. (In an ideal solution, the ions behave completely independently of one another, and their presence does not perturb the interactions between solvent particles in any significant way.) On the other hand, the solubility of copper iodate is not so low that it is difficult to determine the molarities of the ions in the saturated solutions.

In this experiment, the K_{sp} values will be evaluated at several temperatures by measuring the concentration of the Cu^{2+} ions in the saturated solutions. The saturated solutions will be prepared by precipitating $Cu(IO_3)_2$ from mixtures of equal volumes of 0.200 M $CuSO_4$ and 0.400 M KIO_3. Because precipitation will remove Cu^{2+} and IO_3^- ions in a 1:2 ratio from mixtures that initially contained the ions in that ratio, the ratio of copper(II) ions to iodate ions in the saturated solution will also be 1:2. Thus, if $[Cu^{2+}] = s$ in a saturated solution, $[IO_3^-] = 2s$ in that solution. Therefore, K_{sp} will have a value of $4s^3$ for each saturated solution studied.

Variation of K_{sp} with Temperature

If our solutions are ideal and if $\Delta H°$ is reasonably constant, we can calculate important thermodynamic quantities from the dependence of K_{sp} on temperature. We begin with the fundamental expressions:

$$\Delta G° = \Delta H° - T\,\Delta S° \tag{25.3}$$

$$\Delta G° = -RT \ln K \tag{25.4}$$

These may be combined and rearranged as follows:

$$-RT \ln K = \Delta H° - T\,\Delta S° \tag{25.5}$$

$$\ln K = -(\Delta H°/RT) + (\Delta S°/R) \tag{25.6}$$

Note that Equation 25.6 is of the form

$$y = mx + b \tag{25.7}$$

where $y = \ln K$
$x = 1/T$
$m = -\Delta H°/R$
and $b = \Delta S°/R$

A plot of ln K versus $1/T$ should therefore be linear, with a slope equal to $-\Delta H°/R$ and an intercept of $\Delta S°/R$.

Spectrophotometric Determination of Cu^{2+}

Copper(II) ions exist in acidic solution as $Cu(H_2O)_4^{2+}$ ions, which give the solution a characteristic pale-blue color. The intensity of the blue color is increased greatly by the addition of aqueous NH_3, which converts the hydrated copper(II) ions to ammonia (or ammine) complexes. In 0.25 M aqueous ammonia, the predominant complex present is the tetraamminecopper(II) ion, $Cu(NH_3)_4^{2+}$, which shows a broad maximum at 610 nm. If the experimental conditions are carefully controlled, the absorbance of copper–ammonia solutions at 610 nm may be used as a sensitive and rapid method for determining copper concentrations.

To relate absorbance to copper concentration, you must prepare a calibration curve using a series of standard copper–ammonia solutions covering the concentration range 0.002 M to 0.02 M. Such a calibration curve is shown in Figure 19.1. The copper–ammonia solutions used to determine the concentration of Cu^{2+} ions in saturated copper iodate solutions must be prepared by the same procedure used to prepare the calibration curve, with the same NH_3 stock solution. The concentration of copper in a copper-ammonia solution can then be obtained by comparing its absorbance with the calibration curve. The equilibrium concentration of Cu^{2+} in the saturated copper iodate solution can be calculated by multiplying by 8 to account for the dilution of the copper in by ammonia and water.

PROCEDURE

Preparation of the Saturated Solutions

NOTE: These solutions must be prepared at least one laboratory period before the analyses are to be performed.

Clean and dry four 10-mL test tubes and two burets. Add 4.0 mL of stock 0.200 M $CuSO_4$ solution and 4.0 mL of stock 0.400 M KIO_3 solution, respectively, to each of the test tubes. Stir each solution vigorously to start precipitation of excess copper iodate. Place the test tubes in an ice bath (or an ice/water/rock salt bath) if necessary to initiate precipitation. Be sure precipitates have started to grow in each test tube before sealing the tubes and setting them aside so that equilibrium can be established.

Preparation of the Calibration Curve

Clean and rinse five matched spectrophotometer cells. Fill each cell with water and dry the cell on the outside using a soft, absorbent tissue. Inspect each cell to ensure that no dirt, air bubbles, or scratches are in the light path. Insert each cell in turn into the spectrophotometer and measure the absorbance. The cell that has the lowest absorbance should be noted and used as your reference. With that reference cell in the cell compartment, close the cover, and set the light control so that the meter reads 100% *T* (or zero absorbance). Replace the reference with each of the sample cells, close the cover, and read and record its %*T* (or absorbance). The difference between the absorbance of each sample cell and the absorbance of the reference cell should be used as corrections to all the absorbances measured in the sample cell.

Estimate the amount of standard 0.200 M $CuSO_4{\cdot}5H_2O$ solution you will need and put it in a clean, dry beaker. Record the exact concentration of the solution. Using volumetric flasks, pipets, and burets as needed, accurately dilute the $CuSO_4$ solution with water to prepare four solutions in which the $CuSO_4$ concentrations are 0.16 M, 0.080 M, 0.040 M, and 0.016 M. Use a pipet to transfer 1.00 mL of each solution to a separate 10-mL test tube; then add 4.00 mL of 0.5 M NH_3 solution and 3.00 mL of water to each test tube. Mix each solution well and measure its %*T* or absorbance at 610 nm with water in the reference cell. Apply the cell correction. Calculate the concentration of Cu^{2+} in each solution and plot the corrected absorbance versus concentration on graph paper. Draw a smooth curve through the experimental points.

Copper Analysis

Fill a water bath with room-temperature water. Place one of the test tubes containing the copper iodate precipitate and supernatant liquid in the bath. Agitate the contents of the test tube vigorously for five minutes. Then, centrifuge the test tube and use a pipet to transfer 1.00 mL of the clear supernatant liquid to clean, dry, labeled spectrophotometer cell that has had its absorbance correction measured previously.

Add ice to the water bath and stir its contents until all of the ice dissolves and the bath temperature is lowered to approximately 10°C. Then, repeat the procedure described in the previous paragraph with the second test tube containing the products of mixing $CuSO_4$ and KIO_3. After the cooled test tube has been centrifuged, transfer 1 mL of clear supernatant liquid from the second test tube to a second clean, dry, labeled spectrophotometer cell that has had its absorbance correction measured previously.

Add more ice to the water bath and stir its contents until the bath temperature is lowered to 0°C (there should be undissolved ice in the bath). Repeat the process of agitating and centrifuging the copper iodate mixture and transferring 1 mL of clear supernatant liquid from the third test tube to a third clean, dry, labeled spectrophotometer cell that has had its absorbance correction measured previously.

Stir rock salt into the ice/water mixture and lower the bath temperature to approximately –10°C. Place the last of the test tubes in which the copper sulfate and potassium iodate were mixed in this rock salt/ice/water mixture. Agitate the contents of the test tube and centrifuge the tube. Then, transfer 1 mL of clear supernatant liquid from the fourth test tube to a fourth clean, dry, labeled spectrophotometer cell that has had its absorbance correction measured previously.

Add 4.00 mL of the 0.5 M NH_3 solution and 3.00 mL of water to each cell and mix well. Measure the absorbance at 610 nm with water in the reference cell. Then calculate the corrected absorbance of each copper–ammonia solution.

Using the prepared calibration curve, calculate the concentration of $Cu(NH_3)_4^{2+}$ in each of the solutions measured. Next calculate the concentration of Cu^{2+} in each of the saturated copper iodate solutions.

Use the calculated solubility, s, to determine the value of K_{sp} at each temperature. Prepare a table of K_{sp}, ln K_{sp}, T, and $1/T$. Plot ln K_{sp} versus $1/T$ as described in the Principles section and calculate $\Delta H°$ and $\Delta S°$ from slope and intercept values. Finally, calculate $\Delta G°$ for the reaction at the standard temperature, 298 K.

Disposal of Reagents

The solutions used in this experiment may be flushed down the drain after neutralization and dilution.

Questions

1. Comment on the significance of each of the possible sources of error listed.
   a. The saturated solutions may not show ideal behavior.
   b. $\Delta H°$ and $\Delta S°$ may not be independent of temperature.
   c. Obtaining the intercept requires significant extrapolation.
2. Starting with Equation 25.6, show how to arrive at the Clausius-Clapeyron equation:

$$\ln \frac{K_2}{K_1} = -\left(\frac{\Delta H°}{R}\right)\left(\frac{1}{T_2} - \frac{1}{T_1}\right)$$

Date ______ Name ________________________________ Section ______ Desk Number ______

PRE-LAB EXERCISES FOR EXPERIMENT 25

These exercises are to be completed after you have read the experiment but before you come to the laboratory to perform it.

1. The solubility of AgCl in water is 3.97×10^{-6} M at 0°C and 1.91×10^{-4} M at 100°C. Calculate K_{sp} for AgCl at each temperature.

2. Calculate $\Delta H°$ for the formation of a saturated solution of AgCl.

3. Calculate $\Delta S°$ for the formation of a saturated solution of AgCl.

4. Calculate $\Delta G°$ for the formation of a saturated solution of AgCl at 25°C.

5. Calculate the slope of a line for which $x_1 = 1.50$, $y_1 = 3.00$, $x_2 = 1.00$, $y_2 = 3.50$. Which graph—(a), (b), or (c)—corresponds to the data given?

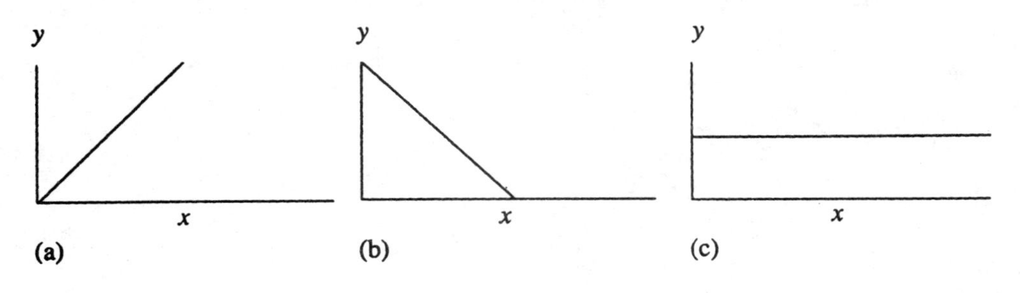

Date ______ Name ________________________________ Section ______ Desk Number ______

SUMMARY REPORT ON EXPERIMENT 25

Calibration Curve

Absorbance correction for sample cell	______	______	______	______
Original concentration of Cu^{2+}	0.16 M	0.080 M	0.040 M	0.016 M
Concentration of Cu^{2+} in copper–ammonia solutions	______	______	______	______
Absorbance of copper–ammonia solutions	______	______	______	______
Corrected absorbance	______	______	______	______

Copper Analysis

Absorbance correction for sample cell	______	______	______	______
Celsius temperature	______	______	______	______
Kelvin temperasture	______	______	______	______
Absorbance of $Cu(NH_3)_4^{2+}$ solution	______	______	______	______
Corrected absorbance of $Cu(NH_3)_4^{2+}$	______	______	______	______
Concentration of $Cu(NH_3)_4^{2+}$ in sample cell	______	______	______	______
Concentration of Cu^{2+} in saturated $Cu(IO_3)_2$	______	______	______	______
K_{sp}	______	______	______	______
$\ln K_{sp}$	______	______	______	______

Procedure for calculating $\Delta H°$ and $\Delta S°$ from graphical data:

Results: $\Delta H°$ ______________

$\Delta S°$ ______________

$\Delta G°_{298}$ ______________

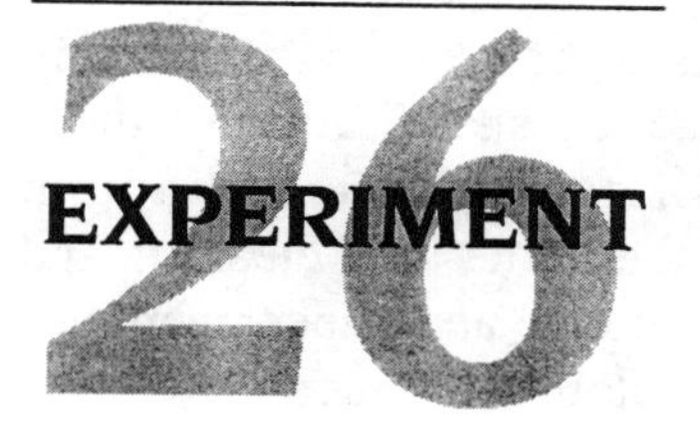

Analysis of Bleach for Hypochlorite Content

Laboratory Time Required

Three hours

Special Equipment and Supplies

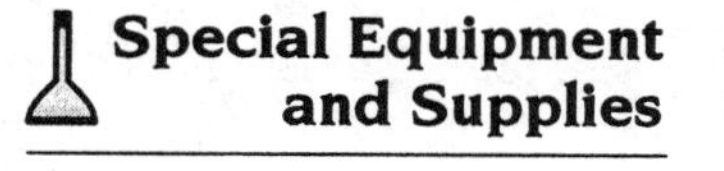

Balance
Burets
Buret clamp
Pipet
Pipet bulb
Volumetric flask, 100-mL

Bleach
Sodium thiosulfate pentahydrate(s)
Standard 0.02 M potassium iodate
Potassium iodide(s)
0.5 M sulfuric acid
0.2% starch
1 M sodium hydroxide

Safety

Most household bleaches are alkaline solutions that contain a strong oxidizing agent, the hypochlorite ion. This combination of caustic and oxidizing properties makes bleach solutions hazardous to the eyes and skin. Exposure of skin to bleach may cause blistering. Serious eye damage may result if bleach gets in the eyes; WEAR GOGGLES WHEN WORKING WITH BLEACH. DO NOT PIPET BY MOUTH. USE THE PIPET BULB. Chlorine gas might be released when bleach is acidified; chlorine is a poison that can cause lung and eye damage. Avoid skin contact with potassium iodate—a strong oxidizer—and sulfuric acid—a corrosive acid that is also a good oxidizing agent.

First Aid

Skin contact with sodium thiosulfate, potassium iodate, potassium iodide and starch solutions: **These chemicals present only minor hazards. Flush the chemical from the skin or eyes with copious amounts of water. No additional attention should be needed.**

Skin or eye contact with bleach, sulfuric acid, or sodium hydroxide: **These solutions present the greatest threat, especially to the eyes. After the chemicals have been thoroughly flushed from the affected areas, consult a doctor immediately if the eyes have been exposed to any of the chemicals or if there is prolonged skin discomfort.**

The chloralkali industry is based on the electrochemical processing of concentrated solutions of sodium chloride, known as **brine**. The electrolysis of brine produces sodium hydroxide and chlorine gas. Reaction and dilution of these chemicals produces bleach, dilute solutions of sodium hypochlorite. Bleach is used in the home as a cleanser and disinfectant. It is also used in the textile and paper industries to remove colored impurities from cloth and paper.

PRINCIPLES

The active ingredient in bleach is sodium hypochlorite, NaOCl, which is formed by the reaction shown in Equation 26.1. Because this reaction is reversed by acid, bleach solutions must be disposed of properly to avoid the formation of chlorine gas.

$$Cl_2(g) + 2Na^+(aq) + 2OH^-(aq) \rightleftharpoons 2Na^+(aq) + Cl^-(aq) + OCl^-(aq) + H_2O \quad (26.1)$$

The reactivity of bleach is due to the presence of chlorine in the +1 oxidation state in the OCl^- ion. Because chlorine is a fairly active nonmetal, it is relatively easy to reduce chlorine from the +1 to the −1 state. This makes the hypochlorite ion a good oxidizing agent.

In the analysis of bleach for hypochlorite content that you will perform, a diluted sample of bleach will be allowed to react with potassium iodide. In the course of this reaction, the iodide ion will be oxidized to iodine while the hypochlorite ion will be reduced to chloride (see Equation 26.2).

$$OCl^- + 2I^- + 2H^+ \longrightarrow I_2 + Cl^- + H_2O \quad (26.2)$$

Iodine is not very soluble in water. The reaction shown in Equation 26.2 is carried out with a large excess of potassium iodide so that the iodine formed is converted to the triiodide ion. Like many ionic compounds, potassium triiodide (KI_3) is very soluble in water (see Equations 26.3 and 26.4).

$$I_2 + I^- \longrightarrow I_3^- \quad (26.3)$$

$$OCl^- + 3I^- + 2H^+ \longrightarrow I_3^- + Cl^- + H_2O \quad (26.4)$$

The analysis of bleach for hypochlorite content will be accomplished by determining how much triiodide ion is produced. This is a routine procedure in which a solution of triiodide ion is titrated with a solution of sodium thiosulfate (see Equation 26.5).

$$I_3^- + 2S_2O_3^{2-} \longrightarrow 3I^- + S_4O_6^{2-} \quad (26.5)$$

The reaction between triiodide and thiosulfate ions is followed visually, with the deep brown color of the triiodide ions fading to yellow as the reaction with thiosulfate converts I_3^- ions to the colorless iodide ions. Starch is added shortly before the end point is reached. The remaining triiodide ions form a blue complex with the starch. This complex is then destroyed by the further addition of $S_2O_3^{2-}$ ions. When all I_3^- ions have been converted to I^-, the titration mixture becomes colorless, signaling the end point.

The sodium thiosulfate solution must be standardized prior to use. This is conveniently accomplished by titrating a known amount of triiodide ion that has been generated by the reaction shown in Equation 26.6. The potassium iodate used in this reaction is a primary standard.

$$IO_3^- + 8I^- + 6H^+ \longrightarrow 3I_3^- + 3H_2O \qquad (26.6)$$

PROCEDURE

Standardization of Thiosulfate Solution

Prepare an approximately 0.08 M thiosulfate solution by dissolving 4 g of $Na_2S_2O_3 \cdot 5H_2O$ (sodium thiosulfate pentahydrate) in 200 mL of distilled water. Be sure the solution is well mixed. Rinse a buret with two small volumes of distilled water, followed by two small volumes of your thiosulfate solution. Then fill the buret with the solution. Record the initial buret reading.

Record the molarity of the standard potassium iodate (KIO_3) solution provided for your use. Rinse a second buret with two small volumes of distilled water, followed by two small volumes of the standard KIO_3 solution. Then fill the second buret with the iodate solution and record the initial buret reading for the second buret. Deliver 20 mL of KIO_3 solution into a clean, 250-mL Erlenmeyer flask and record the final buret reading for the iodate buret. Add 20 mL of distilled water and 2 g of potassium iodide (KI) to the flask. Swirl the flask to dissolve the solid potassium iodide. Continue to swirl the flask while adding 20 mL of 0.5 M H_2SO_4. The mixture in the Erlenmeyer flask should take on the deep red-brown color characteristic of I_3^- ions.

Titrate the I_3^- ions with $S_2O_3^{2-}$, adding 1-mL increments of thiosulfate initially, but reducing the increment size as the color of the titration mixture turns from brown to yellow. Then add 5 mL of 0.2% starch solution. Consult your instructor if the titration mixture does not develop a blue color after starch has been added and the flask has been swirled. Use your wash bottle to rinse all splattered drops from the walls of the flask into the titration mixture. Resume the addition of thiosulfate, adding titrant by drops until the blue color of the starch–iodine complex disappears completely. Usually, only a few additional drops are required. Record the final reading of the thiosulfate buret and calculate the molarity of your thiosulfate solution. Repeat the procedure. Report the molarity obtained in each trial and the average molarity.

Titration of Bleach

Refill your buret with $Na_2S_2O_3$ solution and record the initial reading. Clean a 10-mL volumetric pipet thoroughly. Rinse it twice with distilled water, followed by two small volumes of household bleach. Then use the pipet to deliver 10.00 mL of bleach to a clean (but not necessarily dry) 100-mL volumetric flask.

CAUTION AVOID GETTING BLEACH IN YOUR EYES. IF BLEACH DOES GET IN YOUR EYES, RINSE THEM QUICKLY AND THOROUGHLY AT THE EYEWASH FOUNTAIN. SEEK MEDICAL ATTENTION.

Add 50 mL of distilled water to the bleach and swirl the flask carefully to mix its contents. Dilute the solution with distilled water to the 100-mL mark, then invert the flask several times to promote mixing of the solution.

Dissolve approximately 2 g of KI in a mixture of 20 mL of distilled water and 20 mL of 0.5 M H_2SO_4 in a 250-mL Erlenmeyer flask. Swirl the flask until all solid has dissolved. If the solution turns yellow at this point, discard it. Prepare a new solution, being sure that the flask is clean and that the H_2SO_4 solution is at or below room temperature.

Rinse the pipet with two small portions of distilled water, followed by two small portions of the diluted bleach. Use the rinsed pipet to deliver 10.00 mL of diluted bleach to the flask containing the solution of KI. Swirl the Erlenmeyer flask to allow the contents to mix. Using the procedure for the standardization of the thiosulfate solution, titrate the triiodide ions that result from the reaction of the bleach and the KI solution. Repeat the titration. Report the experimental molarity of the diluted bleach for each titration, and the average molarity. Also report the calculated molarity of commercial bleach.

Disposal of Reagents

The potassium iodate solution should be reduced to potassium iodide before disposal. This may be achieved by following essentially the same procedure as used in the experiment. Dissolve excess KI in the KIO_3 solution. Then add a few mL of 0.5 M H_2SO_4 to acidify the mixture. Mix well. Next add sodium thiosulfate solution until the yellow-brown iodine color has disappeared. Dilute the solution and rinse it down the drain.

The 0.5 M H_2SO_4 solution should be neutralized with aqueous NaOH, and then rinsed down the drain.

All other solutions may be diluted and then poured down the drain with a steady stream of water. Do not dispose of undiluted bleach as it may react with other chemicals in the drainage system to produce poisonous chlorine gas anywhere in the drain pipes.

Questions

1. Most commercial bleaches are 5.25% sodium hypochlorite (by mass). Assume the density of such a bleach is 1.0 g/mL and calculate the molarity of the commercial bleach with respect to hypochlorite ion.
2. How does the molarity obtained in answer to Question 1 compare with your experimental value of the molarity?
3. Mixing bleach with hydrochloric acid can result in the release of noxious chlorine gas. For this reason, many household cleansers, which are very acidic, carry warning labels stating that they should not be mixed with bleach. Why were you able to mix samples of bleach with acid in this experiment and not produce chlorine?

Date ______ Name ________________________________ Section ______ Desk Number ______

PRE-LAB EXERCISES FOR EXPERIMENT 26

These exercises are to be completed after you have read the experiment but before you come to the laboratory to perform it.

1. A 10-mL pipet was used to deliver a sample of commercial bleach to a 100-mL volumetric flask. Distilled water was added and the flask was inverted several times to produce a uniform, 100.00-mL sample of diluted bleach. The pipet was rinsed with diluted bleach and then used to deliver 10.00 mL of diluted bleach to an Erlenmeyer flask. Sufficient quantities of H_2SO_4 and KI were added to convert all OCl^- ions to chloride ions, with consequent conversion of some iodide ions to I_3^- ions. The 10.00 mL of diluted bleach required the addition of 26.37 mL of 0.0512 M thiosulfate for complete reaction. Use this data to compute the molarity of the undiluted bleach with respect to sodium hypochlorite.

2. Use your answer to Question 1 to find the percent NaOCl in the undiluted bleach. State any assumptions you make in performing the calculation.

Date ______ Name ______________________ Section ______ Desk Number ______

SUMMARY REPORT ON EXPERIMENT 26

Molarity of KIO_3 ______

Standardization of Thiosulfate Solution

	Trial 1	*Trial 2*	*Trial 3**	
Final buret reading, KIO_3				
Initial buret reading, KIO_3				
Volume of KIO_3 delivered	20.00	20.00	20.00	20.00
Final buret reading, $Na_2S_2O_3$				
Initial buret reading, $Na_2S_2O_3$				
Volume of $Na_2S_2O_3$ used	29.70	29.50	32.0	31.5
Molarity of $Na_2S_2O_3$				
Average molarity of $Na_2S_2O_3$				

Titration of Hypochlorite

	Trial 1	*Trial 2*	*Trial 3**	
Volume of diluted bleach	10 mL	10 mL	10 ml	10 ml
Final buret reading, $Na_2S_2O_3$	26.55	21.55		
Initial buret reading, $Na_2S_2O_3$	7.5	3.0		
Volume $Na_2S_2O_3$ used	19.05	18.55	18.71	18.79
Molarity of NaOCl in dilute bleach				
Average molarity of NaOCl in dilute bleach				
Molarity of NaOCl in undiluted bleach				
Percent NaOCl by mass in undiluted bleach				

*Optional

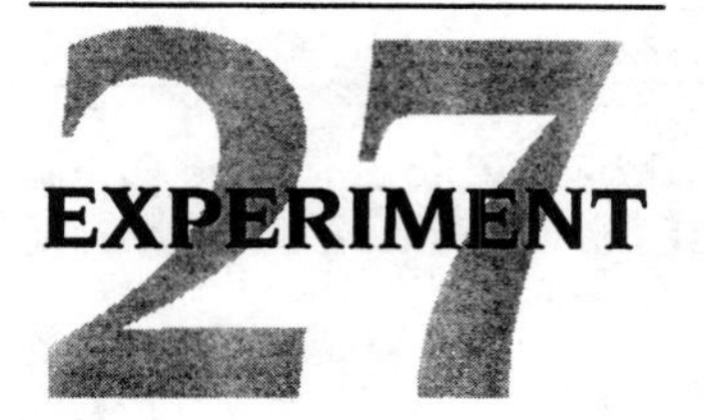

Redox Titration of Ascorbic Acid

Laboratory Time Required

Three hours

Special Equipment and Supplies

Burets
Buret clamp
Pineapple or grapefruit juice
Starch indicator
Sodium thiosulfate pentahydrate
Potassium iodide
0.02 M Potassium iodate (standard)
0.5 M Sulfuric acid

Safety

Spills of sulfuric acid should be flushed away with large amounts of water.

Although it may be tempting, **do not** drink any unused portion of your juice. Drinking is not permitted in the lab. Any unused juice must be discarded.

First Aid

If your skin comes in contact with sulfuric acid, wash the affected area thoroughly. If acid gets in your eyes, flush them with water for at least 20 minutes. Then seek medical attention.

In Experiment 18, a procedure for analyzing pineapple or grapefruit juice by an acid/base titration was described. In this experiment, a second method of analysis, based on the chemical oxidation of ascorbic acid to dehydroascorbic acid, is employed. This method is somewhat more complicated (both in experimental design and in calculation) than the acid/base titration. However, it is also more specific than the acid/base titration and it illustrates the commonly used techniques of iodometric titration and back titration.

PRINCIPLES

Ascorbic acid (vitamin C) is readily oxidized to dehydroascorbic acid (see Equation 27.1). In this experiment, iodine, in the form of the triiodide complex ion, will be used as the oxidizing accent agent.

$$C_6H_8O_6 + I_3^- \longrightarrow C_6H_6O_6 + 3I^- + 2H^+ \tag{27.1}$$

This triiodide ion will be generated by the addition of an excess of potassium iodide to a standard solution of potassium iodate (see Equation 27.2).

$$IO_3^- + 8I^- + 6H^+ \longrightarrow 3\,I_3^- + 3H_2O \tag{27.2}$$

The amount of I_3^- generated will be more than sufficient to react with all of the ascorbic acid in the sample. The exact amount of excess triiodide ion will be determined by titration with sodium thiosulfate, $Na_2S_2O_3$ (see Equation 27.3). This method of generating a known amount of reagent, permitting some of it to react with a sample and, ultimately, determining how much of the reagent is left unreacted, is a standard technique, known as a **back titration**.

$$I_3^- + 2S_2O_3^{2-} \longrightarrow 3I^- + S_4O_6^{2-} \tag{27.3}$$

Although the calculations involved in obtaining an answer from the back titration are more complicated than those involved with the acid/base titration, they are still relatively straightforward. All that is required is that the data from the various titrations be combined with the stoichiometric relations given in Equations 27.1 through 27.3.

PRINCIPLES

A. Standardization of the Sodium Thiosulfate Solution

Prepare an 0.08 M sodium thiosulfate solution for the experiment by dissolving 4 g of $Na_2S_2O_3{\cdot}5H_2O$ in 200 mL of distilled water. Mix the solution well and label the container appropriately. Rinse a clean buret with two small portions of distilled water, followed by two small portions of the thiosulfate solution. Then fill the buret with the solution. Record the initial buret reading.

Rinse another clean buret with two small portions of distilled water, followed by two small portions of standard potassium iodate solution. Then fill the second buret with the iodate solution. Accurately dispense approximately 20 mL of KIO_3 solution into a clean Erlenmeyer flask. Record the initial and final buret readings.

Add 20 mL of 0.5 M sulfuric acid and 2 g of solid potassium iodide to the flask containing the 20 mL of KIO_3 solution. Mix well; the solution should take on a deep reddish-brown color. Titrate immediately with the $S_2O_3^{2-}$ solution. As the color of the triiodide solution fades, add smaller increments of the thiosulfate solution. When the titration mixture is yellow, add 2 mL of starch indicator solution to form the characteristic blue starch–iodine complex.

Continue the titration until the blue color first disappears. Record the final buret reading for the $S_2O_3^{2-}$ buret.

Repeat the titration. Report the molarity of the thiosulfate solution from each trial and the average molarity.

B. Titration of a Sample of Juice

Use a graduated cylinder to measure 20 mL of pineapple or grapefruit juice. Transfer the juice to a clean 250-mL Erlenmeyer flask. Add 20 mL of 0.5 M H_2SO_4 to the flask containing the juice. Swirl the flask.

Fill the burets once again with the appropriate solutions. Add 2 g of potassium iodide and 20 mL of potassium iodate to the Erlenmeyer flask containing the juice and sulfuric acid. Titrate immediately with sodium thiosulfate. Add starch, as before, when the iodine color fades. Continue titrating until the end point has been reached.

Repeat the titration. Report the chemical amount of vitamin C found in each portion of the juice and the molarity of vitamin C in each sample. Determine the number of mg of vitamin C contained in a 6-oz. serving of juice.

If you performed Experiment 18, compare your value for the vitamin C content of the juice (determined via acid/base titration) with the results found in this experiment.

Disposal of Reagents

The potassium iodate solution should be reduced to potassium iodide before disposal. This may be achieved by following essentially the same procedure used in the experiment. Dissolve excess KI in the KIO_3 solution. Then add a few mL of 0.5 M H_2SO_4 to acidify the mixture. Mix well. Add sodium thiosulfate solution until the yellow-brown iodine color has disappeared. Dilute the solution and flush it down the drain.

The 0.5 M H_2SO_4 solution should be neutralized with aqueous NaOH and then flushed down the drain.

All other solutions may be diluted and flushed down the drain.

Questions

If you performed Experiment 18, compare the two methods of analysis. Which was more difficult to carry out? Which seemed more subject to error? In which method did the color and turbidity of the juice cause more interference?

Date ______ Name ______________________ Section ______ Desk Number ______

PRE-LAB EXERCISES FOR EXPERIMENT 27

These exercises are to be completed after you have read the experiment but before you come to the laboratory to perform it.

1. A 20.0-mL portion of pineapple juice was placed in a flask with 20 mL of 0.5 M H_2SO_4. Then 2 g of potassium iodide and 19.75 mL of 0.02001 M potassium iodate were added to the sample. Titration with sodium thiosulfate required the use of 21.04 mL of 0.07965 M $Na_2S_2O_3$. Find the molarity of ascorbic acid in the juice.

2. Find the uncertainty associated with your answer to Exercise 1. Assume the uncertainty associated with reading the buret is ±0.03 mL. Assume an uncertainty of ±0.5 mL associated with using the graduated cylinder. Assume an uncertainty of ±3 in the last place for the molarities of the $Na_2S_2O_3$ and the KIO_3.

Date ______ Name ______________________________ Section ______ Desk Number ______

SUMMARY REPORT ON EXPERIMENT 27

A. Standardization of Sodium Thiosulfate

Molarity of KIO_3 ______

	Trial 1	*Trial 2*	*Trial 3**
Final buret reading, KIO_3	______	______	______
Initial buret reading, KIO_3	______	______	______
Volume of KIO_3 delivered	______	______	______
Final buret reading, $Na_2S_2O_3$	______	______	______
Initial buret reading, $Na_2S_2O_3$	______	______	______
Volume of $Na_2S_2O_3$ used	______	______	______
Calculated molarity of $Na_2S_2O_3$	______	______	______
Average molarity of $Na_2S_2O_3$		______	

B. Titration of a Sample of Juice

	Trial 1	*Trial 2*	*Trial 3**
Volume of juice used	______	______	______
Final buret reading, KIO_3	______	______	______
Initial buret reading, KIO_3	______	______	______
Volume of KIO_3 delivered	______	______	______
Final buret reading, $Na_2S_2O_3$	______	______	______
Initial buret reading, $Na_2S_2O_3$	______	______	______
Volume of $Na_2S_2O_3$ used	______	______	______
Moles of ascorbic acid titrated	______	______	______
Average mg vitamin C per 6-oz. juice		______	
Average mg vitamin C per 6-oz. juice (from Exp.18)		______	

*Optional

EXPERIMENT 28

Determination of the Solubility Product of Copper Iodate[1]

Laboratory Time Required

Three and one-half hours, divided between two periods, for well-prepared students working in teams.

Special Equipment and Supplies

For preparing saturated solutions

Buret	0.200 M $CuSO_4 \cdot 5H_2O$
	0.300 M KIO_3

For analyzing saturated solutions

Spectrophotometer	0.200 M $CuSO_4 \cdot 5H_2O$
Spectrophotometer cells	0.02000 M KIO_3
Buret	0.5 M Aqueous NH_3
Centrifuge	3 M Acetic acid
Mohr pipet	1 M Sodium citrate
Volumetric flask	$Na_2S_2O_3 \cdot 5H_2O$ (s)
	Starch indicator solution
	KI Crystals

Safety

None of the chemicals used in this experiment poses a serious health hazard. However, the KIO_3 is an oxidizer, acetic acid is a weak acid, and aqueous NH_3 is a base. Avoid exposing your skin or eyes to contact with these or other chemicals.

First Aid

Rinse the exposed area thoroughly to flush away chemicals.

In this experiment, solid copper iodate will be precipitated from aqueous solutions prepared by mixing $CuSO_4$ and KIO_3 stock solutions in a number of different volume ratios. The concentrations of Cu^{2+} left in the saturated solutions will be determined by a spectrophotometric method based on the absorption of the $Cu(NH_3)_4^{2+}$ ion. The concentrations of IO_3^- in the saturated solutions will be determined by a volumetric method. These data will be used to calculate the solubility product for copper iodate.

[1]Adapted from an article by J.A. Campbell, *J. Chem. Educ.*, **42** (1965) p.488.

PRINCIPLES

The Solubility Product

The equilibrium between the sparingly soluble salt $Cu(IO_3)_2$ and its saturated solution is represented bv Equation 28.1.

$$Cu(IO_3)_2(s) \rightleftharpoons Cu^{2+}(aq) + 2IO_3^-(aq) \quad (28.1)$$

The equilibrium expression for this reaction is given in Equation 28.2, in which K_{sp} is the solubility product constant for copper iodate and $[Cu^{2+}]$ and $[IO_3^-]$ are the molarities of Cu^{2+} and IO_3^-, respectively, in the saturated solution.

$$K_{sp} = [Cu^{2+}][IO_3^-]^2 \quad (28.2)$$

A reasonable procedure for evaluating K_{sp} would be to measure the concentrations of Cu^{2+} and IO_3^- in a saturated solution independently and then calculate K_{sp} using Equation 28.2. In the saturated solution, however, the concentration of one or both ionic species may be too low for accurate measurement, resulting in an inaccurate value for K_{sp}. This problem can be eliminated by studying solutions in which the equilibrium concentration of either Cu^{2+} or IO_3^- is large enough for accurate measurement. The concentration of the other ionic species can then be accurately calculated using your knowledge of the original composition of the mixture and the stoichiometry of the precipitation reaction.

Spectrophotometric Determination of Cu^{2+}

Copper(II) ions exist in acidic solution as $Cu(H_2O)_4^{2+}$ ions, which give the solution a characteristic pale-blue color. The intensity of the blue color is greatly increased by the addition of aqueous NH_3, which converts the hydrated copper(II) ions to ammonia (or ammine) complexes. Figure 28.1 shows the absorption spectrum of copper(II) in 0.25 M aqueous ammonia. Under these conditions, the predominant complex present is the tetraamminecopper(II) ion, $Cu(NH_3)_4^{2+}$, which shows a broad maximum at 610 nm. By carefully controlling the experimental conditions, the absorbance of copper–ammonia solutions at 610 nm may be used as a sensitive and rapid method for determining copper concentrations.

To relate absorbance to copper concentration, you must prepare a calibration curve using a series of standard copper-ammonia solutions covering the concentration range 0.002 M to 0.02 M. Such a calibration curve is shown in Figure 19.1. The copper–ammonia solutions used to determine the concentration of Cu^{2+} ions in saturated copper iodate solutions must be prepared by the same procedure used to prepare the calibration curve, with the same NH_3 stock solution. The concentration of copper in a copper-ammonia solution can then be obtained by comparing its absorbance with the calibration curve. The equilibrium concentration of Cu^{2+} in the saturated copper iodate solution can be calculated by multiplying by 8 to account for the dilution of the copper by ammonia and water.

Volumetric Determination of Iodate

The concentration of IO_3^- will be determined by the familiar iodometric reaction: reduction of IO_3^- to I_3^- via reaction with excess I^- in acid solution and subsequent

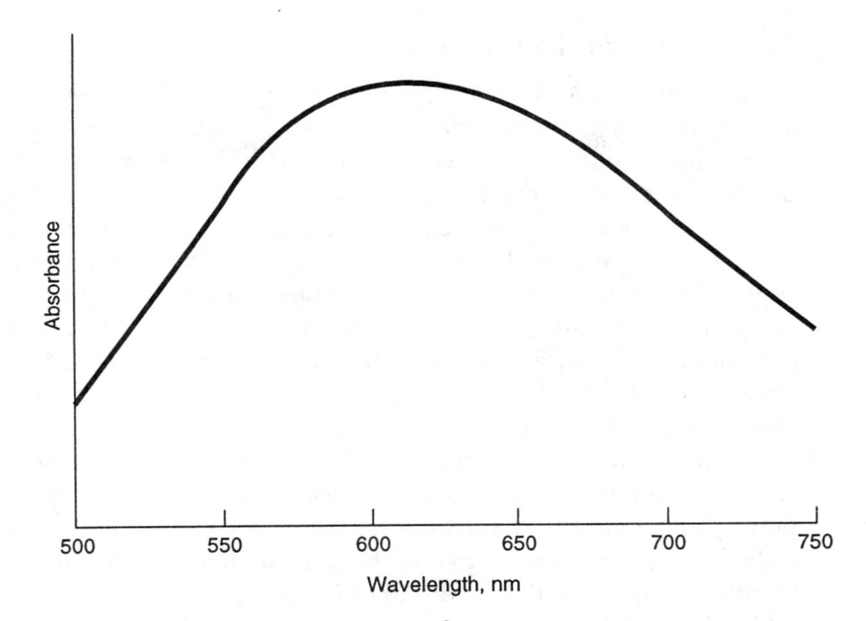

FIGURE 28.1 Absorption spectrum of Cu^{2+} in 0.25 M NH_3.

titration with thiosulfate to the starch end point. You can review the pertinent equations by reading Experiments 26 and 27.

PROCEDURE

Preparation of the Saturated Solutions

NOTE: These solutions must be prepared at least one laboratory period before the analyses are to be performed.

Clean and label five 10-mL test tubes and two burets. Add 3.5 mL, 3.5 mL, 4.0 mL, 4.5 mL, and 5.0 mL volumes of stock 0.200 M $CuSO_4$ solution and 4.5 mL, 4.5 mL, 4.0 mL, 3.5 mL, and 3.0 mL volumes of stock 0.300 M KIO_3 solution, respectively, to the five test tubes. Stir each solution vigorously to start precipitation of excess copper iodate. Place the test tubes in an ice bath (or an ice/water/rock salt bath) if necessary to initiate precipitation. Be sure precipitates have started to grow in each test tube before sealing the tubes and setting them aside so that equilibrium can be established.

Analyses

The analyses should be performed by teams of four students. One student should prepare the copper/ammonia calibration curve. Another should perform the analysis of three mixtures for copper content. A third should prepare and standardize the thiosulfate solution. The fourth student should perform the analysis of two mixtures for iodate content. Team members should share their data and each should calculate values of K_{sp} for copper iodate.

Preparation of the Calibration Curve (Team Member #1)

Clean and rinse four matched spectrophotometer cells. Fill each cell with water and dry the cell on the outside using a soft, absorbent tissue. Inspect each cell to ensure that no dirt, air bubbles, or scratches are in the light path. Insert each cell in turn into the spectrophotometer and meaure the absorbance. The cell that has the lowest absorbance should be noted and used as your reference. With that reference cell in the cell compartment, close the cover, and set the light control so that the meter reads 100% *T* (or zero absorbance). Replace the reference with each of the sample cell, close the cover, and read and record its %*T* (or absorbance). The difference between the absorbance of each sample cell and the absorbance of the reference cell should be used as a correction to each absorbance measured in the sample cells.

Estimate the amount of standard 0.200 M $CuSO_4 \cdot 5H_2O$ solution you will need and put it in a clean, dry beaker. Record the exact concentration of the solution. Using volumetric flasks, pipets, and burets as needed, accurately dilute the $CuSO_4$ solution with water to prepare four solutions in which the $CuSO_4$ concentrations are 0.16 M, 0.080 M, 0.040 M, and 0.016 M. Use a pipet to transfer 1.00 mL of each solution to a separate 10-mL test tube; then add 4.00 mL of 0.5 M NH_3 solution and 3.00 mL of water to each test tube. Mix each solution well and measure its %*T* or absorbance at 610 nm with water in the reference cell. Apply the cell correction. Calculate the concentration of Cu^{2+} in each solution and plot the corrected absorbance versus concentration on graph paper. Draw a smooth curve through the experimental points.

Copper Analysis (Team Member # 2)

Clean and rinse two matched spectrophotometer cells. Fill each cell with water and dry the cell on the outside using a soft, absorbent tissue. Inspect each cell to ensure that no dirt, air bubbles, or scratches are in the light path. Insert each cell in turn into the spectrophotometer and meaure the absorbance. The cell that has the lower absorbance should be noted and used as your reference. With that reference cell in the cell compartment, close the cover, and set the light control so that the meter reads 100% *T* (or zero absorbance). Replace the reference with the sample cell, close the cover, and read and record its %*T* (or absorbance). The difference between the absorbance of the sample cell and the absorbance of the reference cell should be used as a correction to all the absorbances measured in the sample cell.

Centrifuge each $CuSO_4 \cdot KIO_3$ mixture for about 3 minutes using a similar test tube, containing water, as a counterbalance. Set aside the test tubes containing 3.5 mL of $CuSO_4$ for the IO_3^- analysis. Use a pipet to transfer 1.00 mL of the clear supernatant solution from each of the other test tubes to a clean, dry, labeled test tube. Be very careful not to transfer any of the precipitate. Add 4.00 mL of the 0.5 M NH_3 solution and 3.00 mL of water to each test tube and mix well. Measure the absorbance at 610 nm with water in the reference cell. Then calculate the corrected absorbance of each copper-ammonia solution.

Using the prepared calibration curve, calculate the concentration of $Cu(NH_3)_4^{2+}$ in each of the solutions measured. Next calculate the concentration of Cu^{2+} in each of the saturated copper iodate solutions. From the original concentrations of Cu^{2+} and IO_3^- and the concentration of Cu^{2+} at equilibrium, calculate for each mixture:

1. The change observed in the Cu^{2+} concentration.
2. The change in the IO_3^- concentration.
3. The equilibrium concentration of IO_3^-.

Calculate K_{sp} using the results for each mixture. Report the three values for K_{sp} and their average.

Preparation and Standardization of the Thiosulfate Solution (Team Member #3)

Prepare an 0.02 M sodium thiosulfate solution by dissolving 1.5 g of $Na_2S_2O_3 \cdot 5H_2O$ in 300 mL of distilled water. Mix the solution well and label the container appropriately. Rinse a clean buret with two small portions of distilled water, followed by two small portions of the thiosulfate solution. Then fill the buret with the solution. Record the initial buret reading.

Rinse a clean pipet with two small portions of distilled water, followed by two small portions of standard potassium iodate solution. Then fill the pipet with the iodate solution. Accurately dispense 5.00 mL of KIO_3 solution into a clean Erlenmeyer flask.

Add 5 mL of 0.5 M sulfuric acid and 0.5 g of solid potassium iodide to the flask containing the 5.00 mL of KIO_3 solution. Mix well; the solution should take on a deep reddish-brown color. Titrate immediately with the $S_2O_3^{2-}$ solution. As the color of the triiodide solution fades, add smaller increments of the thiosulfate solution. When the titration mixture is yellow, add 0.5 mL of starch indicator solution to form the characteristic blue starch–iodine complex. Continue the titration until the blue color first disappears. Record the final buret reading.

Repeat the titration. Report the molarity of the thiosulfate solution from each trial and the average molarity.

Iodate Analysis (Team Member #4)

Using a pipet, transfer 4.00 mL of clear solution from one of the test tubes set aside for IO_3^- analysis to a 125-mL flask. Add 15 mL of distilled water, 1 mL of 3 M acetic acid, 1 g of potassium iodide crystals, and swirl gently to dissolve the potassium iodide. Add 5 mL of 1 M sodium citrate to prevent the precipitation of CuI; swirl until the solution is clear. Titrate with standardized 0.0200 M $Na_2S_2O_3$ solution until the iodine color fades to pale yellow. Then add 1 mL of fresh starch solution and carefully continue the titration, using the first disappearance of the blue starch–iodine color as the end point.

Calculate the equilibrium concentrations of IO_3^- and Cu^{2+} in the copper iodate solution and use these values to calculate K_{sp}. Repeat the analysis using the remaining solution. Report the two values of K_{sp} and their average.

Disposal of Reagents

The solutions used in this experiment may be flushed down the drain, after neutralization and dilution.

Questions

1. Suggest reasons why identical values of K_{sp} were not obtained for the five solutions that were analyzed.
2. When aqueous NH_3 is added, drop by drop, to solutions containing Cu^{2+} ions, a precipitate forms that redissolves in excess ammonia. Explain.
3. Consider a mixture prepared by mixing 4.0 mL of 0.200 M $CuSO_4$ with 4.0 mL of 0.300 M KIO_3. Which analyses (copper and/or iodate) could be performed on this mixture if K_{sp} were equal to:
   a. 1.0×10^{-2}?
   b. 1.0×10^{-8}?
   c. 1.0×10^{-20}?

Date ______ Name ______________________ Section ______ Desk Number ______

PRE-LAB EXERCICSES FOR EXPERIMENT 28

These exercises are to be completed after you have read the experiment but before you come to the laboratory to perform it.

1. A precipitate of $Cu(IO_3)_2$ was prepared by mixing 3.5 mL of 0.200 M $CuSO_4$ with 4.5 mL of 0.300 M KIO_3. Analysis revealed that the concentration of iodate ions left behind in the saturated solution was 0.0138 M. Find the value of K_{sp} for copper(II) iodate.

2. Explain briefly why the copper and iodate ions do not need to be mixed in a 1:2 ratio. What would be the problems in analyzing mixtures in which one ion is very much in excess?

3. Solid KIO_3 is often used as a primary standard for standardizing $Na_2S_2O_3$ solutions. A typical procedure is to dissolve a weighed sample of KIO_3 in 100.0 mL of solution, add a 10.00 mL aliquot of the resulting solution to KI in dilute acid, and titrate the I_3^- with sodium thiosulfate solution. What mass of KIO_3 is needed to prepare 100.0 mL of solution if a 10.00 mL aliquot produces enough I_3^- to react with 30.00 mL of 0.02 M $Na_2S_2O_3$ solution?

Date ______ Name ____________________ Section ______ Desk Number ______

SUMMARY REPORT ON EXPERIMENT 28

Preparation of the Saturated Solutions

Concentration of the standard $CuSO_4$ solution ____________

Concentration of the standard KIO_3 solution ____________

Analysis of the saturated solutions

Calibration Curve

Absorbance correction for sample cell	______	______	______	______
Original concentration of Cu^{2+}	0.16 M	0.080 M	0.040 M	0.016 M
Concentration of Cu^{2+} in copper–ammonia solutions	______	______	______	______
Absorbance of copper–ammonia solutions	______	______	______	______
Corrected absorbance	______	______	______	______

Copper Analysis—Data

Absorbance correction for sample cell ______ ______ ______

	Solution Number		
	3	4	5
mL of standard $CuSO_4$ used	______	______	______
mL of standard KIO_3 used	______	______	______
Absorbance of $Cu(NH_4)_4^{2+}$ solution	______	______	______
Corrected absorbance of $Cu(NH_3)_4^{2+}$	______	______	______
Concentration of $Cu(NH_3)_4^{2+}$ in sample cell	______	______	______
Concentration of Cu^{2+} in saturated $Cu(IO_3)_2$	______	______	______

Calculated Values and Results

	Solution Number		
	3	4	5
Initial Cu^{2+} concentration	______	______	______
Concentration of Cu^{2+} at equilibrium	______	______	______
Change in the concentration of Cu^{2+}	______	______	______
Initial IO_3^- concentration	______	______	______
Change in the concentration of IO_3^-	______	______	______
Concentration of IO_3^- at equilibrium	______	______	______
Calculated K_{sp} for $Cu(IO_3)_2$	______	______	______
Average value of K_{sp} for $Cu(IO_3)_2$		______	

Date ______ Name ______________________________ Section ______ Desk Number ______

Standardization of Thiosulfate Solution

Concentration of the standard KIO_3 solution ______________

	Trial 1	*Trial 2*	*Trial 3**
Final buret reading, KIO_3	______	______	______
Initial buret reading, KIO_3	______	______	______
Volume of KIO_3 delivered	______	______	______
Final buret reading, $Na_2S_2O_3$	______	______	______
Initial buret reading, $Na_2S_2O_3$	______	______	______
Volume of $Na_2S_2O_3$ used	______	______	______
Calculated molarity of $Na_2S_2O_3$	______	______	______
Average molarity of $Na_2S_2O_3$		______	

*Optional

Iodate Analysis—Data

Concentration of standard $Na_2S_2O_3$ ______________

	Solution Number 1	2
mL of standard $CuSO_4$ used	______	______
mL of standard KIO_3 used	______	______
Final buret reading	______	______
Initial buret reading	______	______
Volume of $Na_2S_2O_3$ required	______	______
Moles of I_3^- titrated	______	______
Concentration of IO_3^- in saturated $Cu(IO_3)_2$	______	______

Calculated Values and Results

	Solution Number 1	2
Initial IO_3^- concentration	______	______
Concentration of IO_3^- at equilibrium	______	______
Change in the concentration of IO_3^-	______	______
Initial Cu^{2+} concentration	______	______
Change in the concentration of Cu^{2+}	______	______
Concentration of Cu^{2+} at equilibrium	______	______
Calculated K_{sp} for $Cu(IO_3)_2$	______	______
Average value of K_{sp} for $Cu(IO_3)_2$	______	
Average value of K_{sp} for $Cu(IO_3)_2$ for all five mixtures	______	

29 EXPERIMENT

Electrochemical Cells

Laboratory Time Required

Two hours. May be combined with Experiment 32.

Special Equipment and Supplies

Hexagonal weighing boats	0.1 M $Zn(NO_3)_2$
Light cardboard	0.1 M $Cu(NO_3)_2$
Double-sided tape	0.1 M $AgNO_3$
Masking tape	0.1 M $Pb(NO_3)_2$
Strips of filter paper (for use as salt bridges)	0.1 M $SnCl_2$
Electrodes (strips of Zn, Cu, Ag, Sn, and Pb foils)	0.1 M KNO_3
Electrical leads	12 M HCl
pH Meter	1.0 M KI
Sandpaper	3.0 M NH_3
Scissors	
Forceps	
Wooden applicator stick	

Safety

Avoid electrical shock when connecting or disconnecting the pH meter power cord.

The Ag^+ and Pb^{2+} ions are **toxic** and their solutions should be handled carefully to avoid skin contact. Wash your hands after sanding and handling the lead electrode. The 12 M hydrochloric acid is corrosive and may cause burns or skin irritation. Avoid breathing HCl fumes. Pipet the solutions (using pipet and bulb) or pour solutions carefully.

First Aid

Remove spilled silver and lead solutions from the skin by washing thoroughly with water.

Electrochemical cells are devices that allow the interconversion of electrical and chemical energy. In voltaic (or galvanic) cells, a spontaneous chemical reaction produces electricity.

Voltaic cells are often used to determine the relative activities of metals. They may also be used to determine equilibrium constants. These applications are explored in this experiment.

PRINCIPLES

Determination of Standard Reduction Potentials

A typical electrochemical cell, such as the one shown in Figure 29.1, consists of two half-cells linked by a wire and a salt bridge. Each half-cell consists of a metal electrode in contact with a solution containing a salt of that metal. One half-cell functions as the anode, where the oxidation reaction ($M \longrightarrow M^{n+} + ne^-$) takes place; the other half-cell functions as the cathode, where the reduction reaction ($M^{n+} + ne^- \longrightarrow M$) takes place. Electrons flow from the anode to the cathode via the wire. The salt bridge allows migration of ions to prevent an imbalance of charge from building up as electrons leave the anode and move to the cathode.

Inserting a voltmeter into the circuit between the half-cells permits a measurement of the voltage—or potential difference—between the half-cells. In general, this voltage is designated by the symbol, $\Delta\varepsilon$. When the solutions in the half-cells are 1 M with respect to the ions involved in the oxidation and reduction reactions, the cell is designated a "standard cell" and its voltage is called a "standard potential," $\Delta\varepsilon^\circ$. Many textbooks and some reference books, such as the *CRC Handbook of Chemistry and Physics*, contain tables of standard reduction potentials, which show the values of ε° for various reduction reactions of the type $M^{n+} + ne^- \longrightarrow M$. The ε°'s in the tables were obtained by assigning a potential of 0.00 V to the reduction reaction: $2H^+(1\ M) + 2e^- \longrightarrow H_2\ (g,\ 1\ atm)$ and measuring the potentials of other standard half-cells coupled to this standard hydrogen electrode.

Whenever two standard half-cells are coupled (joined to create a voltaic cell), the one with the less positive (or more negative) value of ε° functions as the anode. The value of $\Delta\varepsilon^\circ$ for the cell is given by the difference between ε° for the half-cell that functions as the cathode and ε° for the half-cell that functions as the anode. The use of standard reduction potentials to predict the potential of an electrochemical cell is illustrated below (see Equations 29.1–29.5).

Reduction Reaction	ε°	
$Sn^{2+} + 2e^- \rightleftarrows Sn$	–0.14 V	(29.1)
$Al^{3+} + 3e^- \rightleftarrows Al$	–1.66 V	(29.2)

Cathode:	$(Sn^{2+} + 2e^- \longrightarrow Sn) \times 3$	$\varepsilon^\circ_{cathode} = -0.14V$	(29.3)
Anode:	$(Al \longrightarrow Al^{3+} + 3e^-) \times 2$	$-\varepsilon^\circ_{anode} = -(-1.66\ V)$	(29.4)
Cell reaction:	$3Sn^{2+} + 2Al \longrightarrow 3Sn + 2Al^{3+}$	$\Delta\varepsilon^\circ = 1.52\ V$	(29.5)

The electrochemical cells that you will construct in the performance of this experiment will not be standard cells. However, you will be able to use the potentials you observe to obtain standard reduction potentials by the use of the Nernst equation (see Equation 29.6).

$$\Delta\varepsilon = \Delta\varepsilon^\circ - \frac{0.0591}{n} \log Q \qquad (29.6)$$

To use the Nernst equation, you need to know the value of n, which represents the number of electrons transferred in the cell reaction, and the form of Q, which is called the **reaction quotient**. This quotient is the ratio of the concentra-

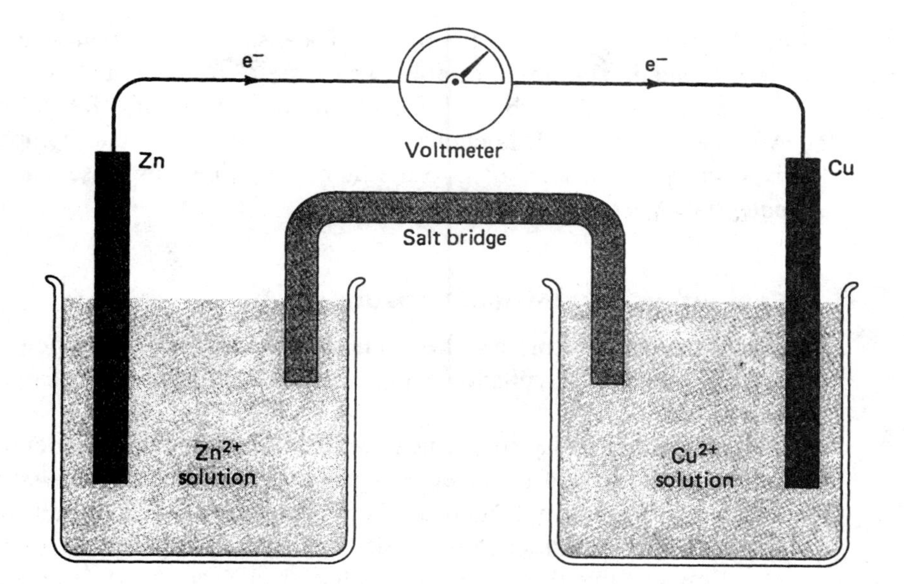

FIGURE 29.1 An electrochemical cell.

tions of ions that are products to the concentrations of ions that are reactants and may be represented as $[P]^a/[R]^b$, where a and b are the coefficients that appear in the equation for the cell reaction. The term 0.0591 is the numerical value of several constants including the temperature (T = 298K) and the Faraday (the charge on a mole of electrons); note that this term has the unit of volts. The relationship between $\Delta\varepsilon$ for a Sn/Sn^{2+} (0.10 M)-Al/Al^{3+} (0.10 M) cell and $\Delta\varepsilon°$ is shown in Equations 29.7 and 29.8. Note that $n = 6$, $a = 2$, and $b = 3$ for this case.

$$\Delta\varepsilon = 1.52\text{ V} - \frac{0.0591}{6}\log\frac{[Al^{3+}]^2}{[Sn^{2+}]^3} \qquad (29.7)$$

$$\Delta\varepsilon = 1.52\text{ V} - 0.0099\text{ V} = 1.51\text{ V} \qquad (29.8)$$

Let us suppose that you prepared an electrochemical cell composed of a tin electrode immersed in 0.01 M $SnCl_2$ and an aluminum electrode immersed in 0.010 M $AlCl_3$. Suppose you found that the aluminum half-cell was the anode and the cell potential was 1.50 V. Let us further assume that you had assigned a value of 1.66 V to the reduction potential of the reaction $Al^{3+}(1.0\text{ M}) + 3e^- \rightarrow Al$. These data would be sufficient for determining the reduction potential for the reaction $Sn^{2+}(1\text{ M}) + 2e^- \longrightarrow Sn$. A method for doing so is illustrated in Equations 29.9–29.11.

$$\Delta\varepsilon = \varepsilon^\circ_{cathode} - \varepsilon^\circ_{anode} - \frac{0.0591}{6}\log\frac{[Al^{3+}]^2}{[Sn^{2+}]^3} \qquad (29.9)$$

$$1.50\text{ V} = \varepsilon^\circ_{cathode} - (-1.66) - \frac{0.0591}{6}\log\frac{[0.01]^2}{[0.01]^3} \qquad (29.10)$$

$$\varepsilon^\circ_{cathode} = -0.14\text{V} \qquad (29.11)$$

In this experiment, you will prepare an apparatus that will permit you to couple the half-cells pairwise, so that you will study each of the half-cells in combination with each of the other half-cells. Using the data you obtain and the Nernst equation, you will be able to evaluate $\mathcal{E}^\circ$ for the equations $M^{2+} + 2e^- \rightleftarrows M$, where M = Cu, Sn, Pb, and Zn, assuming a value of 0.80 V for the reaction $Ag^+ + e^- \rightleftarrows Ag$. This choice of reference will facilitate comparison of your experimental values of $\mathcal{E}^\circ$'s; to the literature values for these standard reduction potentials.

Evaluating Equilibrium Constants

In addition to determining the values of standard reduction potentials, one can use electrochemical cells to evaluate certain equilibrium constants, as illustrated below.

Consider an electrochemical cell that consists of two identical half-cells, each composed of a silver electrode suspended in a solution containing silver ions at some molarity (Figure 29.2). The voltage of the cell will be zero because each half-cell is identical to the other.

Now imagine that some hydrochloric acid (enough to cause the precipitation of silver chloride) is added to one half-cell. A voltage will now be measured in the electrochemical cell because the concentrations of Ag^+ will differ in the two half-cells (Figure 29.3). Such a cell is called a "concentration cell."

In Figure 29.3, the half-cell containing the silver chloride precipitate is designated as the anode, while the half-cell containing 0.10 M silver nitrate is designated as the cathode. Why is this so? A concentration cell will show a non-zero voltage only so long as there is a difference in the concentration of the ions in the two half-cells that comprise the cell. Thus, the spontaneous reaction is the one that tends to equalize the ion concentrations in the two half-cells. The half-cell containing the more concentrated solution becomes the cathode and the concentration of ions decreases as the reaction $M^{n+} + ne^- \longrightarrow M$ occurs. Conversely, the half-cell containing the less concentrated solution becomes the

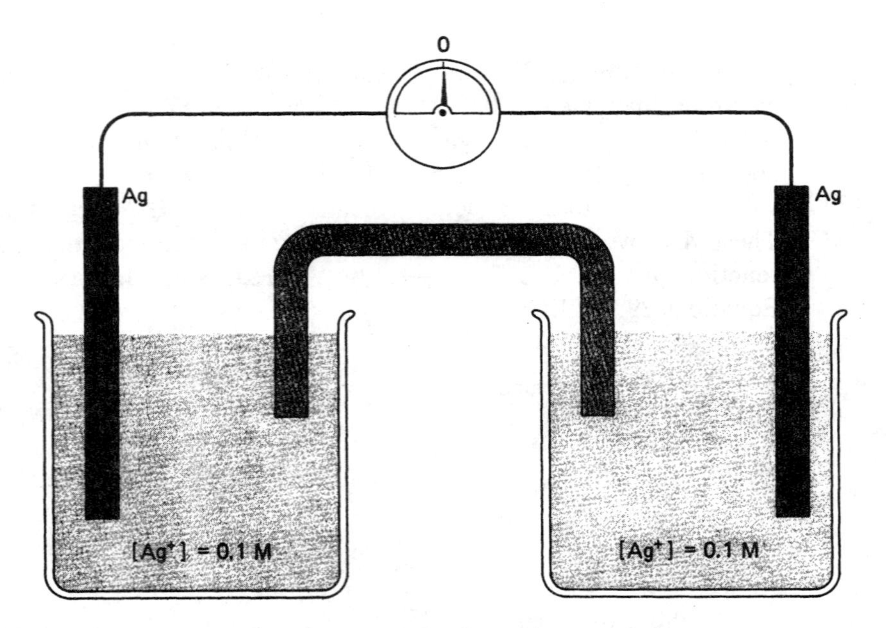

FIGURE 29.2 Linking two identical half-cells produces no voltage.

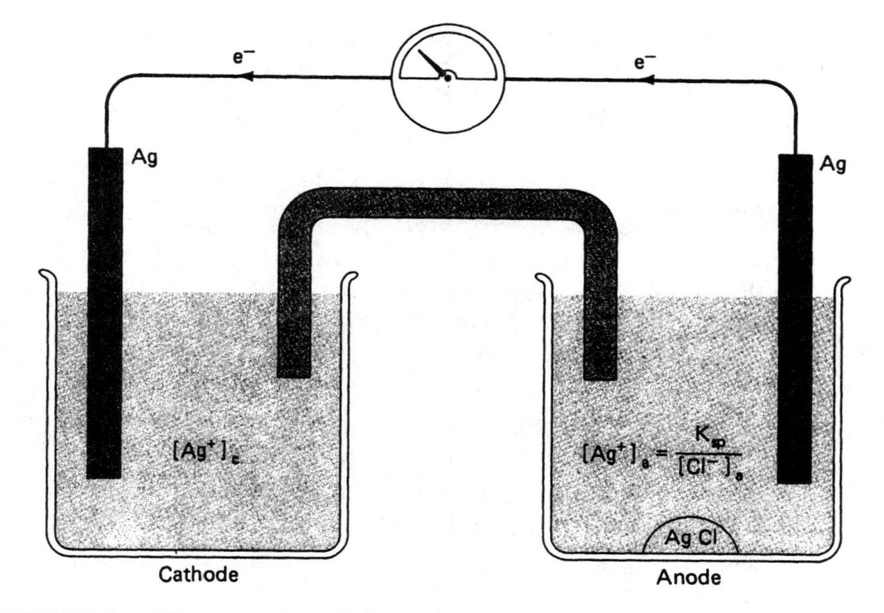

FIGURE 29.3 Direct coupling of silver/silver nitrate and silver/silver chloride half-cells.

anode and the concentration of ions increases via the reaction $M \longrightarrow M^{n+} + ne^-$. The potential of the cell decreases steadily, finally reaching a value of zero when the concentration of ions in both half-cells becomes the same. In the cell pictured in Figure 29.3, the concentration of silver ions is 0.1 M in each half-cell before the addition of chloride ions. Following the addition of chloride ions to one half-cell, the concentration of silver ions in that half-cell decreases because some silver ions are removed from solution as the precipitate of silver chloride forms. This situation is summarized in Equations 29.12–29.15, where X is the value of the silver ion's concentration in the half-cell containing chloride ions.

cathode	$Ag^+(0.10\ M) + e^- \longrightarrow Ag$	(29.12)
anode	$Ag \longrightarrow Ag^+\ (X\ M) + e^-$	(29.13)
overall	$Ag^+\ (0.10\ M) + Ag \longrightarrow Ag + Ag^+(X\ M)$	(29.14)

$$\Delta\mathcal{E} = \Delta\mathcal{E}^\circ - \frac{0.0591\ \text{V}}{1} \log \frac{X}{0.1} \tag{29.15}$$

The value of X may be obtained readily from an experimentally determined value of $\Delta\mathcal{E}$, because $\Delta\mathcal{E}^\circ$ for a concentration cell must equal zero. (Why?) If the cell is constructed with a known excess chloride ion concentration, the value of K_{sp} $(= X[Cl^-])$ can also be obtained.

Arguments similar to those made above can be used to evaluate a number of K_{sp}'s. In this experiment, you may be asked to evaluate K_{sp} for AgCl, AgI, and/or PbI_2. Still other equilibrium constants may be evaluated, provided that the corresponding equilibria result in a change in concentration of a metal ion involved in an oxidation/reduction reaction. For instance, the concentration of silver ions may be reduced by complexing Ag^+ with ammonia (see Equation 29.16).

$$Ag^+(aq) + 2NH_3(aq) \rightleftharpoons Ag(NH_3)_2^+(aq) \tag{29.16}$$

The form of the equilibrium constant, K_f, for this reaction is shown in Equation 29.17.

$$K_f = \frac{[Ag(NH_3)_2^+]}{[Ag^+][NH_3]^2} \qquad (29.17)$$

The value of K_f for $Ag(NH_3)_2^+$ may be obtained by coupling a half-cell containing a mixture of silver nitrate and ammonia to a silver/silver nitrate half-cell. The silver ion concentration, X, is obtained from Equation 29.15. Of course, K_f can be evaluated only if the ammonia is in excess and conditions of mixing are such that $[Ag(NH_3)_2^+]$ and $[NH_3]$ can be evaluated. In this experiment, you may be asked to evaluate K_f for $Ag(NH_3)_2^+$, $Cu(NH_3)_4^{2+}$, and/or $Zn(NH_3)_4^{2+}$.

Although concentration cells are frequently used to evaluate equilibrium constants, it is possible to obtain K values from any voltaic cell in which a half-cell of known potential is coupled to a half-cell in which a chemical reaction limits the concentration of the ion of interest. In this experiment, you will obtain values of K_{sp} for AgCl from a concentration cell and from a cell in which a Ag/AgCl half-cell is coupled to a Zn/Zn^{2+} half-cell.

PROCEDURE

A. Measurement of Standard Reduction Potentials

Crossing three strips of double-sided tape, make an "asterisk" on a cardboard backing. Press a weighing boat firmly onto the middle of the vertical strip of tape. Then press five additional weighing boats onto the cross strips of tape, creating an open honeycomb pattern (Figure 29.4). Place approximately 5 mL of 0.10 M KNO_3 in the center boat. Place approximately 5 mL of 0.10 M $AgNO_3$, 0.10 M $Cu(NO_3)_2$, 0.10 M $Pb(NO_3)_2$, 0.10 M $Zn(NO_3)_2$, and 0.10 M $SnCl_2$ respectively, in each of the outer boats, labeling the boats by showing the symbol

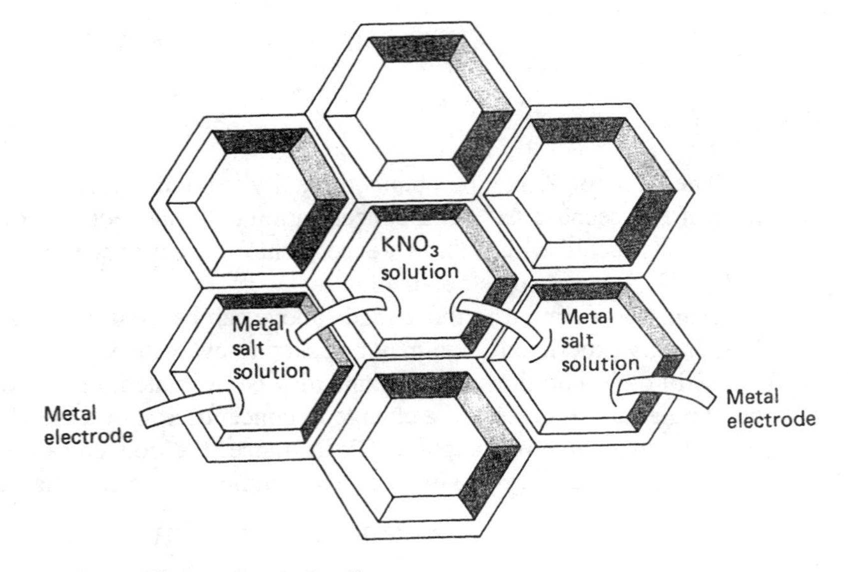

FIGURE 29.4 Electrochemical cells.

for each cation on the cardboard backing to which the weighing boats are attached. Be sure that the contents of the various boats are not allowed to mix.

Obtain Ag, Cu, Pb, Zn, and Sn strips, which will serve as electrodes. Lift each metal electrode with forceps and sand it to remove any oxide coating. Place one end of the electrode in the solution of the corresponding ion (that is, Zn in Zn^{2+}) to complete each half-cell. (Do not handle the electrodes with your fingers; use the forceps at all times.) You may bend the electrodes, if necessary. You may also use a small piece of masking tape to secure the electrodes to the weighing boat edge or the cardboard on which the boats are mounted.

Attach the electrical leads to the pH meter. Position the weighing boat honeycomb so that both electrical leads can reach around it easily. Then secure the cardboard holding the honeycomb to the laboratory bench with masking tape. Plug in the pH meter, with the function selection knob set on standby. Then turn the function selection knob to mV (or + mV) and touch the leads together. Use the calibration knob to set the meter reading to –700 mV (or 0.0 pH).

Dip one end of a strip of filter paper into the solution of zinc nitrate. Place the other end of the filter paper in the solution of potassium nitrate. (You may use a wooden applicator stick, if necessary, to submerge the second end of the paper in the potassium nitrate solution.) The filter paper "bridges" can function effectively as "salt bridges" (conduits for spectator ions) only if the entire length of the filter paper is moistened by the solutions in which the paper is immersed.

Use separate strips of filter paper to link the solutions of copper(II) nitrate, lead nitrate, silver nitrate, and tin(II) chloride to the central boat containing the solution of potassium nitrate. Two strips of filter paper, connecting any two outer boats to the center boat, constitute the salt bridge of your electrochemical cell.

Attach the alligator clip of one of the electrical leads to the zinc electrode. Attach the alligator clip of the other electrical lead to the silver electrode. If the needle of the pH meter goes off scale, indicating a value below –700 mV (or below 0.0 pH), reverse the clips on the electrodes.

Wait 15 seconds, then record the reading on the meter. Use a label (or a piece of masking tape) to mark the electrical lead to which the zinc metal electrode is attached. The lead with the tape will always be the anode in any cell that gives a positive deflection on the pH meter.

Move the alligator clips to new electrodes. Wait 15 seconds before recording the reading on the pH meter. For each cell, note the identity of the electrodes involved, and state which is the anode. Remember that the leads should always be attached to the electrodes in a manner that produces an on-scale reading. Continue until all of the boats have been coupled in pairs.

Convert your readings to volts. Then use the Nernst equation to find the value of $\mathcal{E}^\circ$ for the Zn^{2+}/Zn half-cell, the Cu^{2+}/Cu half-cell, the Sn^{2+}/Sn half-cell, and the Pb^{2+}/Pb half-cell, assigning the Ag^{+}/Ag half-cell an $\mathcal{E}^\circ$ value of 0.80 V. Prepare a table of standard half-cell potentials and compare it to one available in the literature. Comment briefly on the accuracy of your experimental work. Briefly explain any error in your experimental results.

B. Evaluation of Equilibrium Constants

Insert a sixth weighing boat in the open space of your honeycomb. Fill this cell with 5.0 mL of 0.10 M $AgNO_3$ and 0.50 mL of 12 M HCl. Immerse a freshly sanded piece of silver foil into this mixture and connect the sixth boat to the central boat, as usual, with a strip of filter paper. Attach the alligator clips to the two silver foils and wait 15 seconds before recording the reading on your pH meter. Note which half-cell is the anode.

Move the alligator clip from the silver foil that is immersed in silver nitrate to the zinc electrode. Wait 15 seconds before recording the reading on the pH meter. Note which half-cell is the anode.

Remove the weighing boat that contains the silver chloride from the honeycomb. Clean (see Disposal of Reagents) and dry the boat. Place it in the honeycomb once again and fill it, as your instructor directs, using a foil and mixture of solutions specified in Table 29.1. Use the potential of the resulting concentration cell to find the value of your assigned K.

Disposal of Reagents

The $Zn(NO_3)_2$, $Cu(NO_3)_2$, $SnCl_2$ and KNO_3 solutions may be diluted and poured down the drain. Precipitate Ag^+ as AgCl(s) by adding 1 drop of 12 M HCl to the $AgNO_3$ solution. After the precipitate has settled, decant the liquid into a beaker, dilute it with water, and flush the diluted solution down the drain. Transfer the AgCl slurry to the collection bottle labeled "waste AgCl." Dispose of the AgCl generated in the experiment in the same manner.

Decant the supernatant liquid from the half-cell containing AgI; dilute it and flush the diluted solution down the drain. Transfer the AgI slurry to the collection bottle labeled "waste AgI."

Precipitate Pb^{2+} as PbI_2. Decant and dilute the supernatant liquid and flush the diluted solution down the drain. Transfer the slurry of PbI_2 to the collection bottle marked "waste PbI_2." Dispose of the PbI_2 generated in the experiment in the same manner.

TABLE 29.1 Materials Needed for the Evaluation of Equilibrium Constants

K	Electrode	Mixture
K_{sp} for AgI	Ag	5.0 mL of 0.10 M $AgNO_3$ 1.0 mL of 1.0 M KI
K_{sp} for PbI_2	Pb	3.0 mL of 0.10 M $Pb(NO_3)_2$ 3.0 mL of 1.0 M KI
K_f for $Ag(NH_3)_2^+$	Ag	3.0 mL of 0.1 M $AgNO_3$ 3.0 mL of 3.0 M NH_3
K_f for $Cu(NH_3)_4^{2+}$	Cu	3.0 mL of 0.1 M $Cu(NO_3)_2$ 3.0 mL of 3.0 M NH_3
K_f for $Zn(NH_3)_4^{2+}$	Zn	3.0 mL of 0.1 M $Zn(NO_3)_2$ 3.0 mL of 3.0 M NH_3

Date ______ Name ______________________ Section ______ Desk Number ______

PRE-LAB EXERCISES FOR EXPERIMENT 29

Complete these exercises after you have read the experiment but before you come to the laboratory to perform it.

1. An electrochemical cell is constructed by coupling a Cu^{2+}/Cu half-cell with an Ag^{+}/Ag half-cell. The concentrations of Cu^{2+} and Ag^{+} are each 0.010 M. The observed potential, $\Delta\varepsilon$, for the cell is 45 mV, with Cu at the anode. If $\varepsilon^{\circ} = 0.80$ V for the half-reaction $Ag^{+}(aq) + e^{-} \rightleftarrows Ag(s)$, find the value of ε° for the half-reaction: $Cu^{2+}(aq) + 2e^{-} \rightleftarrows Cu(s)$.

2. When 5.0 mL of 0.10 M $AgNO_3$ and 0.50 mL of 12 M HCl are mixed, a precipitate of AgCl forms. Assume the reaction below goes to completion, and find the value of $[Cl^{-}]$ in the mixture.

$$Ag^{+}(aq) + Cl^{-}(aq) \longrightarrow AgCl(s)$$

3. Silver ions are readily complexed by ammonia molecules according to the reaction:

$$Ag^{+} + 2NH_3 \longrightarrow Ag(NH_3)_2^{+}$$

   a. Find the values of $[NH_3]$ and $[Ag(NH_3)_2^{+}]$ when 3.0 mL of 0.10 M $AgNO_3$ and 3.0 mL of 3.0 M NH_3 are mixed. Assume the complexation reaction goes to completion.

b. A cell is constructed in which the anode has a silver wire in contact with a mixture of 3.0 mL of 3.0 M NH_3 and 3.0 mL of 0.10 M $AgNO_3$, while the cathode has a silver wire in contact with 0.10 M $AgNO_3$. The potential of the cell is 505 mV. Find the value of K_f for $Ag(NH_3)_2^+$.

c. The K_f defined here is an "overall" formation constant. It is actually the product of two "stepwise" formation constants corresponding to the reactions:

$$Ag^+ + NH_3 \rightleftharpoons Ag(NH_3)^+$$
$$Ag(NH_3)^+ + NH_3 \rightleftharpoons Ag(NH_3)_2^+$$

If a significant portion of the silver ion is complexed as $Ag(NH_3)^+$, would your calculated value of K_f be too high or too low? Explain your answer briefly.

Date ______ Name ______________________________ Section ______ Desk Number ______

SUMMARY REPORT ON EXPERIMENT 29

A. Determination of Standard Reduction Potentials

Cell	Electrode 1	Electrode 2	Meter Reading	$\Delta\varepsilon$, mV	$\Delta\varepsilon$, V	$\Delta\varepsilon^\circ$, V
1	Ag	Zn(A)				
2	Ag	Cu				
3	Ag	Sn				
4	Ag	Pb				
5	Zn	Cu				
6	Zn	Sn				
7	Zn	Pb				
8	Cu	Sn				
9	Cu	Pb				
10	Sn	Pb				

Be sure to note which electrode is at the anode in each cell. Use Equation 29.6 to determine the value of $\Delta\varepsilon^\circ$ from the value of $\Delta\varepsilon$..

CALCULATION OF STANDARD REDUCTION POTENTIALS

Half-Cell Reduction Reaction	Experimental Value of ε°	Literature Value of ε°
$Ag^+ + e^- \rightleftarrows Ag$	0.80V	0.80V
$Cu^{2+} + 2e^- \rightleftarrows Cu$		
$Pb^{2+} + 2e^- \rightleftarrows Pb$		
$Sn^{2+} + 2e^- \rightleftarrows Sn$		
$Zn^{2+} + 2e^- \rightleftarrows Zn$		

B. Determination of Equilibrium Constants

Evaluation Of K_{sp} For AgCl

Cell	Electrode 1	Electrode 2	Meter Reading	$\Delta\varepsilon$, mV	$\Delta\varepsilon$, V	$\Delta\varepsilon^\circ$, V
1	Ag	Ag				
2	Ag	Zn				

Be sure to note which electrode is at the anode in each cell.

CONCENTRATION OF Cl^- ______________

Cell	Cell Reaction	$\Delta\varepsilon$, V	$[Ag^+]_{6th\ Boat}$	Value of K
1				
2				

Use your data from Part A to determine the value of $\Delta\varepsilon^\circ$ for the cell reaction written *as it occurs in part B*. Recall that $\Delta\varepsilon^\circ = 0$ for any concentration cell.

Evaluation Of Another *K*

Mixture assigned ______

K to be evaluated ______

Meter reading ______

$\Delta\varepsilon$, mV ______

$\Delta\varepsilon$, V ______

Cell reaction ______

$[M^{n+}]_{6th\ Boat}$ ______

Concentration of Cl^-, I^-, or NH_3 ______

Value of *K* ______

Enthalpy of Hydration of Ammonium Chloride

Laboratory Time Required

Three hours

Special Equipment and Supplies

Burets
Buret clamp
Dewar flask calorimeter
Thermometer
Mortar and pestle
pH indicating paper
Balance

Standardized 1 M hydrochloric acid
Aqueous ammonia
Methyl orange indicator
Ammonium chloride
Ice

Safety

Aqueous ammonia is **caustic.** Hydrochloric acid is **corrosive.** Avoid splashing these chemicals on your skin, in your eyes, or in your mouth.

The Dewar flask and thermometer are easily broken, resulting in a possible hazard of cuts from sharp edges. The Dewar flask may implode violently, throwing glass fragments for some distance. Wear safety goggles at all times.

Broken thermometers may contaminate the laboratory with spilled mercury, a poison. Consult your instructor for clean-up procedures.

First Aid

If you have contact with an acid or base, flush the affected area thoroughly with water. See a doctor if your eyes were exposed or if there is continued skin discomfort.

Minor cuts may be treated with antiseptic and covered with an adhesive bandage. Deep or extensive cuts require a doctor's attention.

Calorimeters are used to study the heat absorbed or released by a system that undergoes a physical or a chemical change. A number of reliable procedures have been devised for quantitatively measuring the number of joules lost or gained by the reactants during this change. The ice calorimeter, for example, makes use of the fact that a known number of joules is needed to convert 1 g of ice at 0°C to water at the same temperature. The amount of ice melted, therefore, accurately determines the number of joules released during a reaction. Alternatively, the heat absorbed or released by the reaction system may be used to change the temperature of the calorimeter and its contents. If the temperature change and the heat capacity of the calorimeter and its contents are known, the quantity of heat involved may be calculated easily. In this experiment, the temperature change in a solution calorimeter will be used to study the enthalpy of a neutralization reaction and the enthalpy of solution of a salt.

PRINCIPLES

Chemical reactions are generally accompanied by a gain or loss of energy. This energy is very often in the form of heat. The amount of heat can be measured if the reaction is carried out in a calorimeter (a vessel insulated to restrict the exchange of heat between the reaction system and its surroundings). Thus, the heat evolved by an exothermic reaction would equal the heat absorbed by the calorimeter and its contents. Similarly, the heat absorbed by an endothermic reaction would equal the heat lost by the calorimeter and its contents. These heat changes are represented by the symbol Q. Obviously, Q depends on the amount of reaction that takes place. If we evaluate Q_{reaction} for 1 mole of reaction, Q is then equal to ΔH, the enthalpy change for that reaction.

At times it is not convenient, or even possible, to measure the heat change associated with a reaction directly. In such cases, Hess' Law may be invoked. This principle states that, if a reaction can be regarded as the sum of two or more reactions, then the enthalpy change, ΔH, of the overall reaction is equal to the sum of the enthalpy changes of these contributing steps. The use of Hess' Law to evaluate the heat of formation of carbon monoxide is illustrated below. It is difficult to measure this reaction enthalpy directly because carbon monoxide tends to be oxidized further to carbon dioxide.

Reaction		
$C(\text{graphite}) + O_2(g) \longrightarrow CO_2(g)$	$\Delta H^\circ_{f,\,CO_2}$	$= -394$ kJ/mol
$CO_2(g) \longrightarrow CO(g) + \frac{1}{2}O_2(g)$	$-\Delta H^\circ_{\text{combustion, CO}}$	$=$ 284 kJ/mol
$C(\text{graphite}) + \frac{1}{2}O_2(g) \longrightarrow CO(g)$	$\Delta H^\circ_{f,\,CO}$	$= -110$ kJ/mol

A corollary of Hess' Law entails a cycle of thermochemical equations (a Born-Haber cycle). Because the net change in enthalpy in going around a cycle and returning to the starting point must be zero, the Born-Haber cycle is most often used when the enthalpies of all but one of the reactions in the cycle can be measured directly. The Born-Haber cycle to be investigated in this experiment is shown below.

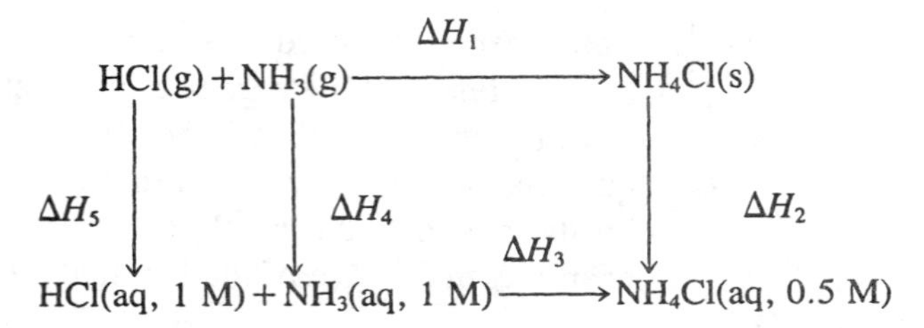

You will use this cycle to find the sum of the enthalpies of hydration of gaseous ammonia and gaseous hydrogen chloride ($\Delta H_4 + \Delta H_5$). You will determine two of the enthalpies experimentally. They are the enthalpy of solution of ammonium chloride (ΔH_2) and the enthalpy of neutralization of aqueous ammonia and aqueous hydrochloric acid (ΔH_3). The last enthalpy (ΔH_1) in the cycle is the enthalpy change for the synthesis of ammonium chloride crystals from gaseous ammonia and hydrogen chloride. You will not determine the value of ΔH_1 experimentally; instead, you will evaluate it from tabulated values of the enthalpies of formation of ammonium chloride crystals, hydrogen chloride gas, and ammonia gas. The standard **enthalpy of formation** of a substance is defined as the enthalpy change involved in the synthesis of a mole of the substance in its standard state from its constituent elements in their standard states. Values of enthalpies of formation are found in textbooks or tabulations of thermodynamic data.

You will use a Dewar flask as a calorimeter to evaluate the various enthalpies for the reactions in the cycle. You must first determine the heat capacity of the calorimeter (the number of joules—(J)—needed to raise the temperature of the calorimeter by 1°C). You will do so by mixing equal volumes of room-temperature water and cool water in the calorimeter and finding the final temperature of the mixture. In this case, the heat lost by the room temperature water will equal the sum of the heat gained by the cool water and the heat gained by the calorimeter. This equality is illustrated in Equation 30.1.

$$C_{H_2O}g_{H_2O}(t_r - t_f) = C_{H_2O}g_{H_2O}(t_f - t_c) + C_{cal}(t_f - t_c) \quad (30.1)$$

where C_{H_2O} is the heat capacity of water in units of J/g-deg,
g_{H_2O} is the mass of cool water or room-temperature water,
t_r is room temperature,
t_c is the initial temperature of cool water,
t_f is the final temperature of the mixture, and
C_{cal} is the heat capacity of the calorimeter in units of J/deg.

Because the calorimeter does not come to thermal equilibrium immediately and because it is not a perfect insulator, the final temperature to be used in evaluating C_{cal} cannot be determined directly. Instead, you will mix the water and monitor the temperature change as a function of time. Plotting a graph of temperature versus time, and extrapolating back to the initial time of mixing, gives the value of t_f (Figure 30.1).

Once the heat capacity of the calorimeter is known, it must be taken into account in evaluating the heats of the reactions taking place in the vessel. The heat associated with the neutralization reaction between HCl(aq) and NH_3(aq) may be obtained from Equation 30.2, where C_{NH_3}, C_{HCl}, and C_{cal}, are the heat capacities of the aqueous ammonia, hydrochloric acid, and calorimeter, respectively; g_{NH_3} and g_{HCl} are the masses of the aqueous ammonia and hydrochloric acid; and Δt is the temperature change that results from the neutralization reaction. A typical plot for determining Δt is shown in Figure 30.2. Note that, for an exothermic reaction, Q_{neut} is negative and the calorimeter will absorb heat, resulting in a temperature increase and a positive value for Q_{cal}.

$$-Q_{neut} = Q_{cal} = C_{NH_3}\, g_{NH_3}\, \Delta t + C_{HCl}\, g_{HCl}\, \Delta t + C_{cal}\, \Delta t \quad (30.2)$$

Similarly, the amount of heat associated with the dissolution of ammonium chloride in water may be obtained from Equation 30.3, where C_{NH_4Cl} and g_{NH_4Cl} are the heat capacity and mass of aqueous ammonium chloride and Δt is the tem-

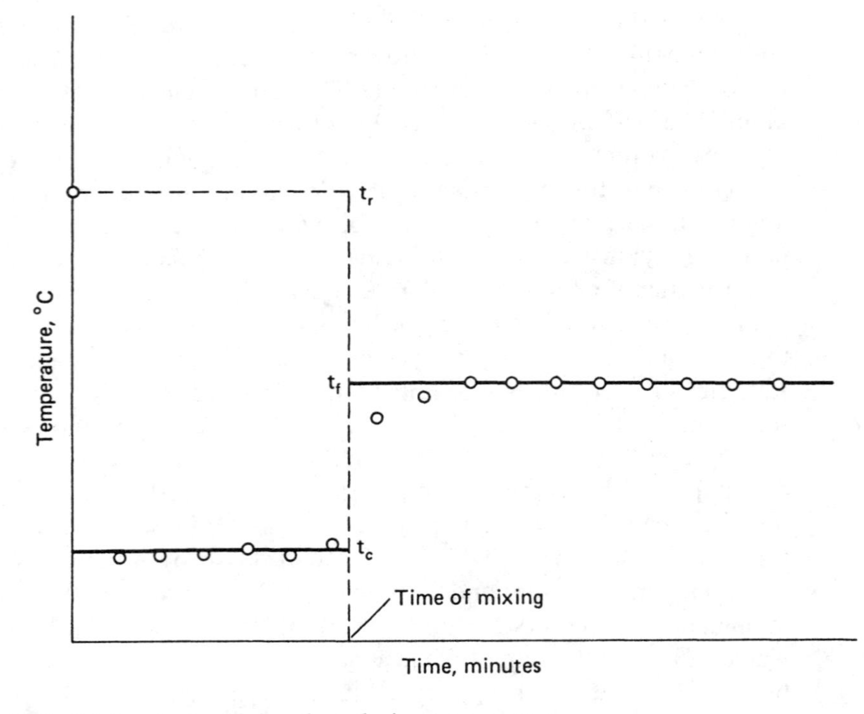

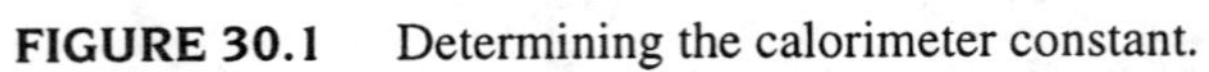
FIGURE 30.1 Determining the calorimeter constant.

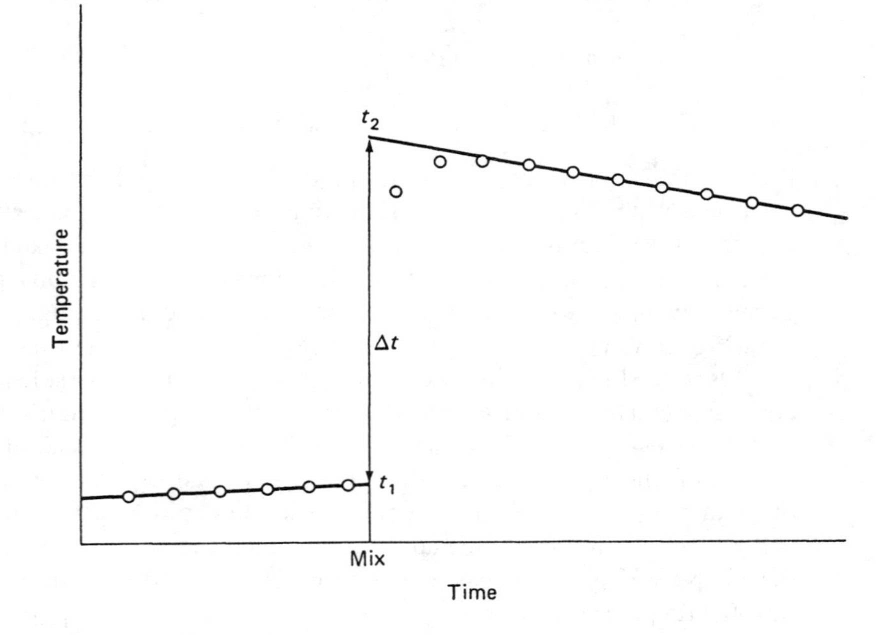

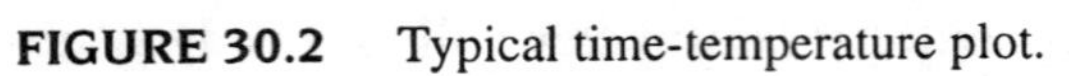
FIGURE 30.2 Typical time-temperature plot.

perature change associated with the dissolution of ammonium chloride. Be careful in determining the sign of Q_{cal}. If the temperature rises, Q_{cal} is positive and the reaction is exothermic (meaning the heat of reaction, $Q_{reaction}$, is negative). If the temperature falls, Q_{cal} is negative and the reaction is endothermic ($Q_{reaction}$ is positive).

$$-Q_{sol} = Q_{cal} = C_{NH_4Cl}\, g_{NH_4Cl}\, \Delta t + C_{cal}\, \Delta t \tag{30.3}$$

Equations 30.2 and 30.3 may be greatly simplified if we make the following assumptions:

1. The heat capacities of all solutions of interest are the same (within 1 to 2%) as pure water at 25°C, 4.180 J/g-deg or 4.167 J/mL-deg, and are not temperature dependent.
2. The initial and final temperature (and therefore Δt) are the same for all solutions in a given reaction.

Making the appropriate changes in Equations 30.2 and 30.3 gives us:

$$-Q_{neut} = (V_{soln}\, C_{H_2O} + C_{cal})\Delta t \tag{30.4}$$

where $V_{soln} = V_{NH_3} + V_{HCl}$ and $C_{H_2O} = 4.167$ J/mL-deg

$$-Q_{sol} = (V_{soln}\, C_{H_2O} + C_{cal})\Delta t \tag{30.5}$$

where $V_{soln} = V_{NH_4Cl}$ and $C_{H_2O} = 4.167$ J/mL-deg.

The heats represented by Q_{neut} and Q_{sol} will have the units of joules. These heats are readily converted to enthalpies, in units of kJ/mole, by applying the appropriate conversion factor and considering the number of moles of reactants involved in the respective reactions.

PROCEDURE

Standardization of Aqueous Ammonia

Record the exact concentration of the standardized HCl. Rinse and fill one buret with the HCl solution. Rinse and fill the other buret with the aqueous ammonia. Record the initial reading of each buret. Dispense a 25-mL aliquot of HCl into a clean (but not necessarily dry) Erlenmeyer flask. Record the final buret reading. Add 25 mL of distilled water and 3 drops of methyl orange indicator to the acid in the flask. Titrate the acid solution with the aqueous ammonia until the indicator color changes from orange to yellow. Record the final reading for the buret containing the aqueous ammonia.

Determining the Heat Capacity of the Calorimeter

Place roughly 250 mL of water in a 400-mL beaker and cool this water to 4°C or 5°C by placing the beaker in an ice bath. Use a clean, dry, 200-mL volumetric flask to transfer 200.00 mL of the cold water to the clean, dry calorimeter. Rinse the volumetric flask with room temperature water and then measure out exactly 200.00 mL of this water. Monitor the temperature of the water in the volumetric flask. Do not proceed until this temperature is constant and equal to room-temperature. Record this temperature. Assemble the calorimeter and stir the cool water, recording its temperature every 30 seconds, for 3 minutes. Add all of the room-temperature water to the cool water in the calorimeter as quickly as possible, while stirring. Record the exact time of mixing. Continue stirring. Record the temperature every 15 seconds for 5 minutes. Prepare a plot, similar to the one shown in Figure 30.1. and calculate the heat capacity of the calorimeter.

Determining the Heat of Neutralization of Ammonia and Hydrochloric Acid

Dry the interior of the calorimeter with a cloth towel or rag.

CAUTION THE DEWAR FLASK MAY IMPLODE VIOLENTLY IF BUMPED OR SCRATCHED.

Rinse the volumetric flask with several small portions of aqueous ammonia. Then fill the flask with NH_3 solution to the mark. Transfer the aqueous ammonia as completely as possible to the calorimeter. Record the temperature of the ammonia solution. Rinse the thermometer with distilled water to remove any base remaining on it. Rinse the volumetric flask with several small portions of distilled water. Then rinse it again with several small portions of hydrochloric acid. Fill the flask to the mark with HCl solution. Check the temperature of the acid in the volumetric flask. Rinse the exterior of the flask with hot or cold water, as necessary, to bring the temperature of the hydrochloric acid to within ±0.1°C of the temperature of the aqueous ammonia. Rinse the thermometer with distilled water to remove any acid remaining on it. Dry the thermometer. Monitor the temperature of the aqueous ammonia, recording its value every 30 seconds for 3 minutes. Add all of the hydrochloric acid to the aqueous ammonia in the calorimeter as quickly as possible, while stirring. Record the exact time of mixing. Continue stirring. Record the temperature every 15 seconds for 5 minutes. Prepare a plot, similar to the one shown in Figure 30.2, and calculate Q_{neut}.

Determining the Heat of Solution of Ammonium Chloride

Using a mortar and pestle, powder 11.0 g of pure, dry NH_4Cl and transfer it to a small (clean, dry) beaker. Weigh the beaker and its contents. Clean and dry the calorimeter **carefully**. Use the volumetric flask twice to add 400.00 mL of distilled water to the calorimeter. Monitor the temperature of the water, recording its value every 30 seconds for 3 minutes. While stirring, add the ammonium chloride to the water in the calorimeter. Record the exact time of mixing. Continue stirring. Record the temperature every 15 seconds for 5 minutes. Weigh the empty beaker. Determine and record the mass of ammonium chloride used. Prepare a plot, similar to the one shown in Figure 30.2, and calculate Q_{sol}.

Use your data to find the values of ΔH_2 (enthalpy of solution of ammonium chloride) and ΔH_3 (enthalpy of neutralization of ammonia and hydrochloric acid). Combine these with the value of ΔH_1, (enthalpy of synthesis of ammonium chloride) to find the sum of ΔH_4 and ΔH_5 (sum of the enthalpies of hydration of hydrogen chloride and ammonia). Note that ΔH_1 is to be calculated using standard enthalpies of formation. Cite your source for these data in your report. Append your plots, calculations, and results to your Summary Report.

Disposal of Reagents

All solutions may be neutralized (if necessary), diluted, and flushed down the drain.

Date ______ Name ________________________ Section ______ Desk Number ______

PRE-LAB EXERCISES FOR EXPERIMENT 30

These exercises are to be completed after you have read the experiment but before you come to the laboratory to perform it.

1. Write the equations corresponding to the enthalpies of formation of hydrogen chloride gas, ammonia gas, and ammonium chloride crystals. Show how these equations can be combined to yield the equation for the synthesis of solid ammonium chloride from gaseous hydrogen chloride and gaseous ammonia.

2. Evaluate ΔH_1 for the Born-Haber cycle shown on p. 316, using the equations shown in answer to Exercise 1. Obtain the necessary values for the enthalpies of formation from appropriate reference works.

PRE-LAB EXERCISES FOR EXPERIMENT

Date ______ Name ______________________________ Section ______ Desk Number ______

SUMMARY REPORT ON EXPERIMENT 30

Standardization of Aqueous Ammonia

Molarity of HCl ______________________

Final buret reading, HCl ______________________

Initial buret reading, HCl ______________________

Volume of HCl used ______________________

Final buret reading, NH_3 ______________________

Initial buret reading, NH_3 ______________________

Volume of NH_3 used ______________________

Determining the Heat Capacity of the Calorimeter

Room temperature ______________________

Temperature of cool water

Time	*Temperature*

Initial time of mixing ______________________

Time-temperature data

Time	*Temperature*	*Time*	*Temperature*

Temperature of cold water ______________________

Temperature of warm water ______________________

Temperature of mixture ______________________

The calculated value of the calorimeter constant C_{cal} ______________________

Date ______ Name ______________________________ Section ______ Desk Number ______

Determining the Heat of Neutralization of Ammonia and Hydrochloric Acid

Temperature of hydrochloric acid ______________________

Temperature of aqueous ammonia

Time	*Temperature*
______	______
______	______
______	______
______	______
______	______
______	______

Initial time of mixing ______________________

Time-temperature data

Time	*Temperature*	*Time*	*Temperature*
______	______	______	______
______	______	______	______
______	______	______	______
______	______	______	______
______	______	______	______
______	______	______	______
______	______	______	______
______	______	______	______
______	______	______	______
______	______	______	______

Determining the Heat of Solution of Ammonium Chloride

Mass of beaker plus NH_4Cl ______________________

Mass of beaker ______________________

Mass of NH_4Cl used ______________________

Temperature of aqueous ammonia

Time	*Temperature*
______	______
______	______
______	______
______	______
______	______
______	______

Initial time of mixing ______________________

Date ______ Name ______________________________ Section ______ Desk Number ______

Time-temperature data

Time	*Temperature*	*Time*	*Temperature*

31 EXPERIMENT

Calorimetric Study of the Decomposition of Hydrogen Peroxide[1]

Laboratory Time Required

Three hours

Special Equipment and Supplies

Calorimeter assembly
Timer
Graduated cylinder, 50- or 100-mL
Buret
Florence flask, 500-mL
Mortar and pestle
Magnetic stirrer

3% H_2O_2 Solution
MnO_2, well powdered
KI crystals
3% ammonium molybdate solution
$Na_2S_2O_3$ (s)
0.02 M KIO_3 standard solution
0.2% xtarch solution
0.5 M H_2SO_4 solution

Safety

The major hazards in this experiment are associated with the calorimeter assembly, consisting of an evacuated Dewar flask and a thermometer. If dropped, bumped, or otherwise treated roughly, the Dewar flask may implode violently, possibly subjecting you to cuts from glass shards. The thermometer is also easily broken, sometimes causing puncture wounds in the process. The toxic vapor from spilled mercury also poses a health hazard.

The hazards associated with concentrated H_2O_2 solutions are discussed in the Principles section. Even 3% H_2O_2 should be treated with caution because of its properties as an oxidizing agent, although its use does not pose a serious hazard. Indeed, it is often used as an antiseptic. The dilute H_2SO_4 can injure your skin and eyes. Eye protection and pipet bulbs should be used when working with this chemical.

First Aid

Flush chemicals from your skin or eyes with a stream of water. Apply pressure to stem rapid bleeding. Seek medical help for serious cuts or burns.

[1]Adapted from the article by D. B. Pattison, J. G. Miller, and W. W. Lucasse. *J. Chem. Educ.*, **20** (1943), p. 319.

In this experiment, you will use a solution calorimeter to study the decomposition of H_2O_2 in the presence of a catalyst. You will then compare the experimental enthalpy of decomposition with the value of ΔH°, calculated using literature values for the enthalpies of formation of H_2O and dilute aqueous H_2O_2. The calculation of ΔH° serves as a practical application of Hess' Law of Constant Heat Summation.

The experiment has three parts: (1) measuring the calorimeter constant C, following the procedure used in Experiment 30, (2) measuring the temperature change of the calorimeter when H_2O_2 is decomposed; and (3) standardizing the H_2O_2 solution.

PROCEDURE

Properties of H_2O_2

Hydrogen peroxide in acidic solution is both a powerful oxidizing agent and a mild reducing agent, as examination of the standard electrode potentials clearly indicates (see Equations 31.1 and 31.2).

$$O_2 + 2H^+ + 2e^- \rightleftarrows H_2O_2 \qquad \Delta\varepsilon^\circ = +0.682V \qquad (31.1)$$

$$H_2O_2 + 2H^+ + 2e^- \rightleftarrows 2H_2O \qquad \Delta\varepsilon^\circ = +1.77\ V \qquad (31.2)$$

Thus, aqueous H_2O_2 is capable of oxidizing Fe^{2+} to Fe^{3+} and I^- to I_2, or of reducing strong oxidizing agents, such as MnO_4^- or $Cr_2O_7^{2-}$, to Mn^{2+} or Cr^{3+}, respectively.

The auto-oxidation-reduction or disproportionation reaction of H_2O_2 is shown in Equation 31.3.

$$H_2O_2 \rightleftarrows H_2O + {}^1/_2\, O_2 \qquad (31.3)$$

Using the standard electrode potentials given above, $\Delta\varepsilon^\circ_{cell}$ for this reaction may easily be shown to be + 1.09 V. Thus, the disproportionation reaction should proceed spontaneously. Fortunately, this reaction is slow at room temperature in the absence of catalysts, as are many of the reactions of hydrogen peroxide. The sluggishness of the uncatalyzed decomposition makes it possible to work safely with dilute H_2O_2 solutions, but concentrated solutions are considerably more hazardous. The accidental contamination of concentrated H_2O_2 hastens the decomposition because many substances that are present in the environment catalyze this reaction.

The Calorimeter Reaction

Approximately 0.3 percent H_2O_2 will be prepared by diluting 50 mL of commercial 3 percent H_2O_2 to 500 mL with distilled water. Part of this solution will be analyzed to determine the exact concentration of H_2O_2. Exactly 400 mL of the dilute solution will be placed in the calorimeter to study the decomposition reaction. The premixing temperature will be recorded at 50-second intervals for several minutes, after which finely powdered MnO_2 will be added to catalyze the decomposition reaction. The time and temperature observations will then be continued until the temperature either stops changing or decreases gradually at a uniform rate. The calorimeter reaction is shown in Equation 31.4.

$$H_2O_2\,(0.\,3\%\ \text{solution}) \longrightarrow H_2O(\ell) + {}^1/_2\,O_2(g) \tag{31.4}$$

Because the solid MnO_2 catalyzes a reaction in solution, this is an example of heterogeneous catalysis. The rate of the reaction depends upon the surface area of the solid exposed to the solution. Therefore, you must use finely powdered MnO_2 and swirl the calorimeter contents rather vigorously to keep the catalyst in suspension. Take care, however, not to splash the solution out through the hole in the rubber stopper, which is provided for the escape of the oxygen gas produced by the reaction (Figure 31.1). Best results will be obtained if you standardize your stirring procedure, swirling the flask for perhaps 30 seconds out of each 50-second interval, and using the remaining 20 seconds to prepare for taking a temperature reading.

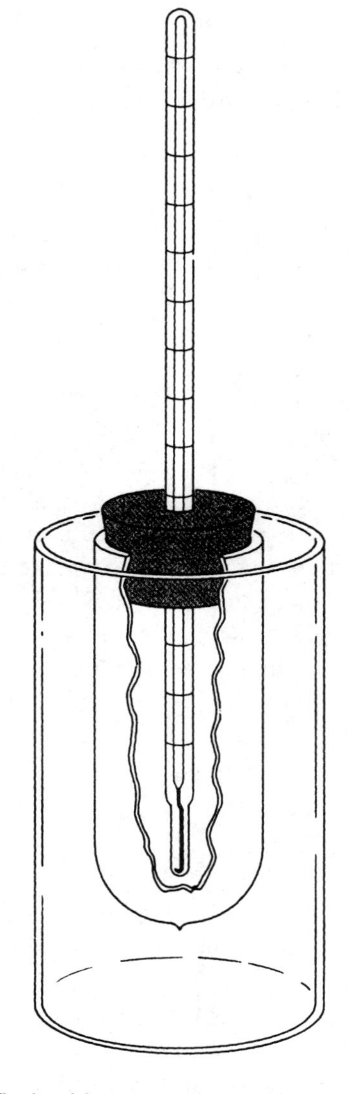

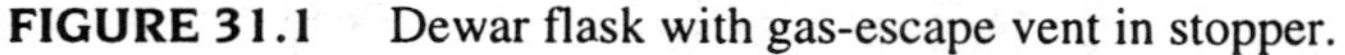

FIGURE 31.1 Dewar flask with gas-escape vent in stopper.

Standardization of the H_2O_2 Solution

The concentration of H_2O_2 in the dilute solution will be determined by adding a 10-mL aliquot to excess KI in dilute H_2SO_4. Ammonium molybdate solution will then be added to catalyze the oxidation of KI by H_2O_2, according to Equation 31.5. The ammonium molybdate serves as a homogeneous catalyst for this reaction.

$$H_2O_2 + 2H^+ + 3I^- = I_3^- + 2H_2O \qquad (31.5)$$

The I_3^- will then be titrated in the customary manner (as in Experiments 26 and 27) with standardized $Na_2S_2O_3$ solution, using starch as an indicator. The titration results will then be used to calculate the concentration of H_2O_2 in the dilute solution.

PROCEDURE

Determining the Calorimeter Constant, C_{cal}

Measure the heat capacity of the calorimeter by mixing warm and cold water, following the procedure outlined in Experiment 30.

Decomposition of Dilute H_2O_2

Add 50 mL of 3% H_2O_2 to approximately 450 mL of distilled water in a Florence flask and mix well. Using a 200-mL volumetric flask twice, measure out 400 mL of the solution and pour it into a clean, well-drained Dewar flask. Weigh out about 2 g of well-powdered MnO_2 on a piece of paper, and set it aside until needed. A mortar and pestle may be used to grind the MnO_2 further if reaction times are inconveniently long using the MnO_2 provided.

Measure the temperature of the solution in the calorimeter to ±0.01°C at 50-second intervals for 300 seconds. Then add the MnO_2, swirl the mixture, and record the mixing time. Continue swirling the mixture according to your standardized procedure and record the temperature at 50-second intervals until the temperature remains constant or starts a gradual decrease. (The temperatures observed may tend to "plateau," remaining constant for several readings and then rising to another plateau. Do not stop taking data too early. Usually, data collection should require no more than 45 minutes.)

Plot the data following the procedure in Experiment 30, and calculate the heat evolved by decomposition of the sample. Do not forget to use the calorimeter constant in your calculations.

Standardization of the H_2O_2 Solution

Before the hydrogen peroxide solution can be standardized, it is necessary to prepare and standardize a sodium thiosulfate solution. Prepare an approximately 0.05 M thiosulfate solution by dissolving 2.4 g of $Na_2S_2O_3{\cdot}5H_2O$ (sodium thiosulfate pentahydrate) in 200 mL of distilled water. Be sure the solution is well mixed. Rinse a buret with two small volumes of distilled water, followed by two small volumes of your thiosulfate solution. Then fill the buret with the solution. Record the initial buret reading.

Record the molarity of the standard potassium iodate (KIO_3) solution provided for your use. Rinse a second buret with two small volumes of distilled water, followed by two small volumes of the standard KIO_3 solution. Then fill the second buret with the iodate solution and record the initial buret reading for the second buret. Deliver 15 mL of KIO_3 solution into a clean, 250-mL Erlenmeyer flask and record the final buret reading for the iodate buret. Add 20 mL of distilled water and 2 g of potassium iodide (KI) to the flask. Swirl the flask to dissolve the solid potassium iodide. Continue to swirl the flask while adding 20 mL of 0.5 M H_2SO_4. The mixture in the Erlenmeyer flask should take on the deep red-brown color characteristic of I_3^- ions.

Titrate the I_3^- ions with $S_2O_3^{2-}$, adding 1-mL increments of thiosulfate initially, but reducing the increment size as the color of the titration mixture turns from brown to yellow. Then add 5 mL of 0.2% starch solution. Consult your instructor if the titration mixture does not develop a blue color after starch has been added and the flask has been swirled. Use your wash bottle to rinse all splattered drops from the walls of the flask into the titration mixture. Resume the addition of thiosulfate, adding titrant by drops until the blue color of the starch–iodine complex disappears completely. Usually, only a few additional drops are required. Record the final reading of the thiosulfate buret and calculate the molarity of your thiosulfate solution. Repeat the procedure. Report the molarity obtained in each trial and the average molarity.

Once the thiosulfate solution has been standardized, it can be used to standardize the hydrogen peroxide solution. Dispense approximately 35 mL of 0.5 M H_2SO_4 solution into a 125-mL flask, add approximately 1 g of solid KI, and swirl the flask to dissolve the solid. Discard the solution if it turns yellow or brown and prepare a new one. The yellow or brown color indicates partial oxidation of the KI by the H_2SO_4 solution, which sometimes occurs when the acid solution is too warm.

Using a pipet, accurately transfer 10.00 mL of the dilute H_2O_2 solution to the flask containing the KI solution. Add 2 drops of 3% ammonium molybdate solution and mix by swirling gently. Add a magnetic stirring bar, and titrate with your standard $Na_2S_2O_3$ solution. When the iodine color fades to pale yellow, add 2 mL of starch solution, and continue titrating to the first disappearance of the blue starch–iodine color. Record the volume and concentration of the $Na_2S_2O_3$ solution used.

Calculate the concentration of the H_2O_2 solution using any necessary equations and your titration data.

Calculate $\Delta H°$ for the decomposition of 1.00 mole of H_2O_2. Show your calculations clearly and be sure that you give ΔH the correct sign.

Disposal of Reagents

All solutions can be neutralized, if necessary, diluted, and flushed down the drain.

Questions

1. Look up the following enthalpies in the U.S. National Bureau of Standards, *Tables of Chemical Thermodynamic Properties or Circular 500*:
   a. ΔH_f°, H_2O_2 (dilute aq)
   b. ΔH_f°, H_2O (ℓ)
   Using these values, calculate the correct value of ΔH° or the decomposition of 1.00 mole of H_2O_2 in dilute solution. Compare the result with your experimental value of ΔH°.
2. Using the value for $\Delta \mathcal{E}^\circ_{cell}$ given in the Principles section, calculate ΔG° and K_{eq} for the reaction

$$H_2O_2\,(1\ M) \longrightarrow H_2O(\ell) + {}^1/_2\,O_2(g,\ 1\ atm)$$

Date ______ Name ______________________________ Section ______ Desk Number ______

PRE-LAB EXERCISES FOR EXPERIMENT 31

These exercises are to be completed after you have read the experiment but before you come to the laboratory to perform it.

1. A student added 10.00 mL of a dilute solution of hydrogen peroxide to a flask containing an acidic solution of potassium iodide. This mixture was titrated with 0.0499 M $Na_2S_2O_3$. The end point was reached after 34.72 mL of the thiosulfate solution had been added. What was the concentration of the hydrogen peroxide solution?

2. The decomposition of the dilute hydrogen peroxide, discussed in Exercise 1, was followed calorimetrically. The decomposition of 400.00 mL of the solution raised the temperature of the system 2.01°C. Assume the calorimeter constant was 196 J/°C and find the value of ΔH for the decomposition of hydrogen peroxide.

Date ______ Name ______________________ Section ______ Desk Number ______

SUMMARY REPORT ON EXPERIMENT 31

Determining the Calorimeter Constant, C_{cal}

Room temperature ______________

Temperature of cool water

Time	Temperature

Initial time of mixing ______________

Time-temperature data

Time	Temperture	Time	Temperature

Temperature of cold water ______________

Temperature of warm water ______________

Temperature of mixture ______________

The calculated value of the calorimeter constant, C_{cal} ______________

Date ______ Name ________________________________ Section ______ Desk Number ______

Decomposition of Dilute H_2O_2

(Note mixing time)

Time	*Temperture*	*Time*	*Temperature*

Date ______ Name ______________________________ Section ______ Desk Number ______

The number of joules Q evolved by decomposition of the sample ____________

Procedure used to calculate Q:

Standardization of the H_2O_2 Solution

	Trial 1	Trial 2	Trial 3*
Final buret reading, KIO_3			
Initial buret reading, KIO_3			
Volume of KIO_3 delivered			
Final buret reading, $Na_2S_2O_3$			
Initial buret reading, $Na_2S_2O_3$			
Volume of $Na_2S_2O_3$ used			
Molarity of $Na_2S_2O_3$			
Average molarity of $Na_2S_2O_3$			

*Optional

Date ______ Name ________________________________ Section ______ Desk Number ______

Titration of the H_2O_2 Solution

Final buret reading, $Na_2S_2O_3$	______	______	______
Initial buret reading, $Na_2S_2O_3$	______	______	______
Volume of standardized $Na_2S_2O_3$ solution used	______	______	______
Calculated concentration of H_2O_2 solution	______	______	______
Average concentration of H_2O_2 solution	______	______	______

Procedure used to calculate concentration of H_2O_2:

Calculated value of ΔH for the decomposition of 1.00 mole of H_2O_2 ______

Procedure used to calculate ΔH:

EXPERIMENT 32 Thermodynamic Prediction of Precipitation Reactions

Laboratory Time Required One hour. May be combined with Experiment 29.

Special Equipment and Supplies Dropper bottles — 0.2 M Solutions containing Ag^+, Ba^{2+}, Na^+, Ca^{2+}, Pb^{2+}, Cl^-, I^-, NO_3^-, and SO_4^{2-} ions

Safety As usual, when chemicals are used in the laboratory, safety glasses should be worn and chemicals should not be ingested.

First Aid **Wash chemicals off your skin and out of your eyes with copious amounts of water.**

The chemistry curriculum has been criticized for being short on descriptive chemistry and long on theory—resulting in students who are often unfamiliar with basic phenomena and uncomfortable with chemical principles. This experiment attempts to overcome both problems by requiring you to work directly with thermodynamic concepts and calculations and to confirm your calculated predictions by direct observation of several precipitation reactions.

PRINCIPLES

A **spontaneous change** is one that occurs by itself, without the exertion of any outside force. A mixture of hydrogen and oxygen gas changes spontaneously (and explosively) into water after being ignited by a spark. Iron rusts spontaneously, albeit slowly, when it is exposed to air and water.

Many spontaneous chemical changes, such as combustion of hydrocarbons, are exothermic. However, there are many examples of endothermic processes that occur spontaneously. These include the melting of ice at ambient temperatures above 0°C and the boiling of water at 100°C (at 760 torr of pressure). In both of these cases, spontaneous change occurs in the direction of a less-ordered state.

Two state functions have been defined to describe the tendency for a change to occur spontaneously. The first of these is ΔH, the enthalpy change. A negative value for ΔH denotes an exothermic process and is a factor that favors spontaneous change. The second state function is ΔS, the entropy change. Entropy is a measure of randomness. A positive value for ΔS denotes that the change will result in a more random (less ordered) system, a factor that also favors spontaneous change.

Very often the direction of spontaneous change is determined by temperature. For instance, water changes spontaneously into ice when placed in a freezer ($t < 0°C$), but ice cubes melt spontaneously when removed from a freezer ($t > 0°C$). In these cases, the signs of ΔH and ΔS work in opposition (e.g., ΔH is favorable when water freezes, but ΔS is unfavorable; ΔH is unfavorable when ice melts, but ΔS is favorable), and temperature is the factor that determines whether ΔH or ΔS will dominate. This information is incorporated into a single state function, the Gibbs free energy, ΔG, defined (for constant temperature systems) in Equation 32. 1.

$$\Delta G = \Delta H - T\Delta S \tag{32.1}$$

It is easy to see that ΔG must be negative if a process is spontaneous because a change for which ΔH is negative and ΔS is positive will surely be spontaneous. Endothermic processes that give positive entropy changes will be spontaneous at high temperatures (where $T\Delta S$ dominates). Exothermic processes that give negative entropy changes will be spontaneous at low temperatures (where ΔH dominates).

Many textbooks give tabulations of values for specific free-energy changes ($\Delta G^\circ_{f,\,298}$), enthalpy changes ($\Delta H^\circ_{f,\,298}$), and entropies (S°_{298}). The ° symbols indicate that the values are being given for changes involving substances in their standard states. The *f* subscript on ΔG and ΔH denotes formation; ΔG°_f and ΔH°_f are the free energy and enthalpy changes, respectively, which are associated with a reaction in which one mole of product is formed from its constituent elements in their standard states. The 298 subscript indicates that all quantities have been corrected to the values they would have if the change were to occur at 298K. Table 32.1 shows how to use tabulated values of $\Delta G^\circ_{f,\,298}$, $\Delta H^\circ_{f,\,298}$, and S°_{298} for carbon, oxygen, and carbon dioxide to calculate the free energy change in the combustion of carbon. Note that you can obtain the standard free energy change for any reaction, $\Delta G^\circ_{rxn,298}$, by subtracting the sum of the $\Delta G^\circ_{f,298}$'s for the reactants from the sum of $\Delta G^\circ_{f,298}$'s for the products (see Equation 32.2).

TABLE 32.1 Calculating the Free Energy Change for the Combustion of Carbon (graphite)

	C(graphite, 298)	+	O_2(g, 298)	$\rightleftarrows$	CO_2(g, 298)
$\Delta H^\circ_{f,\,298}$	0		0		–393.5 kJ/mol
S°_{298}	5.73 J/mol K		205.0 J/mol K		213.6 J/mol K
$\Delta G^\circ_{f,\,298}$	0		0		–394.3 kJ/mol

$\Delta H^\circ_{rxn,298} = \Delta H^\circ_{f,\,298}(CO_2) - \Delta H^\circ_{f,\,298}(O_2) - \Delta H^\circ_{f,\,298}(C)$

$\Delta H^\circ_{rxn,298} = -393.5$ kJ/mol

$\Delta S^\circ_{rxn,298} = S^\circ_{298}(CO_2) - S^\circ_{298}(O_2) - S^\circ_{298}(CO)$

$\Delta S^\circ_{rxn,298} = 2.9$ J/mol K

$\Delta G^\circ_{rxn,298} = \Delta H^\circ_{rxn} - T\Delta S^\circ_{rxn} = -394.4$ kJ/mol

$\Delta G^\circ_{rxn,298} = \Delta G^\circ_{f,\,298}(CO_2) - \Delta G^\circ_{f,\,298}(O_2) - \Delta G^\circ_{f,\,298}(C)$

$\Delta G^\circ_{rxn,298} = -394.3$ kJ/mol

$$\Delta G^\circ_{f,\,298} = \underset{\text{products}}{\Sigma \Delta G^\circ_{f,\,298}} - \underset{\text{reactants}}{\Sigma \Delta G^\circ_{f,\,298}} \qquad (32.2)$$

The combustion of carbon to give carbon dioxide is, of course, a spontaneous process. This is confirmed by the fact that G°_{298} for the combustion process is –394.4 kJ/mol. This means that when one mole of graphite, the most stable form of carbon, is combined with one mole of oxygen, at one atmosphere pressure, to produce one mole of CO_2, at one atmosphere pressure, the free energy of the system decreases by 394.4 kJ. If the reaction is not performed with all materials in their standard states, if the reaction temperature is not 298 K, if the pressure of the either of the gases is not one atmosphere, or if more or less than one mole of carbon is consumed in the reaction, then a value of ΔG_{rxn} will be obtained that will differ from ΔG°_{rxn}. The relationship between ΔG_{rxn} and ΔG°_{rxn} is shown in Equation 32.3.

$$\Delta G_{rxn} = \Delta G^\circ_{rxn} + 2.303\ RT \log Q \qquad (32.3)$$

In Equation 32.3, R is the ideal gas constant (8.314 J/K mol), T is the Kelvin temperature, and Q denotes the reaction quotient. As is customary for evaluating equilibrium constants and reaction quotients, liquids, solids, and solvents are represented by unity; solute concentrations are represented by molarity; and the pressures of gases are given in atmospheres.

In this experiment, you will be calculating the ΔG_{rxn} for a variety of possible precipitation reactions. You will use your calculated values to predict whether a precipitate will form when two solutions are mixed. Then you will actually mix the reagents and attempt to confirm your predictions. You will also calculate ΔG_{rxn} for the precipitation of a few salts at different temperatures. You will place the precipitates in hot or cold baths and attempt to confirm your predictions regarding the change in solubility as a function of changing temperature as well. Tables 32.2 and 32.4 list the data you will use to make your calculations. The following paragraphs explain how to use this data.

The row and column headings of Table 32.2 show the standard free energies

TABLE 32.2 Gibbs Free Energies ($\Delta G^\circ_{f,298}$) for Ions in 1 M Solution and Solids

Anions→ ↓Cations	Cl^- −131.228	I^- −51.57	NO_3^- −108.74	SO_4^{2-} −744.53
Ag^+ 77.107	−109.789	−66.19	−33.41	−618.41
Ba^{2+} −560.77	−1296.32 W2	—	−796.59	−1362.2
Na^+ −261.905	−384.138	−286.06	−367.00	−3646.85 W10
Ca^{2+} −553.58	−748.1	−528.9	−743.07	−1797.28 W2

of formation of the various ions under consideration, with the standard state as a 1 M solution. Thus, $\Delta G^\circ_{f,\,298}$ for Ag^+ is 77.107 kJ/mol and $\Delta G^\circ_{f,\,298}$ for Cl^- is −131.228 kJ/mol. Entries within the body of the table show the standard free energies of formation for the crystalline solids that result from the combinations of the various ions whose rows and columns intersect to create the compound's cell. For instance, $\Delta G^\circ_{f,\,298}$ for AgCl is −109.789 kJ/mol. The "W10" entry in the cell corresponding to sodium sulfate indicates that the most likely precipitate is $Na_2SO_4 \cdot 10H_2O$. When you write the equation for the precipitation of such a hydrated salt, water will appear as a reactant. Therefore, in the calculation of ΔG°_{rxn} you will need to consider $\Delta G^\circ_{f,\,298}$ for water, which has a value of −237.129 kJ/mol.

Because you will not be working with 1 M solutions, the free-energy changes you will be calculating will not be standard free energies; rather, Equation 32.3 will be needed to convert the ΔG°'s to ΔG's. Table 32.3 shows the calculation of ΔG for the possible reaction between Ag^+ ions and Cl^- ions to form the precipitate, silver chloride. Because $\Delta G_{rxn,\,298}$ for the precipitation of AgCl is negative, it is predicted that the precipitate will form when solutions of Ag^+ and Cl^- ions are mixed.

TABLE 32.3 Calculating ΔG for the Precipitation of AgCl

$$Ag^+\,(0.1\text{ M}) + Cl^-\,(0.1\text{ M}) \longrightarrow AgCl\,(s)$$

$$\Delta G^\circ_{rxn,298} = \Delta G^\circ_{f,\,298}(AgCl) - \Delta G^\circ_{f,\,298}(Ag^+) - \Delta G^\circ_{f,\,298}(Cl^-)$$

$$\Delta G^\circ_{rxn,298} = -109.789 - (77.107) - (-131.228)\text{ kJ/mol}$$

$$\Delta G_{rxn,298} = -55.668\text{ kJ/mol}$$

$$\Delta G_{rxn,298} = \Delta G^\circ_{rxn,\,298} + 2.303\,RT\log Q$$

$$\Delta G_{rxn,298} = -55\,.668\text{ kJ/mol} + 2.303\,\frac{(8.314\text{ J/mol K})(298\text{K})}{1000\text{ J/kJ}}\,\log\frac{1}{(0.1)(0.1)}$$

$$\Delta G_{rxn,298} = -44.250\text{ kJ/mol}$$

The $\Delta G°$'s and ΔG's you have calculated so far were evaluated at 298K. There will be times when you may wish to evaluate these functions at other temperatures. This is easily accomplished because the values of $\Delta H°$ and $\Delta S°$ are relatively independent of temperature. Thus, $\Delta G°$ can be evaluated at any temperature by the use of Equation 32.4, where T is the Kelvin temperature and ΔH°_{298} and ΔS°_{298} are the enthalpy and entropy change, respectively, for the reaction.

$$\Delta G^\circ_T = \Delta H^\circ_{298} - T\,\Delta S^\circ_{298} \qquad (32.4)$$

Values of $\Delta H^\circ_{f,\,298}$ and S°_{298} for several ions and crystalline solids are given in Table 32.4. Use the values given to decide whether precipitates would form when equal volumes of 0.2 M solutions of Pb^{2+} and Cl^- are mixed at 273K, 298K, and 373K. Do similar calculations for mixtures of equal volumes of 0.2 M solutions of Ba^{2+} and NO_3^-. Use Equation 32.4 to evaluate ΔG°_T at T = 273K, 298K, and 373K. Use Equation 32.3 to evaluate ΔG_T at those temperatures. An example of this kind of calculation is given in Table 32.5. Because $\Delta G_{rxn,273}$ is negative for the precipitation reaction, it is predicted that a precipitate will form if equal volumes of 0.2 M Pb^{2+} and 0.2 M I^- are mixed and cooled to 273K.

TABLE 32.4 Values of $\Delta H^\circ_{f,298}$ and S°_{298} for Various Ions and Solids

	$\Delta H^\circ_{f,298}$ kJ/mol	S°_{298} J/K mol
Pb^{2+}	− 2	10
Cl^-	−167	56
$PbCl_2$	−359	136
I^-	− 55	111
PbI_2	−175	175
Ba^{2+}	−538	10
NO_3^-	−205	146
$Ba(NO_3)_2$	−992	214

TABLE 32.5 Calculating ΔG for the Precipitation of PbI_2 at 273 K

$$Pb^{2+}\,(0.1\text{ M}) + 2\,I^-\,(0.1\text{ M}) \longrightarrow PbI_2\,(s)$$

$$\Delta H^\circ_{rxn,298} = -175 - (-2 + 2(-55)) = -63\text{ kJ}$$

$$\Delta S^\circ_{rxn,298} = 175 - (10 + 2(111)) = -57\text{ J/K}$$

$$\Delta G^\circ_{rxn,273} = -63000 - 273\,(-57) = -46 \times 10^3\text{ J}$$

$$\Delta G_{rxn,273} = -46 \times 10^3\text{ J} + (2.303)(8.314)(273)\log\,(1/(0.1)(0.1))$$

$$\Delta G_{rxn,273} = -46 \times 10^3\text{ J} + 10.5 \times 10^3\text{ J} = -36 \times 10^3\text{ J}$$

PROCEDURE

Calculate the value of $\Delta G_{rxn,298}$ for each solid that might result when each of the 0.2 M solutions of cations listed in Table 32.2 is mixed with an equal volume of each of the anion solutions listed. Note that mixing equal volumes of 0.2 M solutions of cations and anions will result in 0.1 M solutions after mixing. Record your values of the ΔG's in the upper space of the boxes on the Summary Report sheet. Then mix the solutions and note in the lower spaces whether your observations confirm (C) or deny (D) your predictions. Below the table of results, note the appearance of the precipitates you observe and briefly discuss possible reasons for any discrepancies between your predictions and your observations.

Also, calculate the values of $\Delta G_{rxn,\,T}$ for the precipitation of $PbCl_2$ and $Ba(NO_3)_2$ from solutions that are 0.1 M in lead ions and chloride ions and 0.1 M in barium ions and nitrate ions, respectively, for T = 273K, 298K, and 373K. Mix the appropriate solutions at room temperature. If no precipitate results, cool the mixture in an ice bath. If a precipitate does appear, heat the test tube containing the precipitate and supernatant liquid in a boiling water bath.

Note whether the precipitate dissolves and whether your observations are in accord with your calculations. If they are not, briefly discuss possible reasons for the discrepancies.

Disposal of Reagents

The small quantities of chemicals used in this experiment may be flushed down the drain with copious amounts of water.

Date ______ Name ________________________________ Section ______ Desk Number ______

PRE-LAB EXERICES FOR EXPERIMENT 32

These exercises are to be performed after you have read the experiment but before you come to the laboratory to perform it.

1. Write the chemical equation corresponding to $\Delta G^\circ_{f,\,298}$ for AgCl. How does this equation differ from the net ionic equation that shows AgCl precipitating when solutions of Ag^+ and Cl^- are mixed?

2. The value of $\Delta G^\circ_{f,\,298}$ for Mg^{2+} is –454.8 kJ/mol. The value of $\Delta G^\circ_{f,\,298}$ for $MgCl_2{\cdot}6H_2O$ is –2114.64 kJ/mol. Will a precipitate form when 0.2 M Mg^{2+} is mixed with 0.2 M Cl^-?

SUMMARY REPORT ON EXPERIMENT 32

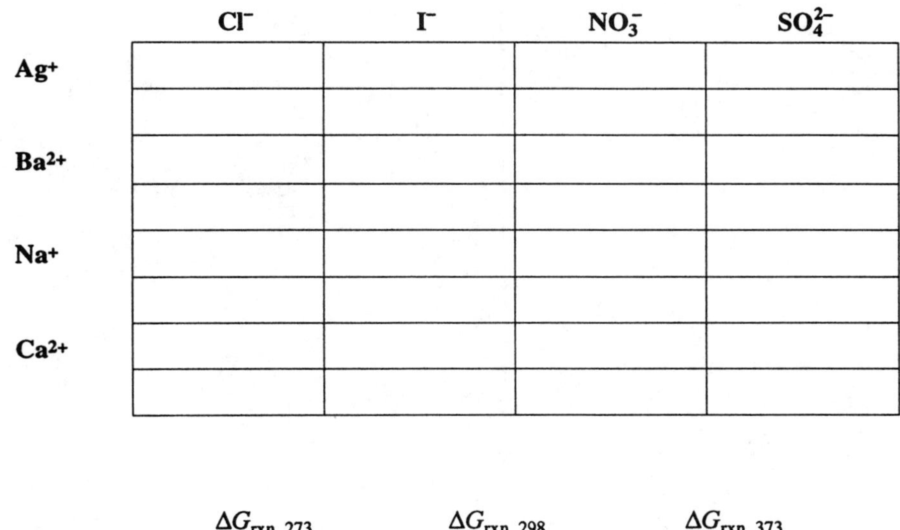

	Cl^-	I^-	NO_3^-	SO_4^{2-}
Ag^+				
Ba^{2+}				
Na^+				
Ca^{2+}				

	$\Delta G_{rxn,\,273}$	$\Delta G_{rxn,\,298}$	$\Delta G_{rxn,\,373}$
$PbCl_2$	______________	______________	______________
$Ba(NO_3)_2$	______________	______________	______________

Observations when 0.2 M Pb^{2+} is mixed with 0.2 M I^-

Observations when 0.2 M Ba^{2+} is mixed with 0.2 M NO_3^-

Kinetic Study of the Reaction Between Ferric and Iodide Ions

Laboratory Time Required

Three hours for well-prepared students, working in teams.

Special Equipment and Supplies

Constant-temperature baths
Burets
Timer
Thermometer, 0.1 subdivisions
Pipets
Pipet bulbs

0.04 M $Fe(NO_3)_3 \cdot 9H_2O$ in 0.15 M HNO_3
0.15 M HNO_3
0.04 M KI
$Na_2S_2O_3 \cdot 5H_2O$ (s)
0.2% Starch solution
$Fe(NH_4)_2(SO_4)_2 \cdot 6H_2O$, solid or 0.002 M solution in 0.15 M HNO_3
0.02 M standard KIO_3
0.5 M H_2SO_4

Safety

None of these solutions is very hazardous, but you should still exercise care to avoid chemical contact with your skin, mouth, or eyes.

First Aid

Thoroughly flush the affected area with water.

Thermodynamics alone does not provide all the information needed to understand a chemical system. It tells us only the direction of spontaneous chemical change, not how fast the reaction proceeds or by which mechanism the reactants are converted to products. You can gain a more complete understanding of a reaction by combining thermodynamics with kinetics, which concerns factors that affect the rate of the reaction. By studying the quantitative dependence of the rate on such factors as reactant concentrations, temperature, and the presence of catalysts, you can write an exact rate expression—or rate law—for the reaction and calculate the activation energy. This information can then be used to postulate a detailed mechanism for the reaction, specifying the individual steps required to convert reactants to products. With kinetic information, you can make practical plans to control the rate of a reaction. This is often important in synthetic chemistry, where the successful synthesis of the desired product may depend on making its rate of formation greater than the rates of competing reactions that lead to other products.

PRINCIPLES

To write the rate expression for a reaction, we must determine, experimentally, how the rate is related to the concentration of each reactant. You can do this easily by varying the initial concentration of one reactant at a time while the concentrations of the other reactants are held constant, observing the effect of each such change on the rate. However, because the rate will probably change as the reactant under study is being used up, you will need to determine the concentrations of all the reactants at the time a rate measurement is made. An elegant solution to this problem is to limit each rate measurement to the initial 1 percent or so of the reaction, so that the reactant concentrations do not have time to change appreciably during the monitoring period. Therefore, you will need to know only the initial concentrations of the reactants in the mixture and not how they change with time. This procedure, called the **initial rate method**, will be used in this experiment.

The Rate Law

Consider the hypothetical reaction represented by Equation 33.1:

$$2A + B \longrightarrow A_2B \tag{33.1}$$

The rate may be related to either the change in product concentration per unit time or the change in reactant concentration per unit time, according to Equation 33.2.

$$\text{rate} = -\frac{1}{2}\frac{d[A]}{dt} = -\frac{d[B]}{dt} = \frac{d[A_2B]}{dt} \tag{33.2}$$

The rate expression or rate law has the general form shown in Equation 33.3,

$$\text{rate} = k[A]^a\,[B]^b \tag{33.3}$$

where k is the specific rate constant for the reaction, and a and b represent the order of the reaction with respect to A and B. It should be emphasized that a and b are determined experimentally and are not deduced from the stoichiometry of the reaction. The order with respect to each reactant may be positive or negative, integral or fractional, or even zero.

Reversible Reactions

The top curve of Figure 33.1 shows how a system initially containing only A and B behaves. The rate of the forward reaction is large at the start but decreases, as predicted from Equation 33.3, as the reactants are converted to A_2B. If the reaction is reversible, as assumed in Figure 33.1, the rate of the reverse reaction is related to the concentration of A_2B. The rate of the reverse reaction is zero at the start, when no A_2B is present, but increases with time as A_2B is formed by the

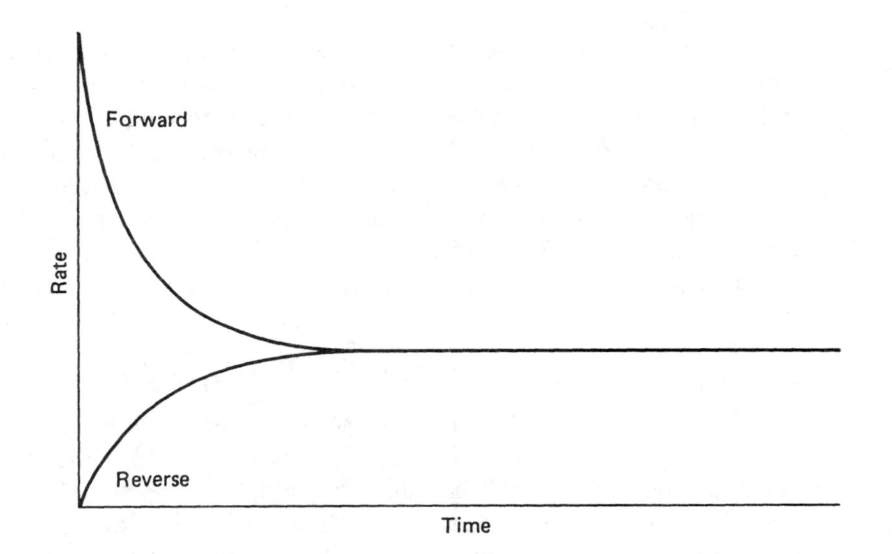

FIGURE 33.1 Change with time of the forward and reverse reaction rates for a reversible reaction.

forward reaction. Eventually, of course, the forward and reverse rates become equal, and the system is then at equilibrium.

We assumed in Equation 33.3 that the rate of formation of product is related to the forward reaction rate, according to the expression shown in Equation 33.4.

$$\frac{d[A_2B]}{dt} = \text{rate}_{(\text{forward})} \tag{33.4}$$

However, if the reaction is reversible, the expression shown in Equation 33.5 applies. The addition of the $\text{rate}_{(\text{reverse})}$ term greatly complicates the calculations, and we therefore try to work under conditions where the reaction is not reversible or where the reverse rate can be set equal to zero. An obvious advantage of the initial rate method is that the reverse rate is negligible under these conditions and, therefore, the reverse reaction can be ignored.

$$\frac{d[A_2B]}{dt} = \text{rate}_{(\text{forward})} - \text{rate}_{(\text{reverse})} \tag{33.5}$$

The Ferric Ion–Iodide Ion System

In this experiment, you will study the oxidation of I^- by Fe^{3+} ions, as shown in Equation 33.6. Note that the triiodide ion, I_3^-, is simply an iodine molecule complexed with an iodide ion. This complexation greatly increases the solubility of the iodine species in water.

$$2Fe^{3+} + 3I^- \longrightarrow 2Fe^{2+} + I_3^- \tag{33.6}$$

The expected rate expression for the reaction of Fe^{3+} and I^- is given in Equation 33.7.

$$\text{rate} = -\frac{1}{2}\frac{d[Fe^{3+}]}{dt} = \frac{d[I_3^-]}{dt} = k[Fe^{3+}]^a[I^-]^b \tag{33.7}$$

Parts A and B of the experiment are concerned with evaluating the exponents *a* and *b*, respectively, in the rate expression. Part C is concerned with the effects of temperature and of other ions on the reaction rate.

You will determine the initial rate by measuring the time, in seconds, required for part (about 4×10^{-5} mole) of the Fe^{3+} to be reduced to Fe^{2+}. You will know that this has occurred because of the presence of starch and a small, constant amount of $S_2O_3^{2-}$ in each mixture. The thiosulfate will react with the triiodide ion produced in the reduction of the Fe^{3+} ion, as shown in Equation 33.8.

$$I_3^- + 2S_2O_3^{2-} \longrightarrow 3I^- + S_4O_6^{2-} \qquad (33.8)$$

As soon as the $S_2O_3^{2-}$ has been consumed, any additional I_3^- formed by the reaction between ferric and iodide ions will react with the starch to give a characteristic blue color. When the blue color first appears, the decrease in the concentration of Fe^{3+} from its initial value is just equal to the initial concentration of $S_2O_3^{2-}$ in the mixture. Thus, the initial rate of the disappearance of iron(III) ions, $\frac{1}{2}\, d[Fe^{3+}]/dt$, is equal to $\frac{1}{2}[S_2O_3^{2-}]_i/\Delta t$, where $[S_2O_3^{2-}]_i$ is the initial concentration of $S_2O_3^{2-}$ and Δt is the time in seconds between mixing and the appearance of the blue color. In order to obtain reasonable reaction times, you will need to use initial rate intervals that allow the Fe^{3+} concentration to decrease slightly (about 4 to 10 percent) from its initial value. To compensate for this change, the average Fe^{3+} concentration during this time interval should be used in place of the initial concentration when plotting your data.

In Part C, the effects of other ions and of temperature will be studied. Ions other than ferric or iodide may affect the rate by varying the ionic strength, serving as catalysts, or serving as inhibitors. Changing the ionic strength affects the activity of the ions reacting and, therefore, changes the rate. This effect will not be studied. Catalysts provide a new and easier path for the reaction and thereby increase the rate. Inhibitors decrease the rate by lowering the concentration or the activity of some species which is important in the reaction mechanism.

The effect of temperature on reaction rate is treated in your text theoretically. Therefore, we will deal with the subject only briefly here. The specific rate constant, *k*, is related to the Kelvin temperature *T* by the expression shown in Equation 33.9, where E_a is the activation energy for the reaction. The activation energy represents the minimum energy required for the reactants to pass over an energy barrier to form products, as shown in Figure 33.2. Because the reactant particles have a statistical distribution of energies, at a given temperature only a fraction of the particles will have sufficient energy to react upon colliding. If the absolute temperature is increased, however, the fraction of the particles having the energy needed to react upon colliding also increases, resulting in a greater rate or a larger value for the specific rate constant.

$$k = Ae^{-E_a/RT} \qquad (33.9)$$

If *A* is a constant in Equation 33.9, then expressions for *k* as a function of *T* may be derived (see Equations 33.10 and 33.11).

$$\ln k = -\frac{E_a}{RT} + \text{constant} \qquad (33.10)$$

$$\Delta \ln k = -\frac{E_a}{R} - \Delta\frac{1}{T} \qquad (33.11)$$

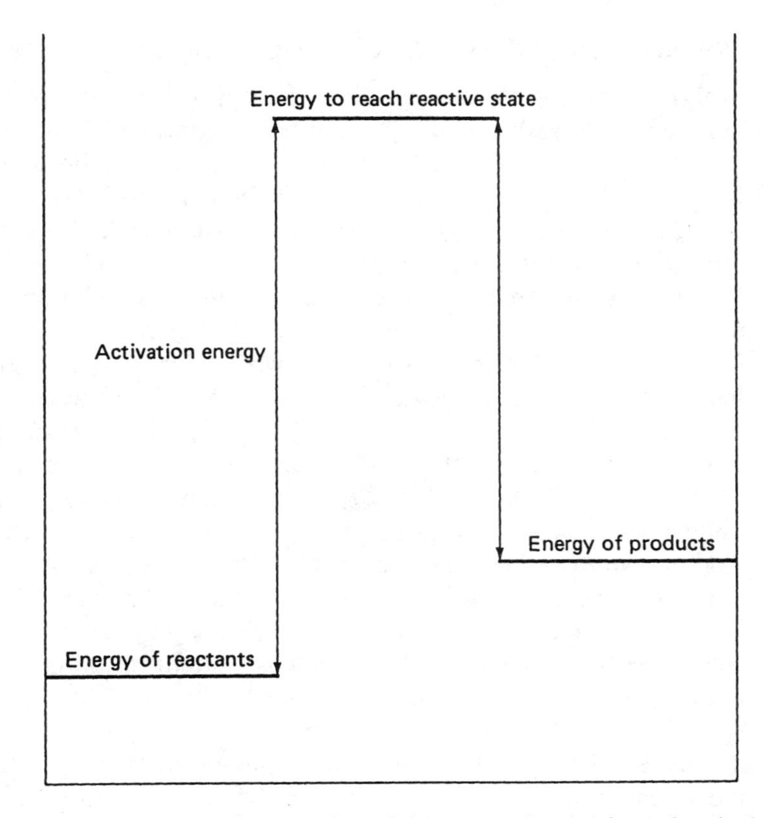

FIGURE 33.2 Schematic representation of the energy barrier for a chemical reaction.

A plot of ln k (or 2.303 log k) versus $1/T$ should therefore be linear, with a slope equal to $-E_a/R$.

PROCEDURE

Many solutions must be used in preparing each reaction mixture to be studied. The thiosulfate solution is somewhat unstable and is best prepared on the day it is to be used. The rate of reaction is quite sensitive to changes of temperature. All of these factors can affect the accuracy of the results obtained in this experiment. Very satisfactory results are obtained efficiently if the class is divided into teams, each with a specific task to perform that will facilitate the work of the individual class members when they begin to collect the data required. One team is responsible for preparing and standardizing the thiosulfate solution. A second team is responsible for filling the water baths and maintaining their temperature at roughly 5°C above room temperature and at roughly 5°C below room temperature. A third team is responsible for preparing work stations with burets containing each of the solutions needed in the reaction mixture. A fourth team is responsible for assembling the special equipment (bottles for storing solutions, timers, thermometers, etc.) needed for the performance of the experiment.

Preparation and Standardization of the Thiosulfate Solution

Weigh out 2.4 g of $Na_2S_2O_3{\cdot}5H_2O$ and dissolve it in approximately 2 L of distilled water. Mix well. Rinse a buret with two small volumes of distilled water, followed by two small volumes of your thiosulfate solution. Then fill the buret with the solution. Record the initial buret reading.

Record the molarity of the standard potassium iodate (KIO_3) solution provided for your use. Rinse a pipet with two small volumes of distilled water, followed by two small volumes of the standard KIO_3 solution. Then use the pipet to deliver 1.00 mL of the standard iodate solution into a clean, 250-mL Erlenmeyer flask. Add 20 mL of distilled water and 20 mL of 0.04 M potassium iodide (KI) to the flask. Swirl the flask to mix the solutions; continue swirling the flask while adding 2 mL of 0.5 M H_2SO_4. The mixture in the Erlenmeyer flask should take on the deep red-brown color characteristic of I_3^- ions.

Titrate the I_3^- ions with $S_2O_3^{2-}$, adding 1-mL increments of thiosulfate initially, but reducing the incremental size as the color of the titration mixture turns from brown to yellow. Then add 5 mL of 0.2% starch solution. Consult your instructor if the titration mixture does not develop a blue color after starch has been added and the flask has been swirled. Use your wash bottle to rinse all splattered drops from the walls of the flask into the titration mixture. Resume the addition of thiosulfate, adding titrant by drops until the blue color of the starch–iodine complex disappears completely. Usually, only a few additional drops are required. Record the final reading of the thiosulfate buret and calculate the molarity of your thiosulfate solution. Repeat the procedure. Report the molarity obtained in each trial and the average molarity.

Give the thiosulfate solution to the team that is setting up the solution stations. Report the concentration of the thiosulfate solution to the class. When the work stations and water baths are set up, individual students, or pairs of students, proceed to the determination of the orders of the reaction and the value of the activation energy.

Maintaining Constant Temperature

It is inevitable that water that has been cooled below room temperature (by the addition of small amounts of ice that melt completely) will absorb heat from the air and begin to warm up. It is also inevitable that water that has been heated a bit above room temperature will begin to lose heat to the surroundings once heating has stopped. Heat transfer with the surroundings can be minimized by insulating the water baths. However, because the reaction rate is very temperature-sensitive, constant vigilance is required to maintain fairly constant bath temperatures. Team members should be assigned to check the bath temperature regularly and should also be prepared to add ice or hot water to the baths as necessary.

Assembling the Work Stations

The number of work stations needed will depend on the size of the class. There should be at least one station for six pairs of students. Refilling is easy if automatic burets are used. If these are not available, be sure there is a good supply of each reagent left at the station so that burets can be refilled as necessary. The starch solution should be dispensed from a graduated cylinder; all other solutions can be dispensed from burets. The solutions needed are identified in Table 33.1.

Glassware

Solutions will need to be kept in a water bath of appropriate temperature for 5–10 minutes before they are mixed. Large beakers and flasks containing small volumes of liquid tend to tip over easily. Your instructor will tell you what measures your class will take to avoid this problem (for instance, vessels might be clamped in place or they might be weighted down in some manner).

Obtaining Data on the Kinetics of the Reaction

Part A. Reaction Order with Respect to Fe^{3+}

Prepare the mixture for Experiment 1 by adding the solutions specified in Table 33.1 to appropriate containers. Briefly swirl the contents of each container and place the containers in a constant temperature water bath set at room temperature. Allow a few minutes for the mixtures to reach temperature equilibrium. Meanwhile, prepare the solutions for Experiment 2 in a second set of containers and place these also in the water bath. By the time you do that, the solution for Experiment 1 should be at bath temperature.

Measure and record the temperature of the Experiment 1 solutions, then simultaneously start the timer and rapidly add the contents of container 1 to container 2. You may temporarily remove the solutions from the water bath when you are doing this. Swirl the solutions until they are well mixed, then return container 2 to the bath. Stop the timer at the first appearance of the blue color. Record the time, *t*, and the temperature (to 0.1°C). Clean and dry the containers, place the solutions for Experiment 3 in them, and return the containers to the water bath. Measure the temperature of the solutions for Experiment 2, mix the solutions, and time the reaction as before. Continue in this manner through Experiment 4.

TABLE 31.1 Reaction Mixtures

	Container 1			Container 2			
Experiment	*0.04 M* Fe^{3+} *mL*	*0.15 M* HNO_3 *mL*	H_2O *mL*	*0.04 M KI mL*	*0.004 M* $S_2O_3^{2-}$ *mL*	*starch mL*	H_2O *mL*
1	10.00	10.00	30.00	15.00	10.00	5.00	20.00
2	15.00	15.00	20.00	15.00	10.00	5.00	20.00
3	20.00	20.00	10.00	15.00	10.00	5.00	20.00
4	25.00	25.00	0.00	15.00	10.00	5.00	20.00
5	10.00	10.00	30.00	12.00	10.00	5.00	23.00
6	10.00	10.00	30.00	18.00	10.00	5.00	17.00
7	10.00	10.00	30.00	21.00	10.00	5.00	14.00

Part B. Reaction Order with Respect to I^-

Repeat the procedures specified in Part A, but use the various reaction mixtures given in Table 33.1 for Experiments 5, 6, and 7.

Part C. Effect of Temperature and of Fe^{2+} Ions

Perform Experiments 8, 9, and 10 by preparing reaction mixtures as in Experiment 1 and measuring the reaction times in water baths at various temperatures. Experiment 8 should be performed at room temperature, Experiment 9 at about 5° higher, and Experiment 10 at a temperature about 5° below room temperature.

If time permits, prepare a solution of 0.002 M in Fe^{2+} by dissolving $Fe(NH_4)_2(SO_4)_2$ in 0.15 M HNO_3. Repeat Experiment 1, substituting the Fe^{2+} solution for the HNO_3 solution.

Calculations

Using the experimental reaction times and the known concentrations of the reagents, calculate the initial rate, $\frac{1}{2}[S_2O_3^{2-}]_i/\Delta t$, for each experiment.

Calculate the initial concentrations of Fe^{3+} and I^- present in each experiment. Approximate $[Fe^{3+}]_{av}$ for each experiment by multiplying each initial Fe^{3+} concentration by 0.93.

Prepare a table showing the following: the initial and average Fe^{3+} concentrations for each experiment, the initial I^- concentration, the initial rate, $\log[Fe^{3+}]_{av}$, $\log[I^-]$, and log (rate).

Using the data for Experiments 1 through 4, plot log(rate) versus $\log[Fe^{3+}]_{av}$. Draw the best straight line through the experimental points, and calculate its slope. This is equal to a, the reaction order for Fe^{3+}.

Using the data from Experiments 1, 5, 6, and 7, plot log (rate) versus $\log[I^-]$. Draw the best straight line and calculate its slope. This is equal to b, the reaction order for I^-.

Round off a and b to integers and write the resulting rate expression. Calculate the initial rates for Experiments 9, 10, and 11. Substitute the measured rates, $[Fe^{3+}]_{av}$ and $[I^-]_i$ in the rate expression to obtain the specific rate constant k at each of the three temperatures. Prepare a table showing k, $\ln k$, T, and $1/T$.

Plot $\ln k$ versus $1/T$. Measure the slope and from it calculate the activation energy, E_a. Comment briefly on the effect of added Fe^{2+} on the reaction rate.

Disposal of Reagents

The acidic solutions (HNO_3 and $Fe(NO_3)_3$) should be neutralized with dilute NaOH solution. The neutralized chemicals may be diluted and poured down the drain. All other solutions may also be diluted and poured down the drain.

Questions

1. For reactions in aqueous solution, would it be easy or difficult to determine the order with respect to H_2O? Explain.
2. Show that multiplying $[Fe^{3+}]_i$ by 0.93 approximates $[Fe^{3+}]_{av}$. Why is it unnecessary to convert $[I^-]_i$ to $[I^-]_{av}$?

PRE-LAB EXERCISES FOR EXPERIMENT 33

These exercises are to be performed after you have read the experiment but before you come to the laboratory to perform it.

1. The reaction between bromate ions and bromide ions occurs according to the equation shown below.

$$BrO_3^-(aq) + 5Br^-(aq) + 6H^+(aq) \longrightarrow 3Br_2(\ell) + 3H_2O(\ell)$$

The following table shows the results of four experiments done on this system. Use these data to determine the order of reaction with respect to each reactant, the overall order of reaction, and the value of the rate constant.

Experiment number	$[BrO_3^-]_i$	$[Br^-]_i$	$[H^+]_i$	Measured initial rate, $-\Delta[BrO_3^-]/\Delta t$ (mol/L s)
1	0.10	0.10	0.10	7.9×10^{-4}
2	0.20	0.10	0.10	1.7×10^{-3}
3	0.20	0.20	0.10	3.1×10^{-3}
4	0.10	0.10	0.20	3.2×10^{-3}

2. Methane gas reacts with diatomic sulfur in the gas phase, producing carbon disulfide and hydrogen sulfide. The rate constant for this reaction is 1.1 L/mol s at 550°C. It is 6.4 L/mol s at 625°C. Use these data to determine the value of the activation energy for the reaction. When k is measured at only two temperatures, Equation 33.11 may be written as shown below.

$$\ln k_2 - \ln k_1 = -\frac{E_a}{R}\left(\frac{1}{T_2} - \frac{1}{T_1}\right)$$

Date ______ Name ______________________________ Section ______ Desk Number ______

SUMMARY REPORT ON EXPERIMENT 33

Concentration of $S_2O_3^{2-}$ ______________________

Experiment Number	*Initial Temperature*	*Reaction Time Δt, sec*
1		
2		
3		
4		
5		
6		
7		
8		
9		
10		
(Fe^{2+} added)		

Experiment Number	*Initial Rate*	*Log Rate*	$[I^-]_i$	*Log* $[I^-]_i$
1				
2				
3				
4				
5				
6				
7				
8				
9				
10				

Experiment Number	$[Fe^{3+}]_i$	$[Fe^{3+}]_{av}$	$Log[Fe^{3+}]_{av}$
1			
2			
3			
4			
5			
6			
7			
8			

Date ______ Name ________________________________ Section ______ Desk Number ______

Experiment Number	$[Fe^{3+}]_i$	$[Fe^{3+}]_{av}$	$Log[Fe^{3+}]_{av}$
9	______	______	______
10	______	______	______

Procedure for calculating *a*, the reaction order with respect to Fe^{3+}:

Procedure for calculating *b*, the reaction order with respect to I^-:

Experimental rate expression for the reaction:

Experiment Number	k	*ln* k	T	*1*/T
8	______	______	______	______
9	______	______	______	______
10	______	______	______	______

Procedure used to calculate *k*:

Experimental value for the activation energy E_a of the reaction ______________

Procedure used to calculate E_a:

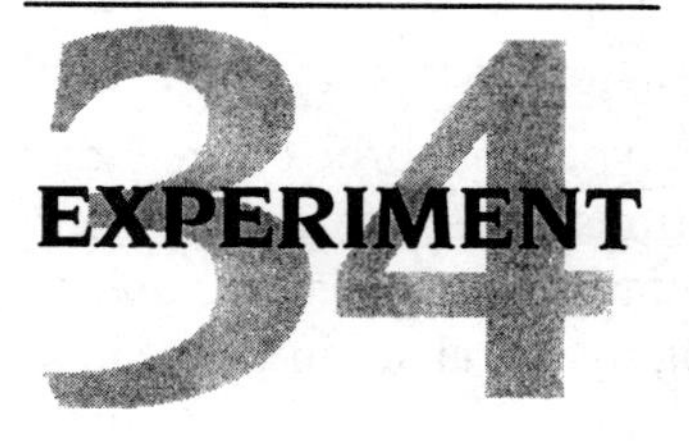

Oxidation of Ethanol by Dichromate

Laboratory Time Required

Three hours

Special Equipment and Supplies

Spectrophotometer	Ethanol
Spectrophotometer cells	0.01667 M Standard potassium dichromate
Timer	3.6 M Sulfuric acid
Buret	3 M NaOH
Pipet	Mossy zinc
Pipet bulb	
Hot water bath	

Safety

Chromium, in all of its oxidation states, is **toxic**. Potassium dichromate ($K_2Cr_2O_7$) is an oxidant and a suspected carcinogen. Solutions of $K_2Cr_2O_7$ are safer to work with than the fine powder because the powder could contaminate the atmosphere with fine $K_2Cr_2O_7$ dust. Solutions can cause a rash or burns to the skin.

Sulfuric acid is **corrosive**. The usual precautions should be observed. Mouth pipetting is not allowed.

First Aid

For acid burns, flush the skin (or eyes, if necessary) thoroughly with water. See a doctor.

For $K_2Cr_2O_7$ on the skin, wash with water for at least 15 minutes. If $K_2Cr_2O_7$ has been ingested, rinse your mouth thoroughly; then, see a doctor.

The order of a reaction with respect to a reactant is generally determined by either the method of initial rates or the integrated rate method. This experiment uses both in the study of the kinetics of the oxidation of ethanol by dichromate ions.

PRINCIPLES

The mild oxidation of alcohols produces ketones, aldehydes, or carboxylic acids, depending on the nature of the alcohol and the conditions under which the reaction occurs. The oxidizing agent is frequently potassium permanganate or potassium dichromate. Each of these oxidizing agents contains a transition metal in a high oxidation state [Mn(VII) in the permanganate ion and Cr(VI) in the dichromate ion]. The extent of reaction between the dichromate ion and ethanol is easy to monitor colorimetrically. It proceeds at a moderate rate, neither too fast nor too slow, and involves a distinct color change as the orange dichromate ion is consumed and the green chromium(III) ion is produced (see Equation 34.1).

$$2Cr_2O_7^{2-} + 16H^+ + 3C_2H_5OH \longrightarrow 4Cr^{3+} + 3HC_2H_3O_2 + 11H_2O \qquad (34.1)$$

The general rate law for this reaction is shown in Equation 34.2.

$$\frac{-d[Cr_2O_7^{2-}]}{dt} = k[Cr_2O_7^{2-}]^a[H^+]^b[C_2H_5OH]^c = 0.5\frac{d[Cr^{3+}]}{dt} \qquad (34.2)$$

The superscripts a, b, and c in Equation34.2 are the orders of the reaction with respect to the dichromate ion, the acid, and the alcohol. The values of a, b, and c must be determined experimentally. They are not necessarily the same as the coefficients of the balanced equation, which represents the overall chemical reaction. The differentials at the extreme ends of Equation 34.2 indicate that the extent of reaction may be monitored by measuring either the rate of disappearance of the dichromate ions or the rate of appearance of the chromium(III) ions.

This experiment will be performed using relatively large excesses of acid and alcohol in comparison to the amount of the dichromate ion. Thus, the concentrations of H^+ and ethanol will remain constant as the limiting agent (the dichromate ion) is consumed. Under theses conditions, the differential rate law is represented as is shown in Equation 34.3 and only the value of a is determined.

$$\frac{-d[Cr_2O_7^{2-}]}{dt} = k'[Cr_2O_7^{2-}]^a \qquad (34.3)$$

The first method that will be used to evaluate a is the method of initial rates. A series of mixtures will be prepared, each of which contains the same amount of acid and alcohol. Each mixture will differ from the others only in the amount of dichromate ion it contains. Thus, the difference in the rates at which two mixtures acquire a green color will depend solely on the difference in their dichromate concentrations, permitting the value of a to be determined (i.e., if the rate of color change remains constant despite the variation in dichromate concentration, the reaction is zero order in dichromate and $a = 0$; if the rate doubles when the dichromate concentration is doubled, the reaction is first order in dichromate and $a = 1$,etc).

The second method for determining the order of the reaction with respect to dichromate is the integrated rate method. This method involves monitoring the concentration of a reactant, as a function of time, as the reaction progresses. Plotting various functions of the concentration (e.g., the concentration itself, the natural log of the concentration, and the inverse of the concentration) reveals the order of the reaction with respect to the reactant under consideration if a straight line plot is obtained. To see why this is so, simply rearrange Equation 34.3 and integrate. (The results of those operations are shown in Equations 34.4 through 34.7.)

$$-\frac{d[Cr_2O_7^{2-}]}{[Cr_2O_7^{2-}]^a} = k'\,dt \qquad (34.4)$$

$$a = 0 \qquad [Cr_2O_7^{2-}] = -k't + [Cr_2O_7^{2-}]_0 \qquad (34.5)$$

$$a = 1 \qquad \ln [Cr_2O_7^{2-}] = -k't + \ln [Cr_2O_7^{2-}]_0 \qquad (34.6)$$

$$a = 2 \qquad \frac{1}{[Cr_2O_7^{2-}]} = k't + \frac{1}{[Cr_2O_7^{2-}]_0} \qquad (34.7)$$

The decrease in concentration of the dichromate ion can be monitored by following the decrease in the absorbance of the reaction mixture at a wavelength at which the dichromate ion absorbs well. Because of Beer's Law, $A = \varepsilon c \ell$, absorbance can be substituted for $[Cr_2O_7^{2-}]$ in Equations 34.5 through 34.7. The value of a is then determined by plotting A, $\ln A$, and $1/A$ versus time and seeing which plot results in a straight line.

In a variation of the integrated rate method, the reaction can be monitored by studying the increase in the absorbance of the chromium(III) ion as the reaction proceeds. The data are collected by monitoring the absorbance at a wavelength at which Cr^{3+} absorbs well. A small complication to using this data is that integrating the differential rate law requires that the rate be expressed in terms of the change in *reactant* concentration. This problem is, however, a minor one. The dichromate concentration will necessarily be zero when the reaction has gone to completion. Therefore, there is some maximum value of the chromium(III) absorbance, A_{max}. As long as some dichromate remains unreacted, A, the absorbance of the chromium(III), will be smaller than A_{max}. Thus, the difference between A_{max} and A is a measure of the amount of dichromate remaining at any time and $(A_{max} - A)$ may be substituted for $[Cr_2O_7^{2-}]$ in Equations 34.5 through 34.7 and in the plots used to determine the order of the reaction with respect to the dichromate ion.

PROCEDURE

Method of Initial Rates

Prepare a series of mixtures as specified in Table 34.1. The acid, dichromate, and water should be mixed first, with the timer being started as soon as the alcohol has been added to the first three components. Stop the timer when the color of the reaction mixture changes from orange to green.

Integrated Rate Method

Place 4.00 mL of water in a clean, dry test tube. Add 2.00 mL of 0.01667 M $Cr_2O_7^{2-}$ and 4.00 mL of 3.6 M H_2SO_4. Mix well and determine the specturm of the diluted solution (review Experiment 19, if necessary). Determine the wavelength to be used in monitoring the rate of disappearance of the dichromate ion.

TABLE 34.1 Mixtures for the Method of Initial Rates.

Mixture	0.01667 M dichromate	3.6 M sulfuric acid	ethanol	water
1	10.00 mL	20.00 mL	20.00 mL	0.00
2	8.00 mL	20.00 mL	20.00 mL	2.00 mL
3	5.00 mL	20.00 mL	20.00 mL	5.00 mL
4	4.00 mL	20.00 mL	20.00 mL	6.00 mL

Place 4.00 mL of water in another clean, dry test tube. Add 2.00 mL of 0.01667 M $Cr_2O_7^{2-}$ and 4.00 mL of 3.6 M H_2SO_4. Mix well and add 4.00 mL of ethanol. Mix once again and heat the mixture in a hot water bath for a few minutes (until the solution turns green). Determine the spectrum of the solution. Determine the wavelength to be used in monitoring the rate of appearance of the chromium(III) ion.

Place 25.00 mL of water in a clean, dry flask. Add 5.00 mL of 0.01667 M $Cr_2O_7^{2-}$ and 10.00 mL of 3.6 M H_2SO_4. Mix well. Add 10.0 mL of alcohol and mix again briefly. Monitor the absorbance of the reaction mixture every 30 seconds for 15 minutes, using the wavelength that is absorbed by the dichromate ion.

Place 25.00 mL of water in a clean, dry flask. Add 5.00 mL of 0.01667 M $Cr_2O_7^{2-}$ and 10.00 mL of 3.6 M H_2SO_4. Mix well. Add 10.00 mL of alcohol and mix again briefly. Monitor the absorbance of the reaction mixture every 30 seconds for 15 minutes, using the wavelength that is absorbed by the chromium(III) ion.

If the absorbance has not become constant after 15 minutes of monitoring has elapsed, heat the mixture for several minutes. Measure the absorbance of the solution that has been heated and use this absorbance as A_{max}.

Discussion

Show the calculations and graphs used to determine the value of *a* by the Method of Initial rates and by the two Integrated Rate Law studies. Discuss whether your values agree and any possible reason for the discrepancies if they don't. What are the advantages and disadvantages of evaluating the order of the reaction by each of the methods used?

Disposal of Reagents

Excess sulfuric acid should be neutralized with 3 M NaOH, diluted, and flushed down the drain.

All solutions containing Cr^{3+} or $Cr_2O_7^{2-}$ must be collected in a large beaker and chemically treated to remove the chromium before the liquid is discarded. Acidify the solution with 3.6 M sulfuric acid (about 5 mL of acid per 100 mL of solution). Then add several pieces of mossy zinc to reduce the $Cr_2O_7^{2-}$ to Cr^{3+}. Stir until the mixture has turned green. Next, decant the green solution into another beaker and carefully add 3 M NaOH until the solution is basic. Allow the $Cr(OH)_3$ to settle and decant the colorless supernatant liquid into another beaker. Dilute this solution and flush it down the drain. Pour the $Cr(OH)_3$ precipitate into a labeled collection bottle. The mossy zinc should be washed, dried with a cloth or paper towel, and saved for reuse.

Date ______ Name ______________________________ Section ______ Desk Number ______

PRE-LAB EXERCISES FOR EXPERIMENT 34

These exercises are to be completed after you have read the experiment but before you come to the laboratory to perform it.
The iodination of acetone was studied by the method of initial rates. Aqueous solutions of acetone, hydrochloric acid, and iodine were mixed in the presence of starch. The diappearance of the characteristic blue color of the complex formed by iodine and starch signaled that the iodine concentration had been reduced to zero. Table 34.2 shows the composition of the various mixtures studied and the amount of time needed for the iodine in the mixtue to be consumed.

Each mixture contained 5 mL of 0.2 % starch and sufficient water to bring the total volume of the mixture to 60.00 mL.

1. How is the rate of the reaction defined in this experiment?

2. What is the order of the reaction with respect to iodine?

3. What is the order of the reaction with respect to acetone? to acid?

4. If the reaction could be carried out so that the concentration of iodine could be monitored as a function of time, which function (concentration, ln concentration, or 1/concentration) would give a straight line when plotted against time?

TABLE 34.2 Initial Rate Study of the Iodination of Acetone

Mixture	10^{-3} M I_2	2.0 M acetone	0.5 M HCl	Time (in seconds)
1	10.00 mL	20.00 mL	20.00 mL	100 sec
2	8.00 mL	20.00 mL	20.00 mL	79 sec
3	6.00 mL	20.00 mL	20.00 mL	61 sec
4	4.00 mL	20.00 mL	20.00 mL	40 sec
5	10.00 mL	10.00 mL	20.00 mL	197 sec
6	10.00 mL	20.00 mL	10.00 mL	203 sec

Date ______ Name ______________________________ Section ______ Desk Number ______

SUMMARY REPORT ON EXPERIMENT 34

Method of Initial Rates

Mixture	*Volume of $Cr_2O_7^{2-}$*	*Time*	*Rate*
______	______	______	______
______	______	______	______
______	______	______	______
______	______	______	______

Integrated Rate Methods

Spectrum of Dichromate Solution

Wavelength	390	400	410	420	430	440	450
% transmittance	____	____	____	____	____	____	____
Absorbance	____	____	____	____	____	____	____

Wavelength	460	470	480	490	500	510	520
% transmittance	____	____	____	____	____	____	____
Absorbance	____	____	____	____	____	____	____

Wavelength	530	540	550	560	570	580	590	600
% transmittance	____	____	____	____	____	____	____	____
Absorbance	____	____	____	____	____	____	____	____

Spectrum of Cr(III) Ion

Wavelength	390	400	410	420	430	440	450
% transmittance	____	____	____	____	____	____	____
Absorbance	____	____	____	____	____	____	____

Wavelength	460	470	480	490	500	510	520
% transmittance	____	____	____	____	____	____	____
Absorbance	____	____	____	____	____	____	____

Wavelength	530	540	550	560	570	580	590	600
% transmittance	____	____	____	____	____	____	____	____
Absorbance	____	____	____	____	____	____	____	____

Best wavelength for monitoring disappearance of dichromate ______________

Best wavelength for monitoring appearance of chromium(III) ion ______________

Absorbance of Dichromate Ion as a Function of Time

Time	*Absorbance*	*Time*	*Absorbance*
______	______	______	______
______	______	______	______
______	______	______	______
______	______	______	______
______	______	______	______

Date ______ Name ______________________________ Section ______ Desk Number ______

Order of the reaction with respect to $Cr_2O_7^{2-}$ ______________

Absorbance of Cr(III) Ion as a Function of Time

A_{max} ______________

Time	*Absorbance*	$A_{max} - A$	*Time*	*Absorbance*	$A_{max} - A$

Date ______ Name ________________________________ Section ______ Desk Number ______

______	______	______	______	______	______
______	______	______	______	______	______
______	______	______	______	______	______
______	______	______	______	______	______

Order of the reaction with respect to $Cr_2O_7^{2-}$ ________________________

EXPERIMENT 35 Atomic Energy Levels and Spectra

Laboratory Time Required Recommended as a one-week, take-home experiment, possibly to be combined with a three-hour, in-class discussion period.

Special Equipment and Supplies None

Safety This experiment does not expose students to chemical hazards.

First Aid **It is not expected that any injuries could result from the performance of this experiment.**

The Bohr model of the atom provides a simple, easy-to-visualize explanation for many phenomena. However, it is strictly correct only for one-electron systems. Atoms that have more than one electron must be described in terms of quantum mechanics. Unfortunately, the more mathematically correct a quantum mechanical calculation becomes, the harder it is to verbalize and visualize. Even people who have learned quantum mechanical terminology may find it easier to think of atoms in terms of a Bohr-type model. This experiment will direct your attention to the successes and shortcomings of the Bohr model in the prediction of the wavelengths of the lines in the emission spectra of the hydrogen atom, the helium ion, and the helium atom.

PRINCIPLES

When a high-energy spark is passed through a sample of gaseous hydrogen, the H_2 molecules absorb energy. As a result, some of the molecules dissociate into H atoms. These free atoms are generally in an excited state when they are formed. As the free atoms return to their ground state, the excess energy that they initially contained is released in the form of light of various wavelengths, producing the characteristic **emission spectrum** of the hydrogen atom.

In the early part of the twentieth century, scientists could not explain why the hydrogen emission spectrum was not a continuous spectrum (i.e., a spectrum containing all wavelengths of light, such as that produced by sunlight passing through a prism). That the emission spectrum of hydrogen is a line spectrum (i.e., one that contains only certain specific wavelengths of light) was a puzzle whose solution eventually led to a radical change in the concept of the structure of the atom.

In 1913, Niels Bohr proposed a model for the hydrogen atom in which the electron was said to orbit the proton in certain "allowed" circular orbits. In the ground state of the atom, the electron moved in the lowest energy (n = 1) orbit. In an excited state, the electron moved in a higher energy (n = 2,3,4 . . .) orbit. Electrons falling from a high-energy level to a lower energy level would emit the radiation corresponding to the difference in energy between those levels. Thus, the model predicted that the emission spectrum of an atom would be a line spectrum.

Bohr's model specified that orbits were allowed only if the angular momentum of the electron in that orbit was evenly divisible by ($\mathrm{h}/2\pi$), where h is Planck's constant, a number that relates the energy (E) and wavelength (λ) of a photon to the speed (c) of light in a vacuum. The angular momentum is the product of the mass (m) and velocity (v) of the electron and the radius (r) of the orbit in which the electron is moving. Planck's constant is defined in Equation 35.1 and the equation for Bohr's orbits is in equation 35.2.

$$h = \frac{E\lambda}{c} \tag{35.1}$$

$$mvr = n\frac{\mathrm{h}}{2\pi} \tag{35.2}$$

Perhaps the most important equation to come from Bohr's model was his expression for determining the energies associated with the allowed orbits of the electron. This expression is given, without derivation, in Equation 35.3, where n designates the number of the orbit whose energy is being calculated; Z is the atomic number of the atom; m is the mass of the electron; e is the charge on the electron; and h is Planck's constant

$$E_n = -\frac{2\pi me^4}{\mathrm{h}^2}\,\frac{Z^2}{n^2} \tag{35.3}$$

The appearance of a negative sign in Equation 35.3 guarantees that all of the orbits are associated with negative energies. This is so because the zero point of energy was chosen to be the situation in which the electron is no longer associated with the nucleus. Thus, the ionization energy of hydrogen is E_1, meaning that E_1 joules of energy must be put into the hydrogen atom to remove the electron from the $n = 1$ orbit and bring it to a distance infinitely separated from the nucleus.

If appropriate values of the various constants are substituted for the symbols in Equation 35.3, we obtain Equation 35.4. It is this equation that you shall use to determine the energies associated with the allowed orbits. Note that the presence of Z in Equations 35.3 and 35.4 allows us to move from our discussion of hydrogen ($Z = 1$) to a discussion of all other elements ($Z > 1$).

$$E_n = -2.178 \times 10^{-18}\text{J}\ \frac{Z^2}{n^2} \qquad (35.4)$$

Suppose an electron changes from the $n = 2$ state of a hydrogen atom to the $n = 1$ orbit, and simultaneously emits a photon of light. What would be the wavelength of this light? Using Equation 35.3, we can easily calculate that $E_1 = -2.178 \times 10^{-18}$ J and $E_2 = -0.5445 \times 10^{-18}$ J, for the hydrogen atom. The energy of the photon that is emitted corresponds to the difference between these energies (i.e., $E_{\text{photon}} = -\Delta E_{\text{atom}} = E_2 - E_1$). Because the electron falls to a lower energy level ($E_1 < E_2$), ΔE_{atom} is negative. Consequently, E_{photon} is positive, as it must be.

$$\lambda_{2\rightarrow 1} = \frac{hc}{E_2 - E_1} = \frac{(6.626 \times 10^{-34})(2.998 \times 10^{8}\ \text{m/s})}{1.634 \times 10^{-18}\text{J}} = 12.15 \times 10^{-8}\,\text{m} \qquad (35.5)$$

The photon emitted when an electron falls from the $n = 2$ level to the $n = 1$ level in the hydrogen atom is predicted to have a wavelength of 121.5×10^{-9} m or 121.5 nm, in excellent agreement with experimental observation. This photon's wavelength is slightly too short for the photon to be observed by the human eye; it is in the ultraviolet region of the spectrum. Light that is visible to the human eye has wavelengths between 400 and 750 nm. (Older texts may quote wavelengths in units of Å, or angstroms. One **angstrom** is 10^{-8} cm or 10^{-1} nm.)

The wavelengths of each of the lines in hydrogen's atomic spectrum is predicted (with better than 99.9 percent accuracy) by the Bohr model. Each can be shown to be associated with a specific transition of hydrogen's electron from a high energy (n_{high}) to a low energy (n_{low}) state.

This is not quite the case for the emission spectrum of the helium atom. Again, the spectrum is a line spectrum. However, its appearance is complicated by the fact that helium has two electrons. Therefore, the application of a spark to a sample of helium has the possibility of producing both excited helium atoms and excited helium ions.

It is not surprising that Equation 35.4 can be used to predict the wavelengths of the lines in the spectrum of the helium ion. This ion is like hydrogen in that it consists of a single electron orbiting a nucleus. The only difference between the hydrogen atom and the helium ion in an electromagnetic sense is that the helium nucleus contains two protons.

However, the helium atom differs greatly from hydrogen, simply because it has two electrons. This complicates matters considerably because the Bohr model is based on a consideration of only one type of energy—the attraction between a positively charged nucleus and a negatively charged electron. When even as few as two electrons are present, their mutual repulsion adds to the total energy of the atom and must be considered.

For any system with two or more electrons, the Bohr model is inadequate. The quantum mechanical picture of the atom, which supplanted the Bohr model, retained Bohr's idea that the electron could have only certain allowed energies. However, it abandoned the concept of fixed orbits in favor of probabilities for locating the electron somewhere in space. If helium's electrons are not forced to orbit its nucleus in a fixed path, it is always possible that one electron will come between the nucleus and the other electron, thereby "shielding" the second electron from the full attraction of the nucleus. Thus, Equation 35.4 may be replaced by Equation 35.6, in which $(Z - \sigma)$ is the effective nuclear charge on the atom and σ is a shielding constant. The values of σ are determined empirically and depend on the ability of one electron to shield the other electrons in the atom.

$$E_n = -2.178 \times 10^{-18}\,\text{J}\ \frac{(Z - \sigma)^2}{n^2} \qquad (35.6)$$

Of course, the quantum mechanical picture of the atom differs from the Bohr model in more than the introduction of the concept of shielding. Quantum mechanical orbitals (probable paths for electrons) are described by other quantum numbers in addition to n. This divides Bohr-like energy levels into sublevels and creates the possibility for more transitions and more lines in the spectrum.

PROCEDURE

Use Equation 35.4 to calculate the energy of each of the five lowest energy levels of the hydrogen atom. Record these energies in Table 35.1.

Use the energies you have calculated for the orbits in Table 35.1 to calculate the energies of the photons emitted as electrons in excited hydrogen atoms fall back to lower energy levels. Also, calculate the wavelengths associated with these photons. Place the results of these calculations in the appropriate spaces in Table 35.2A. (Because the results for the $n = 2 \rightarrow n = 1$ transition were calculated above to illustrate the use of Equations 35.4 and 35.5, they have been entered in the table.)

Complete Table 35.2A, assigning values of n_{high} and n_{low} to the wavelengths actually observed in the spectrum of the hydrogen atom. Once again, the $2 \rightarrow 1$ transition has been entered to illustrate the type of response expected.

Repeat Steps 1 through 3 for the He^+ ion. Place your results in Table 35.2B.

Use the data in Table 35.3 for the $n = 2 \rightarrow n = 1$ transition in the helium atom to determine the value of $(Z - \sigma)$ for the helium atom. Then use your value of $(Z - \sigma)$ to assign the other wavelengths given in Table 35.3 to electronic transitions between energy levels in the helium atom.

Questions

1. What is the physical significance of σ? Would the Bohr model have predicted that electrons in the same orbit could shield each other? Why or why not?
2. Comment briefly on the relationship between predicted and observed wavelengths in the spectra of the hydrogen atom, helium ion, and helium atom.

Date ______ Name ______________________________ Section ______ Desk Number ______

PRE-LAB EXERCISES FOR EXPERIMENT 35

These exercises are to be completed after you have read the experiment but before you come to the laboratory to perform it.

1. Use Equation 35.4 twice to derive a single equation for calculating the energy of transition ($E_{n_{high}} - E_{n_{low}}$) between levels n_{high} and n_{low}.

2. Use the equation derived above to find the value of Z for an ion whose $2 \rightarrow 1$ transition is associated with a wavelength of 13.4 nm.

Date ______ Name ______________________ Section ______ Desk Number ______

SUMMARY REPORT ON EXPERIMENT 35

TABLE 35.1 Prediction of the Energies of the First Five Orbits of the Hydrogen Atom and of the Helium Ion

Hydrogen Atom		Helium Ion	
Orbit	*Predicted Energy*	*Orbit*	*Predicted Energy*
n_5	______	n_5	______
n_4	______	n_4	______
n_3	______	n_3	______
n_2	______	n_2	______
n_1	______	n_1	______

TABLE 35.2A Prediction of Energies and Wavelength for Electronic Transitions in the Hydrogen Atom

n_{low} \ n_{high}		5	4	3	2
1	*E*				1.634×10^{-18}J
	λ				121.5 nm
2	*E*				
	λ				
3	*E*				
	λ				
4	*E*				
	λ				

Observed λ in nm	Probable Transition	Observed λ in nm	Probable Transition
97.2	______	656.2	______
102.6	______	1281.8	______
121.6	______	1875.1	______
433.4	______	4050.0	______
486.1	______		

Date ______ Name ______________________________ Section ______ Desk Number ______

TABLE 35.2B Prediction of Energies and Wavelength for Electronic Transitions in the Helium Ion

n_{low} \ n_{high}		5	4	3	2
1	E				
	λ				
2	E				
	λ				
3	E				
	λ				
4	E				
	λ				

Observed λ in nm	Probable Transition
24.3	______
25.6	______
30.4	______

TABLE 35.3 Assignment of Helium Wavelengths Using $(Z - \sigma)$

Observed Wavelength, nm	Assignment of $n_{high} \rightarrow n_{low}$	Predicted Wavelengths, nm
58.4	$2 \longrightarrow 1$	______
53.7	______	______
52.2	______	______
	$(Z - \sigma) =$ ______	

36 EXPERIMENT Constructing an Alien Periodic Table

Laboratory Time Required Recommended as a one-week, take-home experiment, possibly to be combined with a three-hour, in-class discussion period.

Special Equipment and Supplies Scissors

Safety This experiment does not expose students to chemical hazards.

First Aid **It is not expected that any injuries could result from the performance of this experiment.**

The structure of the Periodic Table is so readily explained in terms of quantum mechanics that it is sometimes difficult for students to appreciate the effort and genius of the scientists who put together its earliest versions. In this experiment, you are challenged to construct a Periodic Table for an alien environment, without the aid of quantum mechanics.

PRINCIPLES

Imagine that you are the science officer on an intergalactic spacecraft, whose mission is to improve the level of scientific research on any planets you may encounter during your voyage. You have landed on Olam, a planet very similar to Earth. Some 50-odd elements are known on Olam. The study of these individual elements consumes most of the time of the Olamite chemists. Having an effective method for classifying these elements in families would give cohesion to Olamite chemical knowledge. Olamite chemists hope that mastering the properties of 10 or 20 families would be simpler than having to worry about the properties and reactions of 54 elements. They have presented you with their accumulated knowledge as shown in Tables 36.1 through 36.4. Can you help them develop an Olamite Periodic Table?

TABLE 36.1 The Olamite Elements

Name	Symbol	Atomic Mass	State*	Type†
Acidium	Ac	27	G (triatomic)	NM
Akumena	Ak	38	S	M
Alstevium	Av	117	S	
Annberin	An	35	S	M
Aquagen	Aq	2	G (triatomic)	NM
Ashkenazin	Az	111	L	SM
Ashon	As	62	S	M
Badgerin	Bd	90	S	M
Boring	Br	120	L	NM
Brooklin	Bk	93	S	M
Chameshan	Ch	80	S	M
Chrisrussium	Cr	84	S	M
Coloran	Cn	75	S	M
Corana	Co	14	S	NM
Devlan	Dv	78	S	M
Doron	Dr	21	S	SM
Fizzon	Fz	31	S	SM
Flintan	Fn	49	S	SM
Flowing	Fl	28	G (diatomic)‡	NM
Gazozite	Gz	25	G (diatomic)	NM
Gemstonan	Gm	9	S	M
Gigantan	Gg	71	S	M
Greening	Gr	57	G (diatomic)‡	NM
Hadashite	Ha	30	G	NM
Halfwanon	Hl	18	S	SM
Jesslynium	Js	16	S	NM
Katjenium	Kt	40	S	M
Kublinium	Kb	103	S	M
Laitan	Lt	33	S	M
Liman	Lm	65	S	M
Lindnicium	Li	100	S	M
Margaran	Mr	76	S	M
Markelin	Mk	12	S	M
Midlanium	Md	106	S	SM

TABLE 36.1 The Olamite Elements (continued)

Name	Symbol	Atomic Mass	State*	Type†
Mikemartiun	Mi	87	S	M
Mishpakton	Ms	109	S	SM
Newairon	Nw	4	G (diatomic)‡	NM
Norskan	Nn	70	S	M
Oldlaceite	Ol	114	S	SM
Pressan	Ps	79	S	M
Printon	Pr	7	S	M
Puzzlite	Pu	122	G	NM
Ruflus	Rf	55	S	NM
Shakorin	Sk	23	S	NM
Shemeshite	Sh	5	G	NM
Simicin	Sm	68	S	M
Techin	Tn	82	S	M
Tennessean	Ts	95	S	M
Ticonium	Tc	42	S	SM
Toldotan	Tl	46	S	NM
Venusite	Vn	52	S	NM
Voyagite	Vo	60	G	NM
Yanatan	Yt	73	S	M
Zinzan	Zz	97	S	M

*G = gas, L = liquid, S = solid
†M = metal, NM = non-metal, SM = semi-metal
‡ = also known in crystalline solid state

TABLE 36.2 Special Characteristics of Some Olamite Elements

1. Hadashite, Puzzlite, Shemeshite, and Voyagite are very unreactive. No compounds of these elements are known to exist on Olam.
2. Aquagen is a triatomic element, as is Acidium.
3. Newairon, Flowing, and Greening all exist as either diatomic gases or crystalline solids. They readily enter into ionic compounds of the form Nw_2X, where X = Fl, Gr, or Br.
4. Badgerin, Brooklin, Tennessean, and Zinzan are rather unreactive metals that are used in Olamite coinage.
5. Halfwanon has a tendency to catenate. It forms a wide variety of compounds such as:
   a. $HlAq_2Nw_2$, $Hl_2Aq_3Nw_3$, $Hl_3Aq_4Nw_4$
   b. $Hl_2Aq_2Nw_2$, $Hl_3Aq_3Nw_3$, $Hl_4Aq_4Nw_4$
   c. Hl_2AqNw, $Hl_3Aq_2Nw_2$

The compounds in set 5a undergo only substitution reactions. The compounds in sets 5b and 5c undergo addition as well as substitution reactions.

TABLE 36.3A Formulas and Properties of Some Olamite Compounds

I. Aquides and Newairides of active metals

Ashon, Fizzon, Printon form M_2Aq, MNw
Gemstonan, Laitan, Liman form MAq, MNw_2
Annberin, Markelin, Simicin form M_2Aq_3, MNw_3
The compounds have crystalline appearance and high melting points.

II. Aquagenide acids

Acidium, Alstevium, and Ruflus form Aq_2X

X	*B.P.**
Ac	200°T
Av	50°T
Rf	0°T

Boring, Flowing, and Greening form Aq*X*

X	*B.P.*
Br	25°T
Fl	115°T
Br	5°T

Gazozite, Oldlaceite, and Venusite form Aq_3X

X	*B.P.*
Gz	50°T
Ol	–70°T
Vn	–120°T

*Temperature is measured in °T (= degrees terran). Room temperature is approximately 100°T. Aqua (Aq_2Ac) is strikingly similar to Earth's water. It freezes at 75°T and boils at 200°T.

TABLE 36.3A Formulas and Properties of Some Olamite Compounds

I. Metals that exhibit variable combining powers*

Metals	*Compounds*
Chameshan	$ChGr_3$, $ChGr_5$, $ChAc_3$
Chrisrussium	$CrGr_3$, $CrGr_4$, $AqCrAc_3$
Coloran	$CnGr_3$, $CnGr_4$, Cn_2Ac_7
Devlan	$DvGr_3$, $DvGr_4$, $AqDvAc_3$
Gigantan	$GgGr_3$, $GgGr_4$, $GgGr_5$
Margaran	$MrGr_3$, $MrGr_4$, $MrAc_3$, $NwMrAc_5$
Mikemartium	$MiGr_3$, $MiGr_5$, $MiAc_4$
Norskan	$NnGr_3$, $NnGr_4$
Pressan	$PsGr_3$, $PsAc_2$
Yanatan	$YtFl_3$, $YtAc_3$

TABLE 36.3A Formulas and Properties of Some Olamite Compounds (*continued*)

II. Formulas and properties of Olamite Greenides and Acidides of the semimetals and certain nonmetals

Compounds	*Properties*
CoGr	Crystalline solid
$CoGr_4$	Waxy solid
JsAc	Solutions are basic
$JsAcGr_3$	Low solubility in aqua; solutions do not conduct electricity
$AqKtAc_3$	Weakly acidic
$TcGr_3$	Crystalline solid
$TcGr_6$	Sublimes at 10°T
As_5Tl	Solutions conduct electricity
$MdAc_3$	Dissolves in both acids and bases
$MsAcGr_2$	Solutions do not conduct electricity
$DrAc_2$	Acidic, electron-deficient compound
SkAc	Electron deficient
$SkGr_2$	Liquid at 100°T
$FnGr_2$	Solid at 100°T, paramagnetic
$FnGr_4$	Not soluble in aqua, diamagnetic, expands valence shell

*Many of the compounds listed are highly colored.

TABLE 36.4 Olamite Electromotive Series

	Element	*Characteristic Reaction*
↑ Metal activity increases	As Fz Pr	Replace Aq from cold aqua* Nw from cold mayin*
	Lm Lt Gm	Replace Aq from hot aqua Nw from hot mayin
	Sm Ak An Mk	Replace Aq, Nw from acids
	Zz Mr	Replace Aq, Nw from acids that are very concentrated
	Li Kb	React only with hot, concentrated acids
	Bd Bk Ts	Acidites of these metals decompose readily to give pure metal

*Mayin, like aqua, is strikingly similar to Earth's water. The formula for mayin is Nw_2Sk.

PROCEDURE

Construct an Olamite Periodic Table based on the hypothetical facts given in Tables 36.1 through 36.4. Begin by cutting apart the tiles shown in Figure 36.1, which may then be ordered into rows and columns in accord with the known properties of the elements (or you may prepare a set of index cards to use in place of the tiles). Recall that the rules governing atomic structure may not be those followed on Earth. Use the report sheet to record your reasoning. Include a detailed Periodic Table in your final report.

Questions

1. Based on your Olamite Periodic Table, predict the nature of the quantum mechanical structure of Olamite atoms.
2. Predict the properties of the 55th Olamite element, tentatively named Erjackium (Er).
3. Is Alstevium most likely to be a metal, nonmetal, or semimetal?
4. Draw the Olamite equivalent of Lewis dot diagrams for Aq_3, Nw_2, Gz_2, Ac_3, and Fl_2. Identify the ions present in the crystalline solid forms of Nw_2 and Fl_2.
5. In what ways was your task in constructing the Olamite Periodic Table harder than Mendeleyev's? In what ways was it easier?
6. An Olamite scientist has claimed the discovery of a new element, Lesrenium (Le). The scientist reports that it has properties between those of Liman and Simicin. Does your proposed Olamite Periodic Table support or refute this claim?
7. An Olamite scientist has claimed the discovery of a new element, Jennerium (Je). It forms JeAq and $JeNw_2$. Where would Jennerium fit in the Olamite Periodic Table? Predict its atomic mass and placement in the Olamite Electromotive Series.

Acidium **Ac** 27	Akumena **Ak** 38	Alstevium **Av** 117	Annberin **An** 35	Aquagen **Aq** 2	Ashkenazin **Az** 111	Ashon **As** 62	Badgerin **Bd** 90	Boring **Br** 120
Brooklin **Bk** 93	Chameshan **Ch** 80	Chrisrussium **Cr** 84	Coloran **Cn** 75	Corana **Co** 14	Devlan **Dv** 78	Doron **Dr** 21	Fizzon **Fz** 31	Flintan **Fn** 49
Flowing **Fl** 28	Gazozite **Gz** 25	Gemstonan **Gm** 9	Gigantan **Gg** 71	Greening **Gr** 57	Hadashite **Ha** 30	Halfwanon **Hl** 18	Jesslynium **Js** 16	Katjenium **Kt** 40
Kublinium **Kb** 103	Laitan **Lt** 33	Liman **Lm** 65	Lindnicium **Li** 100	Margaran **Mr** 76	Markelin **Mk** 12	Midlanium **Md** 106	Mikemartium **Mi** 87	Mishpakton **Ms** 109
Newairon **Nw** 4	Norskan **Nn** 70	Oldlaceite **Ol** 114	Pressan **Ps** 79	Printon **Pr** 7	Puzzlite **Pu** 122	Ruflus **Rf** 55	Shakorin **Sk** 23	Shemeshite **Sh** 5
Simicin **Sm** 68	Techin **Tn** 82	Tennessean **Ts** 95	Ticonium **Tc** 42	Toldotan **Tl** 46	Venusite **Vn** 52	Voyagite **Vo** 60	Yanatan **Yt** 73	Zinazn **Zz** 97

FIGURE 36.1 Names, symbols, and atomic masses of Olamite elements.

Date ______ Name ______________________________ Section ______ Desk Number ______

PRE-LAB EXERCISES FOR EXPERIMENT 36

These exercises are to be completed after you have read the experiment but before you come to the laboratory to perform it.

1. Consult your textbook, other general chemistry textbooks, or a history of chemistry to learn about the historical development of the Periodic Table. Outline briefly:

   a. the forms of the Periodic Law that preceded Mendeleyev's work;

   b. the significant contributions that Mendeleyev made to the construction of the moder Periodic Table.

Date ______ Name ______________________________ Section ______ Desk Number ______

SUMMARY REPORT ON EXPERIMENT 36

Use the grid shown below to construct your Olamite Periodic Table. Note that you may need only some of the spaces provided. Place the symbol and the atomic mass of each element in the appropriate square. It may be desirable to show some elements set off from the main body of the table. Use the reverse side of this sheet to outline the reasoning you used in constructing the table.

Olamite Periodic Table

EXPERIMENT 37 Isomerism in Organic Chemistry

Laboratory Time Required

Three hours. May be performed either as a take-home experiment or in a laboratory where models or computers are available.

Special Equipment and Supplies

Ball-and-stick models

or

Computer and color monitor with appropriate software

Safety

This experiment does not expose students to chemical hazards.

First Aid

It is not expected that any injuries could result from the performance the experiment.

Among the earliest exercises students perform in their introductory chemistry course are those related to determining the empirical and molecular formulas of compounds. Such formulas suffice in working with stoichiometric relationships. However, in many cases, you must go beyond the molecular formula to the structural formula of the compound to understand such fundamental properties of compounds as melting point/boiling-point trends, solubility, and reactivity. Structural formulas are especially important in organic chemistry. This exercise deals with predicting the numbers of isomers that an organic molecule may have and examining the properties of those isomers.

PRINCIPLES

Isomers are molecules that have the same molecular formula yet differ from each other in physical and chemical properties. The differences between isomers arise from the manner in which the molecules are constructed. Some types of isomerism are discussed below.

Molecules that contain the same number and type of atoms, yet differ in the way in which the atoms are grouped together within the molecule, are said to exhibit **structural isomerism**. Ethanol, CH_3CH_2OH, and dimethyl ether, CH_3OCH_3, are structural isomers. Both compounds have the molecular formula, C_2H_6O, but ethanol contains a methyl group (CH_3—), a methylene group (—CH_2—) and an alcohol group (—OH). Dimethyl ether contains two methyl groups, each of whose carbons is bound directly to an oxygen. The presence of the —OH group in ethanol allows the formation of hydrogen bonds both in pure ethanol and in aqueous solutions of ethanol. This leads to a comparatively high boiling point and a high degree of solubility in water for ethanol. Thus, ethanol is a liquid at room temperature and is miscible with water in all proportions. In contrast, dimethyl ether is a gas at room temperature and has limited solubility in water.

Pentane, $CH_3CH_2CH_2CH_2CH_3$, isopentane, $CH_3CH_2\underset{}{\overset{CH_3}{\overset{|}{C}}}H\text{—}CH_3$, and neopentane, $CH_3\text{—}\underset{CH_3}{\underset{|}{\overset{CH_3}{\overset{|}{C}}}}\text{—}CH_3$ are, likewise, structural isomers of the compound C_5H_{12}. Pentane, which is sometimes designated as n-pentane (normal pentane), contains two methyl and three methylene groups. Isopentane has three methyl groups and one methylene group, whereas neopentane contains four methyl groups.

It is sometimes of interest to consider how many isomers of a compound can be obtained by replacing one of the hydrogens with a heteroatom (such as —Cl, —Br, or —I) or a functional group (such as —OH,—NH_2, or $\text{—}\overset{O}{\overset{\|}{C}}OH$). In making this decision, it is helpful to begin by considering how many sets of chemically equivalent hydrogens are contained in the original molecule. For instance, all of the hydrogens in neopentane are chemically equivalent, and only one compound of formula $C_5H_{11}Cl$ can be prepared that is a substituted neopentane. On the other hand, isopentane contains four sets of chemically equivalent hydrogens. The first of these sets contains the six hydrogens of the two methyl groups that are found at the branched end of the isopentane molecule. The second contains the single hydrogen that is attached to the carbon that also bears two methyl groups. The third contains the two hydrogens of the methylene group, and the fourth contains the three hydrogens of the remaining methyl group. Therefore, four isomers of $C_5H_{11}Cl$ can be prepared that are substituted isopentanes. Finally, *n*-pentane contains three sets of chemically equivalent hydrogens (the six methyl hydrogens, the four hydrogens of the methylene groups that are linked directly to the methyl groups, and the two of the methylene group which is in the center of the molecule). Thus, three isomers of $C_5H_{11}Cl$

that are substituted n-pentanes can be prepared. Therefore, there is a total of eight isomers of $C_5H_{11}Cl$, all of which are structural isomers.

Each of the molecules considered above contains only single bonds. Atoms that are linked by single bonds rotate freely, producing various conformations of the molecules in which they are found. In the two extreme conformations of ethane, CH_3CH_3, the hydrogens of the two methyl groups may be staggered or they may be eclipsed. These conformations are shown as Newman projections as in Figure 37.1. When viewing the Newman projection, you are sighting along the C—C bond. The solid lines represent the hydrogens on the front carbon, while the broken lines represent the hydrogens on the rear carbon.

Although the staggered conformation of ethane may be somewhat more stable than the eclipsed form, the low energy barrier to rotation about the C—C single bond causes the two forms to be interconvertible. Staggered and eclipsed ethane are not isomers that can be isolated as separate forms of the compound.

In contrast, rotation about a C═C double bond is restricted. Geometric isomers are characterized by molecules that contain the same functional groups and differ in the manner in which these groups are distributed about a double bond. For example, cis-2-butene and trans-2-butene both have the formula C_4H_8. Each contains two carbon atoms united by a double bond. Each of these carbons, in both isomers, is attached to a methyl group and a hydrogen. However, in cis-2-butene the methyl groups both lie on the same side of the double bond, whereas in trans-2-butene, the methyl groups lie on opposite sides of the double bond. (Another form of C_4H_8 is 1-butene, in which a carbon bearing two hydrogens is double bonded to a carbon bearing one hydrogen and an ethyl group. There is only one form of 1-butene, and it is a structural isomer of the 2-butenes.)

Each 2-butene contains two sets of chemically equivalent hydrogens (i.e., the six hydrogens of the methyl groups and the two hydrogens that are not in the methyl groups). The l-butene contains four sets of equivalent hydrogens. Thus, there are eight distinct isomers of a monosubstituted butene, such as C_4H_7Cl. (Another compound of formula C_4H_8 is 2-methylpropene, which is a structural isomer of the butenes.)

Another form of C_4H_8 contains four methylene groups, bound together in a ring. This compound is called cyclobutane. Rotation is restricted in ring systems just as it is in systems containing double bonds. Thus, disubstituted cyclobutanes, such as $C_4H_6Cl_2$, exhibit geometric isomerism if the two heteroatoms are not attached to the same carbon. There are, in fact, five geometrical and structural isomers of $C_4H_6Cl_2$ that contain a four-carbon ring system. These include

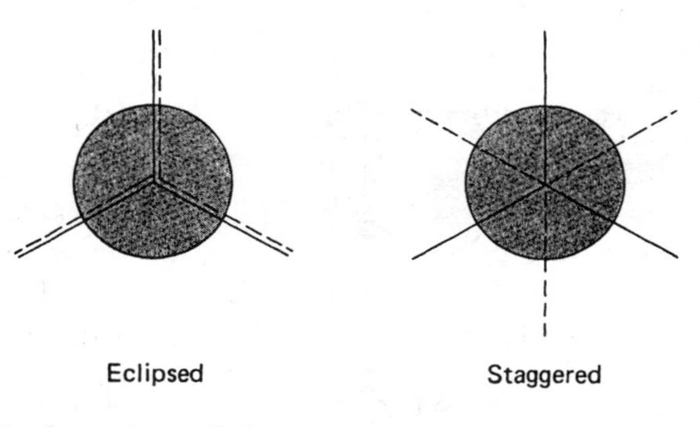

FIGURE 37.1 Conformations of ethane.

two pairs of geometric isomers. (There are also eight isomers of $C_4H_6Cl_2$ that contain a three-carbon ring.)

There are also compounds in which heteroatoms are inserted into the ring system of the molecule. Compounds of this type include tetrahydrofuran (C_4H_8O), and imidazolidine ($C_3H_8N_2$). The structures of these molecules are shown in Figure 37.2.

Rotation is, of course, restricted in all ring systems. Therefore, disubstituted tetrahydrofurans (or imidazolidines or any other nonplanar rings) may exhibit geometric isomerism.

Heteroatoms may also form multiple bonds to carbon. In the carbonyl group, a carbon atom is linked to an oxygen atom by a double bond. Because oxygen forms only two bonds, there is no possibility of geometric isomerism about a C═O bond. However, structural isomers can, and do, exist. For instance, C_3H_6O is the molecular formula for both acetone ($CH_3\overset{\overset{O}{\|}}{C}CH_3$) and propionaldehyde ($CH_3CH_2\overset{\overset{O}{\|}}{C}H$). Ketones, such as acetone, differ from aldehydes, such as propionaldehyde, in several ways. One is that aldehydes can be oxidized to give carboxylic acids, while ketones generally resist oxidation. The acid that can be made from propionaldehyde is $CH_3CH_2\overset{\overset{O}{\|}}{C}OH$ propionic acid.

Even when there is free rotation about all of a molecule's bonds, the relative orientation of the various functional groups may be important. Molecules that are nonsuperimposable mirror images of each other are called **optical isomers**. These isomers will have identical physical properties, except that the two isomers will rotate plane polarized light in opposite directions. The isomers will also have identical chemical properties, except in **stereoselective reactions**, which are reactions whose outcomes depend on the relative orientation in space of the reactants' various functional groups.

The simplest species that will exhibit optical activity (or chirality) is an **asymmetric carbon**—a carbon that is linked to four different groups by four single bonds. Chlorobromoiodomethanol, shown in Figure 37.3, contains such a carbon.

Suppose you were to grasp this molecule (ClBrICOH) by its C—Cl bond. A head-on look at the two mirror images would then yield the Newman projections shown in Figure 37.4.

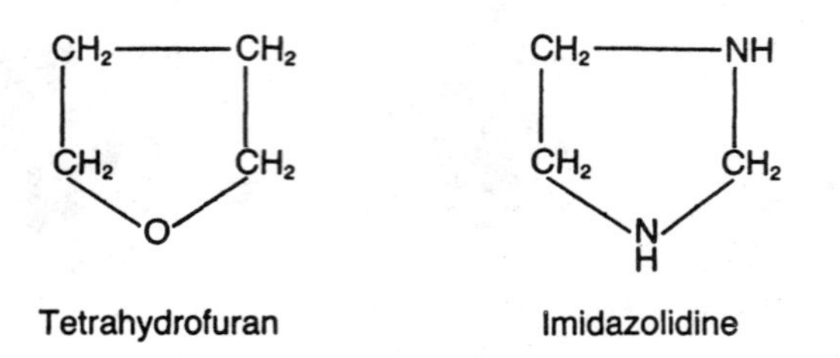

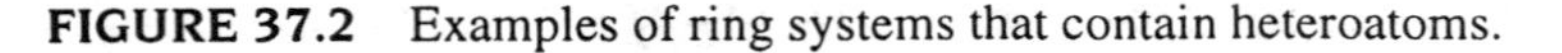
FIGURE 37.2 Examples of ring systems that contain heteroatoms.

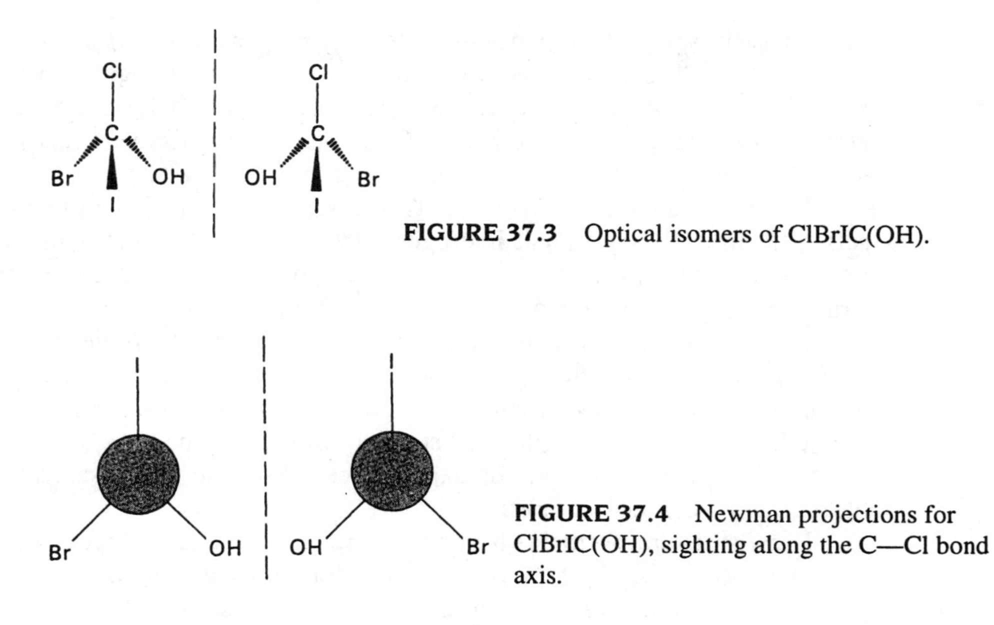

FIGURE 37.3 Optical isomers of ClBrIC(OH).

FIGURE 37.4 Newman projections for ClBrIC(OH), sighting along the C—Cl bond axis.

No amount of rotation about the C—Cl axis can change the Br—OH—I order of substituents in one isomer to the OH—Br—I order in the other. If this molecule were to react with an enzyme which, for instance, needed the —Br to be on the left of the —OH when the —Cl was below the rest of the molecule, one isomer could react while the other could not. Many reactions that occur in the body are, in fact, stereoselective, and chirality plays an important role in determining the biological activity of many chemicals, such as sugars and amino acids.

Like geometric (*cis/trans*) isomers, optical isomers differ from one another in the spatial arrangement of their substituent groups. Both geometric and optical isomers are sometimes classed together as **stereoisomers**; this indicates that the isomers differ from each other by the arrangement of their atoms in space.

PROCEDURE

The Summary Report Sheet has spaces in which you will draw structures and answer questions concerning them. You may wish to construct ball-and-stick models of the compounds to be considered or you may wish to view the structures on a computer monitor. Your instructor will inform you if a computer program is available that would meet your needs.

Begin by considering the simplest hydrocarbon, methane, CH_4. In the days before X-ray diffraction studies were possible, the structure of methane was deduced from a consideration of the number of isomers its disubstituted forms would have. If CH_4 were square planar, how many sets of chemically equivalent hydrogens would it have? How many sets would be present in CH_3X? How many isomers of CH_2X_2 would exist? Answer the same questions for tetrahedral CH_4. Sketch the structures called for on the Summary Report Sheets. What is the correct geometry for CH_4 (only one form of CH_2X_2 has ever been found)?

Now consider the staggered conformation of C_2H_6 (ethane). How many sets of chemically equivalent hydrogens are there in ethane? Draw a Newman projection for each isomer of a monosubstituted ethane (e.g., C_2H_5Cl) in its staggered conformation. How many sets of chemically equivalent hydrogens does C_2H_5Cl have? Draw a Newman projection for each isomer of a C_2H_4ClBr, a disubstituted ethane, in its staggered conformation. Circle any pairs of isomers that would be optically active.

Draw structural diagrams of the two isomers of C_4H_{10}. Note the number of sets of chemically equivalent hydrogens for each isomer. Draw structural diagrams of all the isomers of C_4H_9OH (i.e., the four-carbon alcohols). Also, draw structural diagrams of all ethers that have the formula $C_4H_{10}O$.

Draw structural formulas for all molecules having the molecular formula C_3H_6. Include both compounds containing ring systems and those containing a double bond. Then draw structural formulas for all molecules with the formula $C_3H_4Cl_2$ (which represents the disubstituted forms of all the C_3H_6's). Circle pairs of geometric isomers. Would you expect three-membered ring systems to be stable? Explain your answer briefly.

Draw structural formulas for an aldehyde, a ketone, and a cyclic compound with the formula $C_5H_{10}O$. Draw structural formulas for derivatives of each of your $C_5H_{10}O$ molecules in which one hydrogen on the fourth carbon and one hydrogen on the fifth carbon have been replaced by —OH groups. (Numbering begins on the carbon of the C=O group in aldehydes, at the carbon nearest the C=O group in ketones, and at a carbon adjacent to oxygen in a cyclic compound in which oxygen has been incorporated into the ring.) If you expect the derivatives to be optically active, mark their chiral centers with asterisks and draw the structural formulas of their mirror images. How many optical isomers would there be for a derivative of a $C_5H_{10}O$ aldehyde or ketone in which each carbon, except that of the carbonyl group, bears a hydroxyl group? Such a compound, with the molecular formula, $C_5(H_2O)_5$, is classified as a carbohydrate. More specifically it is a sugar. It would be called an aldopentose if it were derived from an aldehyde, or a ketopentose if it were derived from a ketone. The *pent-* root indicates five carbons are present while the *-ose* ending (as in glucose, fructose, and sucrose) designates a sugar.

Date ______ Name ______________________________ Section ______ Desk Number ______

PRE-LAB EXERCISES FOR EXPERIMENT 37

These exercises are to be completed after you have read the experiment but before you Come to the laboratory to peform it.

1. Draw Newman projections for propane, $CH_3CH_2CH_3$, in which the methylene hydrogens are

a. staggered
b. eclipsed

with respect to the hydrogens on the methyl groups.

2. Draw structural diagrams for all the isomers of $C_4H_6Cl_2$.

Date ______ Name ______________________________ Section ______ Desk Number ______

SUMMARY REPORT ON EXPERIMENT 37

1. Methane

No. of sets of chemically equivalent H's, if square planar ______

No. of sets of chemically equivalent H's in square planar CH_3X ______

No. of isomers of CH_2X_2 ______

No. of sets of chemically equivalent H's, if tetrahedral ______

No. of sets of chemically equivalent H's in tetrahedral CH_3X ______

No. of isomers of CH_2X_2 ______

Correct geometry for CH_4 ______

2. Staggered ethane

No. of sets of chemically equivalent H's in ethane ______

Structure(s) of C_2H_5Cl

Structure(s) of C_2H_4ClBr

3. Butane

Structures of C_4H_{10}

No. of sets of chemically equivalent H's in structure 1 ______

No. of sets of chemically equivalent H's in structure 2 ______

Structures of C_4H_9OH

Structures of ethers with the formula $C_4H_{10}O$

Date ______ Name ______________________________ Section ______ Desk Number ______

4. Propene and cyclopropane
Structures of C_3H_6

Structures of $C_3H_4Cl_2$

Comments on the stability of three-membered ring system

____

5. $C_5H_{10}O$

Ketone

Aldehyde

Ring

Substituted ketone

Substituted aldehyde

Substituted ring

No. of optical isomers of aldopentose ______________

No. of optical isomers of ketopentose ______________

EXPERIMENT 38 Properties of Aspirin

Laboratory Time Required

Three hours

Special Equipment and Supplies

Two burets, 50-mL
pH Meter
Pasteur pipets
Mortar and pestle

pH 4 buffer
pH 7 buffer
0.1 M HCl
0.1 M NaOH
Phenolphthalein indicator
Aspirin tablets
Methanol
Toluene
5% $NaHCO_3$ (aq)
0.1 M $FeCl_3$
Potassium hydrogen phthalate (KHP)

Safety

Bases, such as sodium hydroxide, can cause skin burns and are especially hazardous to the eyes. Acids, such as hydrochloric acid, can also cause skin burns. Do not remove any of the aspirin tablets from the lab. Tablets that have been in the lab may have been contaminated and should not be ingested.

First Aid

Following skin contact with either sodium hydroxide or hydrochloric acid, wash the area thoroughly with water. Should either solution get in the eyes, rinse the eyes thoroughly with water. At least 20 minutes of flushing with water is recommended. Then seek medical attention.

Aspirin is a substance that is familiar to everyone; it is found in virtually every house. In this experiment, this everyday product is studied via titration and back titration, and qualitative analysis of aspirin's functional groups.

PRINCIPLES

It has long been known that organic compounds called salicylates have medicinal activity. They have antipyretic activity, lowering body temperature when it is elevated, but have little effect if body temperature is normal. More important, the salicylates are mildly analgesic, relieving headaches, neuralgia, and rheumatism. The principal and most widely used derivative is the compound called aspirin (acetylsalicylic acid). There is little doubt that it is one of the most widely used, commercially available pharmaceutical drugs in the world today.

In this experiment, you will study the properties of acetylsalicylic acid, the active ingredient in aspirin. You will test aspirin tablets for the presence or absence of certain functional groups and determine the molecular mass and the dissociation constant for acetylsalicylic acid in aqueous solution by titration with sodium hydroxide (which will give you the molecular mass) and back titration with hydrochloric acid (which will give you the dissociation constant).

Salicylic acid, the starting material for the synthesis of aspirin, can be viewed as a benzene ring (C_6H_6) in which two hydrogens have been replaced by functional groups (substituents that have distinctive characteristics). In salicylic acid, one substituent on the ring is an —OH (hydroxyl) group and the other is a $-\underset{\underset{\text{O}}{\|}}{\text{C}}\text{OH}$ (carboxylic acid) group.

When a hydroxyl group is attached to a carbon atom that is not a part of a benzene ring, the compound containing the hydroxyl group is classified as an alcohol. Thus, CH_3OH is called methanol or methyl alcohol and $C_6H_5CH_2OH$ is called benzyl alcohol. When the —OH group is bonded to a carbon atom that is part of a benzene ring (as is the case with salicylic acid) the substituted compound is classified as a phenol.

Phenols are not called alcohols (although both contain —OH groups) because they are weakly acidic. This is a result of the presence of the aromatic benzene ring. Aqueous solutions of iron(III) chloride develop a characteristic color in the presence of phenolic compounds.

Many organic acids, such as formic acid, acetic acid, and benzoic acid, contain carboxylic acid groups. As the name implies, organic acids are acidic—more so than phenols—less so than typical inorganic acids. Treatment of a carboxylic acid with aqueous sodium hydrogen carbonate produces the fizzing characteristic of the release of carbon dioxide gas that is formed when carbonates are acidified. Phenols are not acidic enough to cause the release of CO_2 from sodium hydrogen carbonate.

Performing tests with solutions of iron(III) chloride and sodium hydrogen carbonate on aspirin and on salicylic acid will provide you with clues for determining the structure of acetylsalicylic acid. Further information will be provided by performing solubility tests on aspirin and on salicylic acid. Nonpolar materials, such as hydrocarbons and carbon dioxide, are generally relatively insoluble in water and more readily soluble in nonpolar solvents such as toluene, itself a benzene derivative. The presence of polar functional groups, especially those that can form hydrogen bonds with water, such as —OH and $-\underset{\underset{\text{O}}{\|}}{\text{C}}\text{OH}$, greatly enhance the solubility of an organic compound in water.

Thus, methanol is miscible with water in all proportions. However, dimethyl ether, which can be regarded as a molecule in which the O—H bond has been replaced by an O—C bond, has a much lower solubility in water.

The molecular mass of acetylsalicylic acid is to be determined via titration of a known mass of aspirin with a base of known concentration. The pK_a of acetylsalicylic acid is then determined by back titration. At the basic end point, the acetylsalicylic acid exists as the acetylsalicylate ion. If the amount of acid used in the back titration is set equal to half the amount of base used in neutralizing the aspirin, then the solution will contain an equal amount of acid and conjugate base. The pH of the solution at this point is equal to the pK_a for the acid. A few equations may help to illustrate this point. Equation 38.1 shows the dissociation of the weak acid HA.

$$HA + H_2O \rightleftharpoons H_3O^+ + A^- \qquad (38.1)$$

The dissociation constant, K_a, of a weak acid is given in Equation 38.2. In Equation 38.2, $[A^-]$ is the concentration of the conjugate base and $[HA]$ is the concentration of the acid.

$$K_a = \frac{[H_3O^+][A^-]}{[HA]} \qquad (38.2)$$

Examination of Equation 38.2 reveals that $K_a = [H_3O^+]$ when $[A^-] = [HA]$. Equation 38.3 results from the definition, $pH = -\log_{10}[H_3O^+]$ and $pK_a = -\log_{10}K_a$.

$$pK = pH \text{ (when } [HA] = [A^-]) \qquad (38.3)$$

Thus, you can determine the pK_a of acetylsalicylic acid by following the back titration procedure and measuring the pH of the aspirin solution with a pH meter.

PROCEDURE

Determination of Solubility

Test the solubility of salicylic acid in each of the solvents listed on the Summary Report Sheet by placing a pea-sized sample of salicylic acid in 2 mL of the solvent. Shake the test tube to encourage the mixing of acid and solvent. Report your observation on the Summary Report Sheet and note whether salicylic acid is soluble, slightly soluble, or insoluble in each solvent listed. Repeat the tests, using a pea-sized sample of crushed aspirin tablet. Retain your mixtures of salicylic acid or aspirin in the various solvents for further testing.

Determination of the Presence or Absence of Functional Groups

Add 1 drop of aqueous $FeCl_3$ to the test tubes containing salicylic acid and aspirin in methanol. Record your observations on the Summary Report Sheet.

Add 2 mL of aqueous $NaHCO_3$ to the test tubes containing salicylic acid and aspirin in cold water. Record your observations on the Summary Report Sheet.

Determination of Molecular Mass and pK_a

Standardize the sodium hydroxide solution provided via titration of 0.6 g samples of potassium hydrogen phthalate (KHP), weighed to the nearest 0.1 mg on the analytical balance (see Experiment 8). Record your data on the Summary Report Sheet.

Use the standardized NaOH to titrate a 0.5 g sample of crushed aspirin (weighed to the nearest 0.1 mg on the analytical balance). Record your data on the Summary Report Sheet.

Standardize the hydrochloric acid provided via titration with the standardized sodium hydroxide. Record your data on the Summary Report Sheet. Determine the volume of the HCl solution which contains one-half the number of moles of NaOH used to titrate the aspirin sample, to the phenolphthalein end point. Add this volume to the aspirin/sodium hydroxide titration mixture. Swirl the flask containing the titration mixture to ensure complete mixing of the titration mixture and the added HCl. Follow your instructor's directions in setting up a pH meter and measuring the pH of the mixture of aspirin, NaOH, and HCl. Record your data on the Summary Report Sheet.

Use the data you have accumulated to predict the structural formula of acetylsalicylic acid.

Disposal of Reagents

Excess salicylic acid and crushed aspirin may be discarded in the solid waste containers. All solutions and suspensions may be flushed down the sink with copious amounts of water.

Questions

1. What evidence is there that the aspirin tablets are not pure acetylsalicylic acid?
2. Suppose that a typical aspirin tablet is 85% acetylsalicylic acid. Would ignoring this fact make your experimental molecular mass of acetylsalicylic acid higher or lower than the true value, or would it have no effect on the accuracy of the molecular mass determination? Explain your answer, briefly.
3. Suppose that a typical aspirin tablet is 85% acetylsalicylic acid. Would ignoring this fact make your experimental value for the K_a of acetylsalicylic acid higher or lower than the true value, or would it have no effect on the accuracy of the K_a determination? Explain your answer, briefly.

Date ______ Name ______________________________ Section ______ Desk Number ______

PRE-LAB EXERCISES FOR EXPERIMENT 38

These exercises are to be completed after you have read the experiment but before you come to the laboratory to perform it.

1. It took 25.67 mL of 0.1022 M NaOH to titrate a 0.5134 g sample of an organic acid to the phenolphthalein end point. What is the molecular mass of the acid?

2. How many mL of 0.1134 M HCl should be added to the titration mixture considered in question 1 to bring it back to the halfway point in the titration of the organic acid and the sodium hydroxide?

3. The pH at the halfway point of the titration of the organic acid was 5.13. What is the value of K_a for the organic acid?

Date ______ Name ______________________ Section ______ Desk Number ______

SUMMARY REPORT ON EXPERIMENT 38

Solubility Tests

Solvent	*Salicylic acid*	*Aspirin*
methanol	______	______
toluene	______	______
cold water	______	______
hot water	______	______

Functional groups

Test reagent		
$FeCl_3$	______	______
$NaHCO_3$	______	______

Standardization of NaOH

	Trial 1	*Trial 2*	*Trial 3**
Mass of container + KHP	______	______	______
Mass of container	______	______	______
Mass of KHP	______	______	______
Final buret reading, NaOH	______	______	______
Initial buret reading, NaOH	______	______	______
Volume used, NaOH	______	______	______
Molarity of NaOH solution	______	______	______
Average molarity of NaOH solution		______	

Standardization of HCl

	Trial 1	*Trial 2*	*Trial 3**
Final buret reading, HCl	______	______	______
Initial buret reading, HCl	______	______	______
Volume delivered, HCl	______	______	______
Final buret reading, NaOH	______	______	______
Initial buret reading, NaOH	______	______	______
Volume used, NaOH	______	______	______
Molarity of HCl solution	______	______	______
Average molarity of HCl solution		______	

*Optional

Date ______ Name ______________________________ Section ______ Desk Number ______

Determination of Molecular Mass and pK_a

	Trial 1	*Trial 2*	*Trial 3**
Mass of container + aspirin	________	________	________
Mass of container	________	________	________
Mass of aspirin	________	________	________
Final buret reading, NaOH	________	________	________
Initial buret reading, NaOH	________	________	________
Volume used, NaOH	________	________	________
Volume of HCl needed to react with half of NaOH	________	________	________
Volume of HCl added	________	________	________
pH of solution after HCl addition	________	________	________
pK_a of aspirin	________	________	________
K_a of aspirin	________	________	________
Average K_a of aspirin		________	
Moles of NaOH used to titrate aspirin	________	________	________
Molecular mass of aspirin	________	________	________
Average molecular mass of aspirin		________	

*Optional

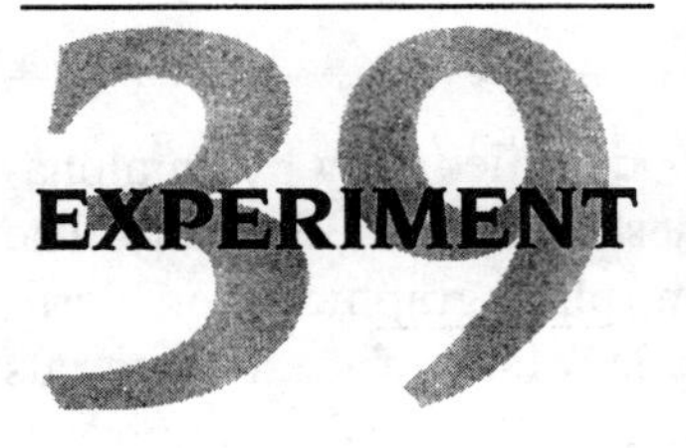

Qualitative Analysis of Household Chemicals

Laboratory Time Required

Three hours

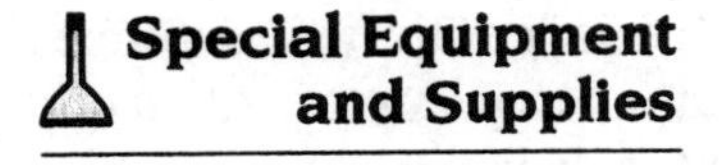

Special Equipment and Supplies

	Unknowns	Reagents
Balance	Photographic fixer	Acetic acid
Thermometer	Cornstarch	Ammonia
Conductivity tester	Chalk	Iodine
Labels	Baking soda	Phenolphthalein
	Washing soda	Rock salt
	Table salt	Ice
	Sugar	
	Epsom salt	
	Alum	
	Sour salt	

Safety

Do not remove any of the unknowns or reagents from the laboratory for home use.

First Aid

For exposure to any of the materials used in this experiment, thoroughly flush the skin or eyes with water. Chemicals may be removed from the stomach by drinking large amounts of water and inducing vomiting.

Several years ago, the American Chemical Society offered a prize to anyone who could find a substance that was not a chemical. The point, of course, was that all matter, whether "natural" or "artificial," is composed of atoms and molecules. Although most college students accept this truism readily, they still often view the chemicals they work with in the laboratory as something different from the materials that abound in the "real world." This experiment, in which the "active ingredient" in many household products will be identified, is an attempt to bridge the gap between the chemistry lab and the household environment.

PRINCIPLES

You will be provided with a number of white solids, identified only by an alphanumeric code. Your job will be to identify each of these unknowns correctly. The possibilities are table salt, sugar, epsom salt, alum, photographic fixer, cornstarch, sour salt, chalk, baking soda, and washing soda. Many of these materials are likely to be found in the typical home. Others, while not as common, are readily available in pharmacies, building supply stores, or camera shops, and may be purchased without any special license or prescription.

Suggestions for identifying the unknowns using chemical tests are made below. Note that many of the test reagents are themselves ingredients in products commonly used in the home. For instance, phenolphthalein is a weak acid, solid in pure form, which appears pink in alkaline (basic) solution. Phenolphthalein has a laxative effect and, for many years, was the active ingredient in Feen-A-Mint® and Epso® tablets. (Recent concerns about possible toxicity have led to reformulation of these products.) Acetic acid ($HC_2H_3O_2$) is a weak acid, with a characteristic sour taste; vinegar is essentially a 1 M solution of acetic acid. Ammonia is a gas in pure form; its aqueous solutions are alkaline and are commonly used as household cleansers. Pure iodine is a solid that sublimes (passes directly into the gas phase) at slightly elevated temperatures. Alcoholic solutions of iodine (tinctures of iodine) are often used as disinfectants.

The information given below should enable you to devise a plan for identifying the unknowns. The formulas given are those of the main ingredient in each household item (of course, the materials used in the home are generally not pure and there is no single formula for the complete mixture).

Insoluble in Water

Chalk (main ingredient: $CaCO_3$, calcium carbonate) and cornstarch (main ingredient: a polymer of glucose, $(C_6H_{10}O_5)_x$), are insoluble in water. Like all carbonates, chalk will fizz when treated with acid. Starch will turn blue when treated with iodine.

Soluble in Water

Form Precipitates when Treated with Ammonia

All of the other unknowns are soluble in water. Epsom salt (main ingredient: $MgSO_4 \cdot 7H_2O$, magnesium sulfate heptahydrate) dissolves with a noticeable cooling effect. Addition of aqueous ammonia to a solution of epsom salt produces "milk of magnesia," a suspension of magnesium hydroxide that has a milky appearance.

Alum (main ingredient: $NH_4Al(SO_4)_2 \cdot 12H_2O$, ammonium aluminum sulfate dodecahydrate) is used as an astringent and as a pickling agent. Addition of aqueous ammonia to a solution of alum produces a gelatinous precipitate of aluminum hydroxide. The other water-soluble unknowns do not form precipitates when treated with ammonia.

No Precipitate when Treated with Ammonia

Washing soda (main ingredient: Na_2CO_3, sodium carbonate) dissolves in water to produce solutions basic enough to turn phenolphthalein pink. Baking soda (main ingredient: $NaHCO_3$, sodium hydrogen carbonate) also has basic properties, although its solutions are so weakly alkaline that phenolphthalein remains colorless when added to them. Both washing soda and baking soda, like other carbonates, will fizz when treated with acetic acid.

Sour salt (main ingredient: $C_6H_8O_7$, citric acid) is often used in the preparation of "sweet and sour" meats. Solutions of sour salt are sufficiently acidic to remove the pink color of solutions of washing soda that contain phenolphthalein. Adding sour salt to solutions of washing soda may also cause fizzing.

Photographic fixer (main ingredient: $Na_2S_2O_3 \cdot 5H_2O$, sodium thiosulfate pentahydrate) is soluble in water and is capable of reducing I_2 molecules to I^- ions. It will decolorize brown iodine solutions and also remove the blue color of the iodine-starch complex.

No Reaction with Test Reagents

Table salt (main ingredient: NaCl, sodium chloride), like epsom salt, alum, fixer, washing soda, and baking soda, produces ions when dissolved in water. Thus, solutions of table salt will conduct electricity. These solutions also have very low freezing points. A solution of 4 g of table salt in 10 mL of water will not freeze in an ice/rock salt/water bath (bath temperature: –10°C) but a solution of 4 g of table sugar (main ingredient: $C_{12}H_{22}O_{11}$, sucrose) in 10 mL of water will freeze when placed in such a bath. In addition, sugar molecules do not dissociate in aqueous solution, so solutions of sucrose do not conduct electricity. Neither sugar nor salt solutions react with acetic acid, ammonia, phenolphthalein, or iodine.

PROCEDURE

Devise a scheme for separating the unknowns into groups and then identifying the members of each group. You may, if you wish, base your procedure on the four groups discussed in the Principles section or you may try an alternate approach, such as trying to separate the electrolytes (those whose solutions conduct electricity) from the nonelectrolytes (those whose solutions do not conduct electricity) or you may try separating those materials that fizz when treated with acid from those that do not. No matter where you start, you will find it helpful to follow the suggestions given below.

1. In determining solubility, use only a pea-sized (or smaller) quantity of solid, placed in a test tube. Add roughly 10 mL of water and mix well. You need just enough solid to be seen easily if it does not dissolve, but not enough to form a saturated solution if it is moderately soluble. A slight cloudiness may be due to a trace of insoluble filler and should not lead to a conclusion of "insoluble" if the major portion of the sample dissolves.
2. Do not add anything to the containers of unknowns or test reagents. Do not use spatulas from one container in another.
3. To determine if an unknown will fizz when treated with acid, place a pea-sized quantity of solid in a test tube and add a few drops of acetic acid.
4. When testing materials with phenolphthalein or iodine, add only 2–3 drops of reagent.

5. Remember that reagents may interfere with one another. For instance, if you have several solutions and add iodine to each, all of the solutions, except the one containing the photographic fixer, will turn brown. Adding phenolphthalein (in an effort to determine basicity) would then be futile; the brown color would mask any pink that might develop. Use fresh samples whenever it seems necessary.
6. Your instructor will show you how to test solutions for conductivity.
7. Be sure to record the identity code of each unknown and any data collected that would help you to identify the unknown.

Disposal of Chemicals

The household chemicals that are insoluble in water may be discarded in the containers designated for solid, nonhazardous waste. All soluble household chemicals and test reagents may be diluted and flushed down the sink with a voluminous stream of water.

Date ______ Name ______________________________ Section ______ Desk Number ______

PRE-LAB EXERCISE FOR EXPERIMENT 39

This exercise is to be completed after you have read the experiment but before you come to the laboratory to perform it.

Complete the table shown below. The first line is completed as a guide.

	Soluble in H_2O?	Solution conducts electricity?	Reacts with vinegar?	Reacts with phenol-phthalein?	Precipitate with NH_3?	Solution of 4 g/10 mL freezes?	Reacts with I_2?
Table salt	yes	yes	no	no	no	no	no
Sugar							
Epsom salt							
Alum							
Photographic fixer							
Cornstarch							
Chalk							
Baking soda							
Washing soda							
Sour salt							

Date ______ Name ______________________________ Section ______ Desk Number ______

SUMMARY REPORT ON EXPERIMENT 39

Unknown	Test Reagent	Observation

Date ______ Name ________________________________ Section ______ Desk Number ______

Use this table to summarize your data.

Unknown	Soluble in H_2O?	Solution conducts electricity?	Reacts with vinegar?	Reacts with phenol-phthalein?	Precipitate with NH_3?	Solution of 4 g/10 mL freezes?	Reacts with I_2?	Identity of unknown

EXPERIMENT 40

Qualitative Analysis of the Acid Chloride Group

Laboratory Time Required

One and one-half hours. Can be combined with a discussion to prepare for Experiments 41, 42, and 43.

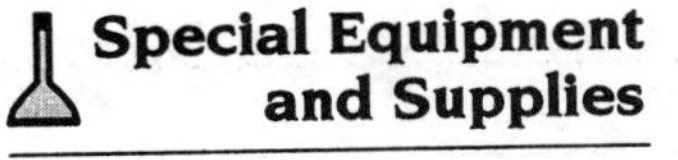

Special Equipment and Supplies

	Cation test solutions	*Reagents*
Labels	0.2 M $AgNO_3$	12 M HCl
Bunsen burner or hot plate	0.2 M $Hg_2(NO_3)_2$	1 M K_2CrO_4
Centrifuge	0.2 M $Pb(NO_3)_2$	6 M NH_3
		6 M HNO_3

If Experiments 41 and 42 are to be performed:

0.2 M $BaCl_2$
0.2 M $Al(NO_3)_3$
0.2 M $Fe(NO_3)_3$
0.2 M $Cu(NO_3)_2$
0.2 M $Ni(NO_3)_2$
0.2 M $Mn(NO_3)_2$
0.2 M KNO_3

Safety

Exercise great care when working with concentrated acids and bases. When dilute solutions must be prepared from these reagents (as will be necessary in this experiment), always add the acid or base to water (never put a drop of water into a concentrated solution of acid or base). The addition of water to the concentrated reagent may cause splattering.

When centrifuging mixtures, always use pairs of test tubes to help keep the centrifuge in balance. Wait for the centrifuge to come to a complete stop, after it is turned off, before reaching in to get your test tubes out. **Never touch a spinning centrifuge.**

First Aid

For skin contact with any reagents, thoroughly wash the affected area with water. Flush your eyes with water for at least 20 minutes if any reagents splash into them. See a doctor if you have skin or eye contact with concentrated acids or bases.

The classical qualitative analysis scheme assigns common cations to five groups. Group I contains those ions whose chloride salts are insoluble in dilute HCl and is therefore known as the acid chloride group. Groups II and III are precipitated as sulfides by H_2S in acidic or basic solution, respectively. Group III is also known as the basic sulfide group. It includes cations that precipitate as hydroxides or as sulfides. Group IV includes the elements that form insoluble carbonates. Group V includes the cations whose salts are generally soluble. Although this traditional scheme is still widely applied, alternative procedures that do not utilize H_2S have been devised for the analysis of selected cations, although none deals with as many as the classical scheme.

In this experiment, you will study the properties and separation procedures for the conventional acid chloride group. In subsequent experiments, you will learn to separate and identify cations using such reagents as H_2SO_4, NaOH, and NH_3, but not H_2S, for separations. When you have mastered these principles, you will demonstrate your skill by analyzing unknown samples.

PRINCIPLES

Although, in principle, a specific test could be devised for each of the common cations and anions, it is obvious that the analysis of a mixture would then require that all these tests be performed in order to verify the presence or absence of each ion under consideration. It is more efficient to divide the ions into smaller groups, the members of which share a distinctive chemical property. A single chemical test can then determine the presence or absence of each group of ions. Then only the groups found to be present need be tested further to determine which group members are present. Another advantage of this procedure is that the chemical properties of the elements, both their similarities and their differences, are presented in a systematic manner.

As in the usual analytical scheme, we will begin by separating the acid chloride group (Group I). There are few ions in this group, and dilute HCl is the only reagent required for the separation. The distinctive property of the ions in this group is the low solubility of their chloride salts in water. These ions therefore precipitate as chlorides when dilute HCl is added. The group is composed of the Ag^+, Pb^{2+}, and Hg_2^{2+} ions because these are the only cations normally encountered in aqueous solution that form insoluble chlorides. We will disregard other ions, such as Cu^+, which form insoluble chlorides but are, under most conditions, unstable in aqueous solution. We will also ignore ions such as Tl^+, which are stable but relatively rare.

After the insoluble chlorides have been precipitated, various properties of the individual ions may be used to separate each ion for positive identification. Properties such as change in solubility with temperature, formation of complex ions, and oxidation or reduction of the cations are used to separate and identify the ions of this group. Lead chloride, for example, differs from AgCl and Hg_2Cl_2 in that it is moderately soluble in hot water. Therefore, $PbCl_2$ can easily be separated from the other acid chlorides by treatment of the precipitate with hot water.

Once $PbCl_2$ has been removed from the precipitate, the remaining solid is treated with aqueous ammonia. Silver chloride dissolves in this medium because of the formation of the $Ag(NH_3)_2^+$ complex ion, whereas mercury(I) chloride disproportionates to give a mixture of elemental mercury and mercury(II) amide chloride. Thus, the behaviors of AgCl and Hg_2Cl_2 in the presence of ammonia are used as the basis for separating and identifying the remaining members of the acid chloride group.

In this experiment, you will perform preliminary experiments on solutions known to contain Ag^+, Hg_2^{2+}, and Pb^{2+}, respectively. After you have done the preliminary tests, you will be given an unknown for analysis. This unknown may contain Ag^+, Hg_2^{2+}, and/or Pb^{2+} ions. If you will be performing Experiments 41 and 42 , you will also be given the opportunity to see the behavior of Ag^+ and Hg_2^{2+} in the presence of the other ions in the analysis scheme (Ba^{2+}, Cu^{2+}, Ni^{2+}, Al^{3+}, Mn^{2+}, Fe^{3+}, and K^+). This will help you when you are working with the general unknown in Experiment 42. The Pb^{2+} ion is not included in the general unknown because the relatively high solubility of $PbCl_2$ means that not all the lead ions are separated when the acid chloride group is precipitated, and lead ions remain in solution, interfering with tests for ions in other groups.

PROCEDURE

Preliminary Experiments

Place 5 drops of 0.2 M $AgNO_3$ solution, 5 drops of 0.2 M $Hg_2(NO_3)_2$ solution, and 5 drops of 0.2 M $Pb(NO_3)_2$ solution, respectively, in separate, labeled test tubes. To each test tube, add 5 drops of water and 1 drop of 1 M HCl solution. Mix well.

Centrifuge the tubes. Then add 1 drop more of HCl to test for complete precipitation. Centrifuge again. Repeat the process of adding HCl and centrifuging until no further precipitation is observed.

Carefully decant the liquid from each test tube, retaining any solid in the test tube. Add 1 mL of water to each test tube, mix, and then heat briefly in a water bath. Which solid dissolves? Add 5 drops of 1 M K_2CrO_4 to the test tube that no longer contains a precipitate.

Add 1 mL of concentrated NH_3 to the test tubes containing AgCl and the Hg_2Cl_2. Record your observations.

Carefully add 2 M HNO_3, drop by drop, while mixing, to the test tube containing silver ions and ammonia, continuing until the solution indicates acidity on litmus paper. Record your observations.

Analysis of Unknown Solution

Obtain an unknown solution from your instructor and note its appearance on the Summary Report Sheet. Mix well and centrifuge the tube containing your unknown, using a test tube filled with water as a counterbalance. If a precipitate is present following centrifugation, decant the liquid from the solid and save both portions for subsequent analysis. Figure 40.1 gives a brief outline of the analytical scheme. More complete details are given below.

Transfer 1 mL of the liquid portion of your unknown to a test tube and add 1 drop of 1 M HCl to this portion of your unknown. Mix the contents well by

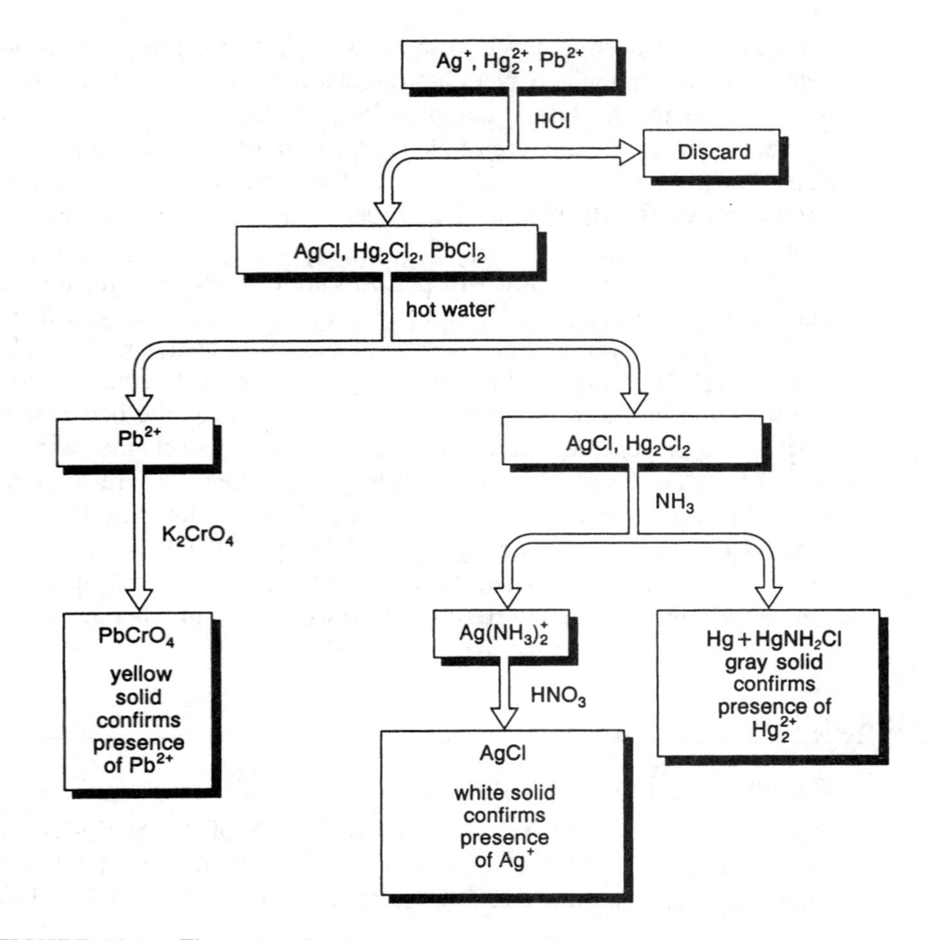

FIGURE 40.1 Flow chart for the analysis of the acid chloride group.

tapping the test tube with your forefinger. If a precipitate does not form immediately, add another drop or two of HCl. After the precipitate forms, centrifuge the test tube. Add another drop of HCl to test for complete precipitation. Centrifuge the test tube. Continue the process of adding a drop of acid and centrifuging until no further precipitation is observed. Discard the centrifugate (the liquid portion) after the final addition of acid.

Add 1 mL of distilled water to the test tube containing the precipitate and heat the tube in a hot-water bath for 5 minutes. Mix well. Then centrifuge the mixture. Transfer the centrifugate to a separate test tube, retaining any solid that remains in the original tube. Add 5 drops of 1 M K_2CrO_4, to the centrifugate. The formation of a yellow precipitate confirms the presence of Pb^{2+}.

If any acid chloride precipitate remains after treatment with hot water, add 1 mL of 1.5 M NH_3 to this solid. Stir the mixture and centrifuge. If the solid turns gray, the presence of Hg_2^{2+} is confirmed. (If the presence of Ag^+ is subsequently confirmed, a trace amount of black residue at this point might be metallic silver, formed by photochemical decomposition of AgCl.)

Decant the ammonia-containing centrifugate from the solid. Cautiously add 2 M HNO_3, drop by drop, to the centrifugate, while stirring, until the mixture is acid to litmus paper. Formation of a white precipitate confirms the presence of Ag^+.

If a solid was present in your original unknown, and you have found fewer than three ions in the liquid portion of the unknown, repeat the tests described above, substituting some of the solid for the acid chloride precipitate.

Disposal of Reagents

Because small quantities of chemicals are used in this experiment, solutions may be diluted and washed down the drain. Precipitates may be deposited in crocks (as solids) or resuspended in water and also flushed down the drain.

Date ______ Name ______________________________ Section ______ Desk Number ______

PRE-LAB EXERCISES FOR EXPERIMENT 40

These exercises are to be completed after you have read the experiment but before you come to the laboratory to perform it.

1. Explain, in detail, how you would go about preparing 2 M HNO_3 from 6 M HNO_3.

2. Mixing a solution containing Ag^+, Hg_2^{2+}, and/or Pb^{2+} ions with an aqueous solution of any soluble chloride salt will produce a white precipitate. Why is it preferable to use HCl rather than $CuCl_2$ or KCl as the precipitating agent for the acid chloride group?

Date ______ Name ______________________ Section ______ Desk Number ______

SUMMARY REPORT ON EXPERIMENT 40

Pb^{2+}

Observations **Equations**

Hg_2^{2+}

Observations **Equations**

Ag^+

Observations **Equations**

Date ______ Name ______________________________ Section ______ Desk Number ______

Analysis of Unknown Solution

As you perform each step in the unknown analysis, record your procedure and observations by building onto the flow chart started below.

At the bottom of the flow diagram, list the cations found to be definitely present in the unknown.

Ag^{+}, Hg_2^{2+}, Pb^{2+}

Qualitative Analysis: Preliminary Tests for Ba^{2+}, Al^{3+}, Fe^{3+}, Cu^{2+}, Ni^{2+}, Mn^{2+}, and K^+

Laboratory Time Required

Three hours

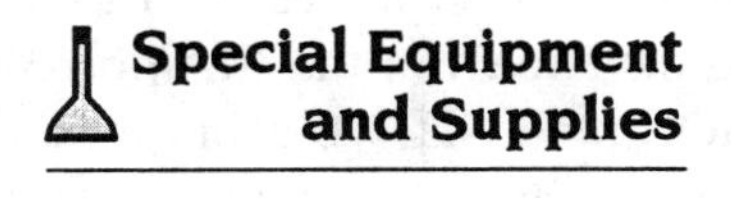

	Cation test solutions	*Reagents*
Labels	0.2 M $BaCl_2$	12 M HCl
Bunsen burner	0.2 M $Al(NO_3)_3$	3 M H_2SO_4
Litmus paper (blue and red)	0.2 M $Fe(NO_3)_3$	6 M HNO_3
Centrifuge	0.2 M $Cu(NO_3)_2$	3 M $HC_2H_3O_2$
Test wires	0.2 M $Ni(NO_3)_2$	6 M NH_3
Cobalt blue glass	0.2 M $Mn(NO_3)_2$	6 M NaOH
Sandpaper	0.2 M KNO_3	$NaBiO_3$ (s)
		DMG (s)
		Iron nail
		NaCl (s)
		0.2 M KSCN

Safety

Exercise great care when working with concentrated acids and bases. When dilute solutions must be prepared from these reagents (as will be the case in this experiment), always add the acid or base to water (never put a drop of water into a concentrated solution of acid or base). The addition of water to the concentrated reagent may cause splattering.

When centrifuging mixtures, always use pairs of test tubes to help keep the centrifuge in balance. Wait for the centrifuge to come to a complete stop, after it is turned off, before reaching in to get your test tubes out. Never touch a spinning centrifuge.

First Aid

For skin contact with any reagents, thoroughly wash the affected area with water. Flush your eyes with water for at least 20 minutes if any reagents splash into them. See a doctor if you have skin or eye contact with concentrated acids or bases.

The analysis of the acid chloride group (Ag^+, Hg_2^{2+}, and Pb^{2+}) has been discussed in Experiment 40. This experiment deals with Ba^{2+}, Al^{3+}, Fe^{3+}, Cu^{2+}, Ni^{2+}, Mn^{2+}, and K^+. The Ba^{2+} will be separated as $BaSO_4$. The other ions, except for K^+, will be precipitated as hydroxides and then treated with reagents that will effect the separation or identification of each ion. The K^+, which represents the soluble group, will be identified via a flame test. Experiments 41 and 42 are intended for classes that will spend more than one week on qualitative analysis.

PRINCIPLES

Once the acid chloride group cations have been separated from a general unknown, many analytical schemes call for the separation of an acid sulfide and a basic sulfide group. The members of the first of these groups are precipitated as sulfides whereas the members of the second group are precipitated as a mixture of hydroxides and sulfides. The precipitating agent has usually been hydrogen sulfide, H_2S, produced by the acidification of iron sulfide or by hydrolysis of thioacetamide, CH_3CSNH_2.

There are difficulties associated with the use of these reagents. Hydrogen sulfide is poisonous and has a vile odor. Thioacetamide has been classified as a possible carcinogen. Therefore, we have chosen to deviate from the traditional scheme and rely on less toxic reagents (H_2SO_4, NaOH, and NH_3) to achieve the necessary separations of ions in our analytical scheme. Our scheme also offers the advantage of using relatively straightforward reactions to achieve separation and identification, whereas the chemistry used in more traditional schemes is often too complicated for beginning students to comprehend.

In our analytical scheme, barium ions will be precipitated as barium sulfate, following the precipitation of the acid chloride group. The precipitating agent will be sulfuric acid. The net ionic equation for the reaction is shown in Equation 41.1.

$$Ba^{2+} + SO_4^{2-} \longrightarrow BaSO_4 \downarrow \qquad (41.1)$$

We have chosen a single ion, Ba^{2+}, to represent the sulfate group because inclusion of Pb^{2+} and Sr^{2+}, which also form insoluble sulfates, would excessively complicate the separation and identification scheme. Should $PbSO_4$, $SrSO_4$, and $BaSO_4$ be precipitated together, the similarity of their properties would make it difficult to separate and confirm the presence of the individual cations. Indeed, traditional schemes use precipitants, such as HCl, H_2S, Na_2CO_3, and $Na_2C_2O_4$ to remove Pb^{2+} and Sr^{2+} before the precipitation of Ba^{2+} as $BaSO_4$. Therefore, we include only Ba^{2+} in our analytical scheme. Its presence in an unknown would be indicated by the formation of a white precipitate upon the addition of sulfuric acid and confirmed by a flame test. The use of the flame test is also a feature of many traditional analytical schemes.

In contrast to the single ion precipitated by sulfuric acid, most of the remaining ions in our scheme are precipitated by sodium hydroxide. The net ionic equations involved here are shown in Equations 41.2 and 41.3.

$$M^{2+} + 2OH^- \longrightarrow M(OH)_2 \downarrow \qquad M = Ni, Cu, Mn \qquad (41.2)$$

$$M^{2+} + 3OH^- \longrightarrow M(OH)_3 \downarrow \qquad M = Al, Fe \qquad (41.3)$$

The group of hydroxide precipitates will be treated with a variety of reagents that will allow the identification of each of the five ions (Fe^{3+}, Al^{3+}, Ni^{2+}, Cu^{2+}, and

Mn^{2+}) that have insoluble hydroxides. This section of our analytical scheme closely resembles the traditional scheme.

Following the precipitation of the hydroxides, additional sodium hydroxide is used to separate $Al(OH)_3$ from the other compounds in the precipitate. Only $Al(OH)_3$ is amphoteric, which means that it dissolves in both acids and bases. The reaction of $Al(OH)_3$ with an acid is a typical neutralization reaction. The reaction of $Al(OH)_3$ with hydroxide ions is shown in Equation 41.4.

$$Al(OH)_3 + OH^- \longrightarrow Al(OH)_4^- \qquad (41.4)$$

The presence of Al^{3+} in the unknown is confirmed if the aluminum hydroxide precipitate can be regenerated from the solution containing $Al(OH)_4^-$ complex ions. This is achieved by adding a controlled amount of acid and then base to the solution (see Equations 40.5 and 40.6).

$$Al(OH)_4^- + 4H^+ \longrightarrow Al^{3+} + 4H_2O \qquad (41.5)$$

$$Al^{3+} + 3OH^- \longrightarrow Al(OH)_3\downarrow \qquad (41.6)$$

After $Al(OH)_3$ has been dissolved in excess NaOH and the liquid phase poured off, the remaining solid metal hydroxides are treated with ammonia. This causes the $Cu(OH)_2$ and $Ni(OH)_2$ to dissolve and leaves $Mn(OH)_2$ and $Fe(OH)_3$ unchanged. The reactions between ammonia and $Cu(OH)_2$ and $Ni(OH)_2$ are represented in Equations 41.7 and 41.8.

$$Cu(OH)_2 + 4NH_3 \longrightarrow Cu(NH_3)_4^{2+} + 2OH^- \qquad (41.7)$$

$$Ni(OH)_2 + 6NH_3 \longrightarrow Ni(NH_3)_6^{2+} + 2OH^- \qquad (41.8)$$

The solution containing the ammine complexes, $Cu(NH_3)_4^{2+}$ and $Ni(NH_3)_6^{2+}$, is treated with acetic acid and divided into two parts. A few grains of table salt and a clean iron nail are added to one portion. A coating of metallic copper will plate out on the nail if Cu^{2+} ions are present in the solution. The net ionic equations for the confirmatory tests for Cu^{2+} are shown in Equations 41.9 and 41.10.

$$Cu(NH_3)_4^{2+} + 4H^+ \longrightarrow Cu^{2+} + 4NH_4^+ \qquad (41.9)$$

$$Cu^{2+} + Fe \longrightarrow Cu + Fe^{2+} \qquad (41.10)$$

The second portion of the solution that contains Ni^{2+} and Cu^{2+} is treated with a few grains of dimethylglyoxime, which has the formula

$$\begin{array}{l} CH_3{-}C{=}N{-}O{-}H \\ \quad\;\;\;| \\ CH_3{-}C{=}N{-}O{-}H \end{array}$$

and is usually represented by the abbreviation DMG. The appearance of a red precipitate confirms the presence of nickel ions in the unknown. The net ionic equations for the confirmatory tests for Ni^{2+} are shown in Equations 41.11 and 41.12.

$$Ni(NH_3)_6^{2+} + 6H^+ \longrightarrow Ni^{2+} + 6NH_4^+ \qquad (41.11)$$

$$Ni^{2+} + 2DMG \longrightarrow Ni(DMG)_2\downarrow + 2H^+ \qquad (41.12)$$

After the original metal hydroxide precipitate has been treated with excess sodium hydroxide and ammonia, any solid remaining would be $Fe(OH)_3$ and/or $Mn(OH)_2$. These hydroxides are then dissolved in nitric acid. The resulting solution is divided into two portions and KSCN is added to the first portion. If the solution becomes blood-red in color, the presence of Fe^{3+} in the unknown is confirmed. The net ionic equations for the confirmatory tests for iron(III) ions are shown in Equations 41.13 and 41.14.

$$Fe(OH)_3 + 3H^+ \longrightarrow Fe^{3+} + 3H_2O \qquad (41.13)$$

$$Fe^{3+} + SCN^- \longrightarrow Fe(SCN)^{2+} \qquad (41.14)$$

A few grains of sodium bismuthate, $NaBiO_3$, are added to the second portion of the solution containing Fe^{3+} and Mn^{2+}. If no color change is observed immediately, the solution is heated gently in a beaker of hot water for a few moments. The development of a pink or purple color, resulting from the formation of permanganate ions, confirms the presence of manganese(II) ions in the unknown. The net ionic equations for the confirmatory tests for Mn^{2+} are given in Equations 41.15 and 41.16.

$$Mn(OH)_2 + 2H^+ \longrightarrow Mn^{2+} + 2H_2O \qquad (41.15)$$

$$2Mn^{2+} + 5BiO_3^- + 14H^+ \longrightarrow 2MnO_4^- + 5Bi^{3+} + 7H_2O \qquad (41.16)$$

The remaining ion in our analytical scheme is K^+, representing the soluble group, which in traditional analytical schemes may also include Na^+ and NH_4^+. The presence of potassium ions is confirmed by observing a fleeting violet color in a flame test. This color is often obscured by the very strong yellow/orange color imparted to the flame by sodium ions. However, this difficulty can be overcome by observing the flame through a piece of cobalt blue glass.

In this experiment, you will test solutions known to contain Ba^{2+}, Al^{3+}, Cu^{2+}, Ni^{2+}, Fe^{3+}, Mn^{2+}, and K^+, respectively. In performing these preliminary tests, you will mix a solution containing the ion of interest with the reagents that would be added to the general unknown in Experiment 42. In other words, when testing a solution of Al^{3+}, you will treat it with HCl (to simulate the removal of Ag^+ and Hg_2^{2+}), H_2SO_4 (to simulate the removal of Ba^{2+}), NaOH (which will precipitate the aluminum as $Al(OH)_3$), and HNO_3, followed by excess NaOH (the confirmatory reagents for Al^{3+}). In this manner, you will be able to see if testing for any of the early ions in the analytical scheme will interfere with tests for later ions.

The performance of these tests is intended to help you with the analysis of the general unknown in Experiment 42. Be sure to take careful notes on color changes; the formation or dissolution of precipitates; the color, quantity, and type of precipitates; and on the appearance of flame tests. Be sure to mix the test tube contents well after each addition of reagent. To test a solution for acidity or basicity, add a drop of acid or base to the test solution, stir the mixture, and touch the stirring rod to a piece of moistened litmus paper that has been draped over a watch glass.

PROCEDURE

Test for Ba^{2+}

Place 5 drops of 0.2 M $BaCl_2$ in a test tube. Add 5 drops of distilled water. Then add one drop of 1 M HCl. Do you observe a reaction? Next add 3 drops of 3 M H_2SO_4. What do you observe? Centrifuge the test tube. Add 1 drop more of

3 M H_2SO_4 to test for complete precipitation. Continue centrifuging and testing the centrifugate until all Ba^{2+} has been removed from the solution. Note the appearance of the $BaSO_4$ precipitate.

Clean a test wire by dipping it in concentrated HCl and heating it in the burner flame. Repeat the process of dipping and heating until the wire does not impart any color to the flame. Decant the centrifugate from the test tube containing the $BaSO_4$ precipitate. Add 5 drops of concentrated HCl to the precipitate. Stir the mixture of $BaSO_4$ and HCl and dip the test wire into the resulting suspension. Bring the loop of the test wire to the edge of the outer cone of the burner flame, at the height of the top of the inner cone. Observe the color imparted to the flame. Record your observations.

Test for Al^{3+}

Place 5 drops of 0.2 M $Al(NO_3)_3$ in another test tube. Add 5 drops of distilled water, followed by 1 drop of 1 M HCl and 1 drop of 3 M H_2SO_4. Next, add 2 M NaOH, drop by drop, until the solution is just basic. Centrifuge the test tube. Add 1 drop more of 2 M NaOH to test for complete precipitation. Continue adding NaOH, centrifuging after the addition of each drop. Note the appearance of the $Al(OH)_3$ precipitate and record the number of drops of NaOH that must be added before the precipitate appears and before it dissolves again. After the precipitate has dissolved completely, add 6 M HNO_3 until the solution is just acid to litmus. Then add 2 M NaOH, dropwise, until the precipitate reappears. Record the number of drops of NaOH that must be added before the precipitate reappears.

Test for Cu^{2+}

Place 5 drops of 0.2 M $Cu(NO_3)_2$ in a test tube. Add 5 drops of water, followed by 1 drop of 1 M HCl and 1 drop of 3 M H_2SO_4. Next, add 2 M NaOH, by drops, until the solution is just basic. Centrifuge the test tube. Add 1 more drop of NaOH to test for complete precipitation. Continue adding NaOH, centrifuging after the addition of each drop, until precipitation is complete. Discard the centrifugate. Note the appearance of the $Cu(OH)_2$ precipitate. Add 1 mL of 3 M NH_3 and stir the mixture. Add more NH_3, if necessary, to dissolve the precipitate completely. Note the appearance of the solution. Add 3 M $HC_2H_3O_2$, dropwise, until the solution is nearly colorless. Then add a few grains of NaCl. Gently sand an iron nail and place it in the solution. Leave the nail in the solution, undisturbed, for 15–30 minutes. Record your observations.

Repeat the test for copper(II) starting with 5 drops of 0.2 M $Cu(NO_3)_2$ and 5 drops of 0.2 M $Ni(NO_3)_2$, instead of 5 drops of 0.2 M $Cu(NO_3)_2$ only. This will permit you to see what a positive Cu^{2+} test looks like in the presence of Ni^{2+} (Recall that our analytical scheme will not separate Cu^{2+} from Ni^{2+}.)

Repeat the test for copper starting with 5 drops of 0.2 M $Ni(NO_3)_2$ instead of 5 drops of 0.2 M $Cu(NO_3)_2$. Observations from this test should help you to avoid reporting a "false positive" if your unknown has Ni^{2+} ions but no Cu^{2+} ions.

Test for Ni^{2+}

Place 5 drops of 0.2 M $Ni(NO_3)_2$ in a test tube. Add 5 drops of distilled water, followed by 1 drop of 1 M HCl and 1 drop of 3 M H_2SO_4. Next, add 2 M NaOH drop by drop, until the solution is just basic to litmus. Centrifuge the tube. Add 1 drop more of 2 M NaOH to test for complete precipitation. Continue adding NaOH, centrifuging after the addition of each drop, until precipitation is complete. Discard the centrifugate. Note the appearance of the $Ni(OH)_2$ precipitate. Add 1 mL of 3 M NH_3 and stir the mixture. Add more NH_3, if necessary, to dissolve the precipitate completely. Note the appearance of the solution. Add 3 M $HC_2H_3O_2$, by drops, until the solution is nearly colorless. Then add a few grains of DMG. Note the appearance of the mixture.

Repeat the test for nickel(II) starting with 5 drops of 0.2 M $Cu(NO_3)_2$ and 5 drops of 0.2 M $Ni(NO_3)_2$, instead of 5 drops of 0.2 M $Ni(NO_3)_2$ only. Then, repeat the test starting with 5 drops of 0.2 M $Cu(NO_3)_2$ in place of $Ni(NO_3)_2$.

Test for Mn^{2+}

Place 5 drops of 0.2 M $Mn(NO_3)_2$ in a test tube. Add 5 drops of distilled water, 1 drop of 1 M HCl, and 1 drop of 3 M H_2SO_4. Then add 2 M NaOH by drops until the solution is just basic to litmus. Centrifuge the test tube. Add 1 drop more of 2 M NaOH to test for complete precipitation. Continue centrifuging and adding NaOH until precipitation is complete. Discard the centrifugate. Add 1 mL of 3 M NH_3 and stir the mixture. Next, add 1 mL of 6 M HNO_3, and stir the mixture. Add more HNO_3, if necessary, to dissolve the precipitate completely. Then, add a few grains of $NaBiO_3$. If no color change is seen immediately, heat the solution gently in a beaker of hot water for a few minutes. Note the final color of the mixture.

Repeat the test for manganese(II) ions starting with 5 drops of 0.2 M $Mn(NO_3)_2$ and 5 drops of $Fe(NO_3)_3$, instead of 5 drops of $Mn(NO_3)_2$ only. Then repeat the test starting with 5 drops of 0.2 M $Fe(NO_3)_3$ in place of $Mn(NO_3)_2$.

Test for Fe^{3+}

Place 5 drops of 0.2 M $Fe(NO_3)_3$ in a test tube. Add 5 drops of distilled water, 1 drop of 1 M HCl, and 1 drop of 3 M H_2SO_4. Then add 2 M NaOH, by drops, until the solution is just basic to litmus. Centrifuge the test tube. Add 1 drop more of NaOH to test for complete precipitation. Continue centrifuging and adding NaOH until precipitation is complete. Discard the centrifugate. Note the appearance of the $Fe(OH)_3$ precipitate. Add 1 mL of 3 M NH_3 and stir the mixture. Next, add 1 mL of 6 M HNO_3, and stir the mixture. Add more HNO_3, if necessary, to dissolve the precipitate completely. Then add 5 drops of 0.2 M KSCN. Note the final color of the mixture.

Repeat the test for iron(III) ions starting with 5 drops of 0.2 M $Mn(NO_3)_2$ and 5 drops of 0.2 M $Fe(NO_3)_3$, instead of 5 drops of $Fe(NO_3)_3$ only. Then, repeat the test starting with 5 drops of 0.2 M $Mn(NO_3)_2$, in place of $Fe(NO_3)_3$.

Test for K^+

Clean a test wire by dipping it in concentrated HCl and heating it in the burner flame. Repeat the process of dipping and heating until the wire does not impart any color to the flame. Place 5 drops of 0.2 M KNO_3 in a clean test tube and dip the test wire into this solution. Bring the loop of the test wire to the edge of the outer cone of the burner flame, at the height of the top of the inner cone. Observe the color imparted to the flame. Record your observations.

Next, place 5 drops of 0.2 M KNO_3 in a clean test tube. Add 5 drops of distilled water, 1 drop of 1 M HCl, and 1 drop of 3 M H_2SO_4. Then add 2 M NaOH, by drops, until the solution is basic to litmus. Repeat the flame test. Record your observations. Repeat the flame test once again, observing the flame through a piece of cobalt blue glass. Record your observations once again.

Disposal of Reagents

Because small quantities of chemicals are used in this experiment, solutions may be diluted and washed down the drain. Precipitates may be deposited in crocks (as solids) or resuspended in water and also flushed down the drain.

Date ______ Name ______________________________ Section ______ Desk Number ______

PRE-LAB EXERCISES FOR EXPERIMENT 41

These exercises are to be completed after you have read the experiment but before you come to the laboratory to perform it.

1. Write the equations that show the confirmatory tests for each ion.

2. Prepare an outline showing the tests to be done on each ion in this experiment. The first outline is prepared for you as an example.

   Ba^{2+}: Add water, HCl, H_2SO_4
   Observe color of $BaSO_4$ precipitate
   Add HCl to precipitate and perform flame test
   Note color imparted to flame

Date ______ Name ______________________________ Section ______ Desk Number ______

SUMMARY REPORT ON EXPERIMENT 41

Ba^{2+}	Al^{3+}
Cu^{2+}	Ni^{2+}
Mn^{2+}	Fe^{3+}
K^+	

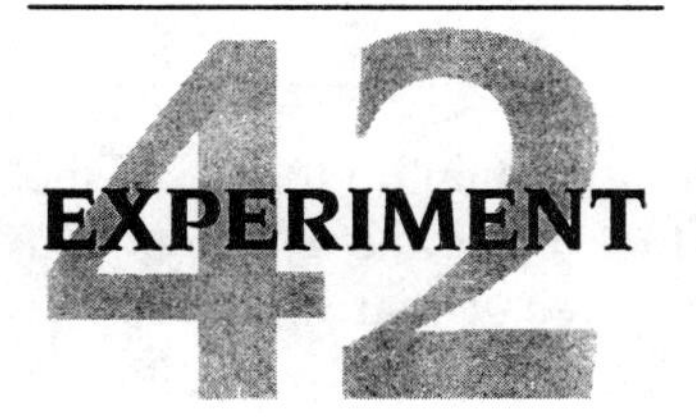

EXPERIMENT 42 Qualitative Analysis of an Unknown

Laboratory Time Required

Three hours

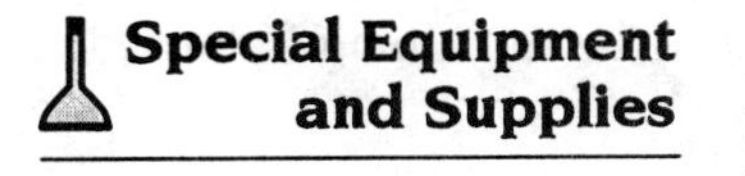

Special Equipment and Supplies

	Unknowns	*Reagents*
Labels	1 M $AgNO_3$	12 M HCl
Bunsen burner	1 M $Hg_2(NO_3)_2$	3 M H_2SO_4
Litmus paper (blue and red)	1 M $Al\ (NO_3)_3$	6 M HNO_3
Centrifuge	1 M $BaCl_2$	3 M $HC_2H_3O_2$
Test wires	1 M $Fe(NO_3)_3$	6 M NH_3
Cobalt blue glass	1 M $Cu(NO_3)_2$	6 M NaOH
Sandpaper	1 M $Mn(NO_3)_2$	0.2 M KSCN
	1 M $Ni(NO_3)_2$	$NaBiO_3$ (s)
	1 M KNO_3	DMG (s)
		NaCl (s)
		Iron nail

Safety

When diluting concentrated acids and bases, as will be necessary in the performance of this experiment, always add the acid or base to water (never put a drop of water into a concentrated solution of acid or base). The addition of water to the concentrated reagents may cause splattering of caustic or corrosive material.

When centrifuging mixtures, always balance the test tubes. Never reach into a spinning centrifuge.

First Aid

If you have skin contact with any reagents, wash the affected area with copious amounts of water. Flush the eyes with water for at least 20 minutes if any reagents splash into the eyes. See a doctor if you have skin or eye contact with concentrated acids or bases.

In Experiments 40 and 41, you performed preliminary experiments on 10 cations. In this experiment, you will try your hand at finding up to 5 of 9 cations in a general unknown.

PRINCIPLES

In this experiment, you will analyze a solution that may contain up to five of the following ions: Ag^+, Hg_2^{2+}, Ba^{2+}, Al^{3+}, Cu^{2+}, Ni^{2+}, Mn^{2+}, Fe^{3+}, and K^+. You will begin by separating the acid chloride group. You may recall that this group is composed of the Ag^+, Pb^{2+}, and Hg_2^{2+} ions. Dilute HCl will be the precipitating agent.

The first step in an analysis of the acid chloride precipitate is to treat it with hot water, which separates the moderately soluble $PbCl_2$ from AgCl and Hg_2Cl_2. Lead chloride is, in fact, so soluble in water that when it is present in a general unknown only a fraction of the Pb^{2+} ions precipitate as $PbCl_2$. The presence of the remaining lead ions greatly complicates the analysis of later groups. Thus, we have chosen to omit lead from our analytical scheme when a general unknown is considered.

After the separation of Ag^+ and Hg_2^{2+} from the general unknown, you will precipitate the Ba^{2+} ion as $BaSO_4$; the precipitating agent will be H_2SO_4. Next, NaOH will be added to precipitate $Al(OH)_3$, $Cu(OH)_2$, $Ni(OH)_2$, $Mn(OH)_2$, and $Fe(OH)_3$. The $Al(OH)_3$ will be dissolved in excess sodium hydroxide. The $Cu(OH)_2$ and $Ni(OH)_2$ will be dissolved in ammonia. The remaining hydroxides will be dissolved in nitric acid.

One ion, K^+, will not form a precipitate. Its presence will be detected via a flame test. A flame test will also be used to confirm the presence of barium ions in the $BaSO_4$ precipitate.

PROCEDURE

Obtain an unknown from your instructor and observe its appearance. If a solid is present, anions from one or more salts used in preparing the unknown may have formed an insoluble compound with one or more of the cations for analysis. If no additional precipitation is observed when HCl is added to precipitate the acid chloride group, or when H_2SO_4 is added to precipitate $BaSO_4$, and the unknown contains a solid, be sure to test the solid as if it were the expected precipitate. Further instructions are given below. Figures 42.1 and 42.2 give a brief outline of the analytical scheme.

Separation and Analysis of the Acid Chloride Group

Mix your unknown well. If a solid was initially present, centrifuge the tube, using a tube containing water as a counterbalance. Prepare labeled test tubes to store: (1) a 1-mL sample of your unknown to be used in the analysis; (2) the remainder of the liquid portion of your unknown; and (3) any solid that was initially present.

Begin the analysis by adding 1 drop of 1 M HCl to the 1-mL portion of the unknown. If no precipitation occurs and no solid was initially present in your unknown, the acid chloride group is absent. Proceed to the test for Ba^{2+}.

If no precipitation occurs when HCl is added to the unknown and a solid was initially present; it may contain AgCl and/or Hg_2Cl_2 and/or $BaSO_4$. Treat a portion of the solid with ammonia as if it were the acid chloride precipitate (see below).

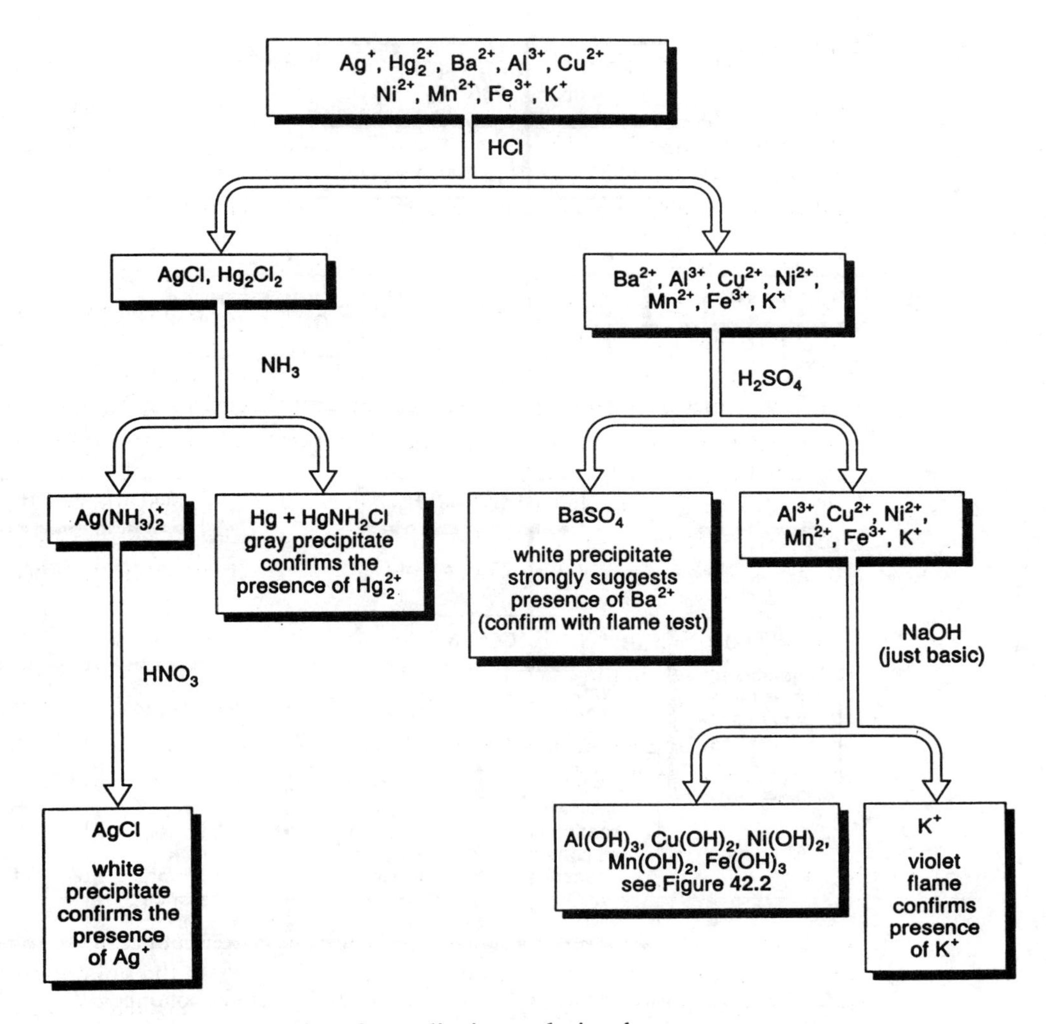

FIGURE 42.1 Flow chart for qualitative analysis scheme.

If a precipitate does form when 1 drop of 1 M HCl is added to 1 mL of unknown solution, centrifuge the tube. Then add 1 drop more of HCl to test for complete precipitation. Repeat the process by adding HCl and centrifuging until no further precipitation is observed.

Transfer the centrifugate to a labeled test tube and save it for subsequent analysis. Add 1 mL of 3 M NH_3 to the solid that remains after the centrifugate has been decanted and stir the mixture. If the solid turns gray, the presence of Hg_2^{2+} is confirmed. (If the presence of Ag^+ is confirmed later, a trace amount of black residue at this point may be metallic silver, produced by the photochemical decomposition of AgCl.)

Centrifuge the tube containing the ammonia and discard any solid. Cautiously add 2 M HNO_3 drop by drop, while mixing, to the test tube containing the ammonia, continuing until the solution indicates acidity on litmus paper. The formation of a white precipitate at this point confirms the presence of Ag^+.

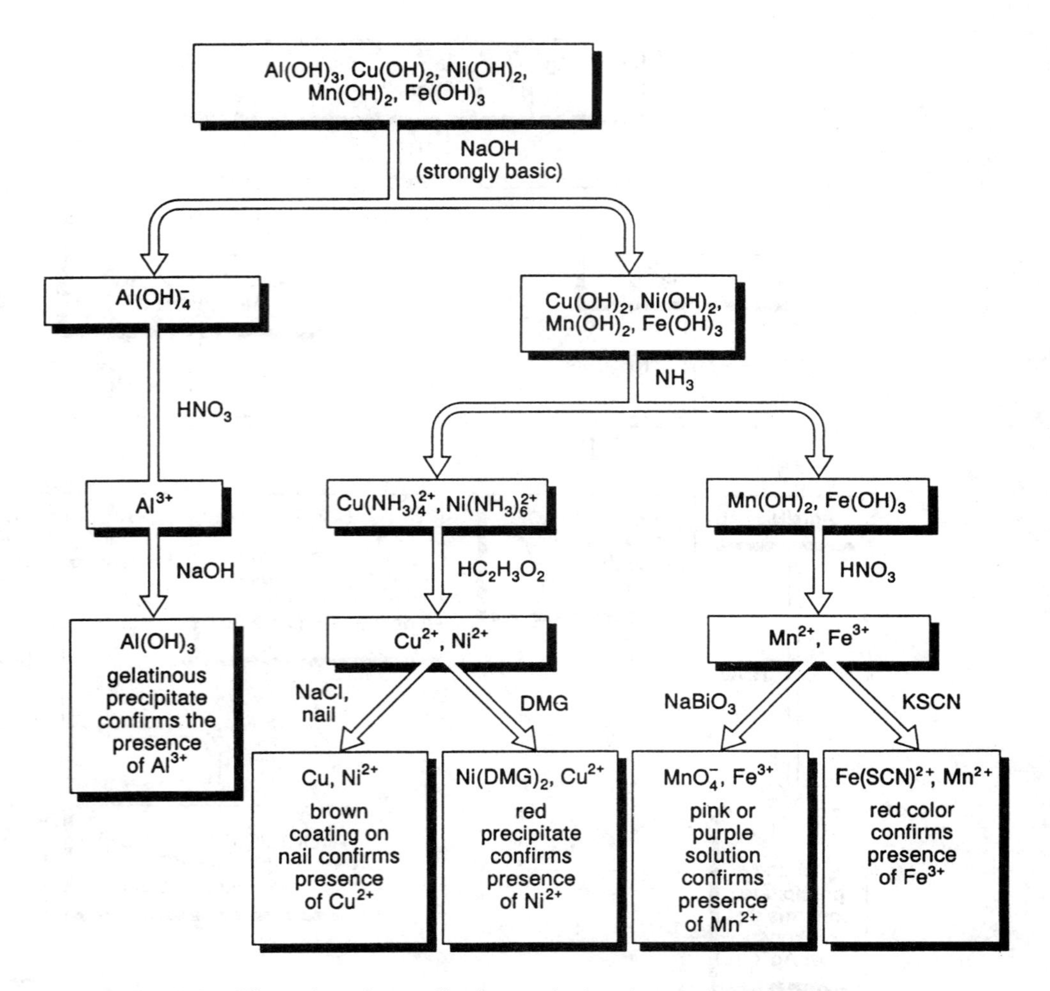

FIGURE 42.2 Flow chart for qualitative analysis scheme. (continued)

Separation and Analysis of $BaSO_4$

Add 3 drops of 3 M H_2SO_4 to the centrifugate from which the original separation of the acid chloride group was attempted. If no precipitation occurs, and no solid was present in the original unknown, Ba^{2+} is absent. Proceed to the separation of the hydroxide group.

If no precipitation occurs, and a solid was present in the original unknown, it may contain $BaSO_4$. Proceed to the flame test for $BaSO_4$ (see below).

If precipitation does occur when 3 drops of 3 M H_2SO_4 are added to the centrifugate, centrifuge the test tube. Add 1 drop more of H_2SO_4 to test for complete precipitation. Continue adding H_2SO_4 and centrifuging until no further precipitation is observed. Transfer the centrifugate to a labeled test tube and save it for subsequent analysis.

Clean a test wire by dipping it in concentrated HCl and heating it in the burner flame. Repeat the process of dipping and heating until the wire does not impart a color to the flame. Add 5 drops of concentrated HCl to the suspected $BaSO_4$ precipitate. Stir the mixture of $BaSO_4$ and HCl and dip the test wire into the resulting suspension. Bring the loop of the test wire to the edge of the outer cone of the burner flame, at the height of the inner cone. The appearance of a green flame confirms the presence of Ba^{2+}.

Separation and Analysis of the Insoluble Hydroxides

Add 2 M NaOH, dropwise, to the centrifugate from which the separation of $BaSO_4$ was attempted, until the solution is just basic to litmus paper. Centrifuge the test tube. Transfer the centrifugate to a labeled test tube and save it for subsequent tests to detect the Al^{3+} and K^+ ions.

Wash the precipitate three times with 5 drop portions of 3 M NaOH. Centrifuge after each addition of a 5-drop portion. Save these washings and the precipitate in separate test tubes. Test for the presence of Al^{3+} in the washings by adding 6 M HNO_3, by drops, until the solution is just acid to litmus paper. Then, add 2 M NaOH, again by drops, until a precipitate appears or until the solution is just basic to litmus paper. If a gelatinous precipitate appears, the presence of Al^{3+} is confirmed. If Al^{3+} is not found in the washings, test one-half of the centrifugate from which the metal hydroxides precipitated in the same manner as the washings were treated. Retain the other half of the centrifugate in a labeled test tube for the K^+ test.

Add 1 mL of 6 M NH_3 to the metal hydroxide precipitate and stir the mixture. Centrifuge the test tube. Transfer the centrifugate to a labeled test tube and retain any remaining precipitate for subsequent testing.

Add 3 M $HC_2H_3O_2$ to the centrifugate until the distinctive colors of the $Cu(NH_3)_4^{2+}$ and $Ni(NH_3)_6^{2+}$ complexes are destroyed. Divide the acidified centrifugate into two portions. Add a few grains of NaCl to one portion. Lightly sand an iron nail and place the nail into that portion of the acidified centrifugate as well. Leave the nail in the solution, undisturbed, for 15–30 minutes. The formation of a brown coating on the nail confirms the presence of Cu^{2+}.

Add a few grains of DMG to the other portion of the acidified centrifugate. The formation of a red precipitate confirms the presence of Ni^{2+}.

If any metal hydroxide precipitate remains following the treatment with excess sodium hydroxide and ammonia, dissolve it in 6 M HNO_3. Divide the resulting solution in two parts. Add 5 drops of 0.2 M KSCN to one portion. The formation of a blood-red solution confirms the presence of Fe^{3+}.

Add a few grains of $NaBiO_3$ to the other portion. If no color change is seen, heat the solution gently in a water bath for a few minutes. The formation of a pink or purple solution confirms the presence of Mn^{2+}.

Test for the Presence of K^+

Clean a test wire by dipping it in concentrated HCl and heating it in the burner flame. Repeat the process of dipping and heating until the wire does not impart any color to the flame. Place 5 drops of the centrifugate from which you attempted to separate the metal hydroxides in a test tube and dip the test wire into this solution. Bring the loop of the test wire to the edge of the outer cone of the burner flame, at the height of the top of the inner cone. Observe the color imparted to the flame through cobalt blue glass, which will mask the sodium flame. If a fleeting violet flame is seen, the presence of K^+ is confirmed.

Disposal of Reagents

Because small quantities of chemicals are used in this experiment, solutions can be diluted and washed down the drain. Precipitates may be deposited in crocks (as solids) or suspended in water and flushed down the drain.

Date ______ Name ________________________________ Section ______ Desk Number ______

PRE-LAB EXERCISE FOR EXPERIMENT 42

This exercise is to be completed after you have read the experiment but before you come to the laboratory to perform it.

Prepare a bibliography of at least five qualitative analysis texts and other books that you might use in writing your report for this experiment.

Date ______ Name ________________________________ Section ______ Desk Number ______

SUMMARY REPORT ON EXPERIMENT 42

As you perform each step in the analysis of your unknown, record your procedure and observations by building onto the chart started below. Continue on the next page if necessary.

At the bottom of the flow chart, list the cations found to be definitely present in the unknown.

Unknown no. ____________ (Brief description)

Add 1 M HCl

Does a white precipitate form?

Appearance of the centrifugate

EXPERIMENT 43

Analysis of a Soluble Salt

Laboratory Time Required

One to three hours, depending on the number of salts to be analyzed.

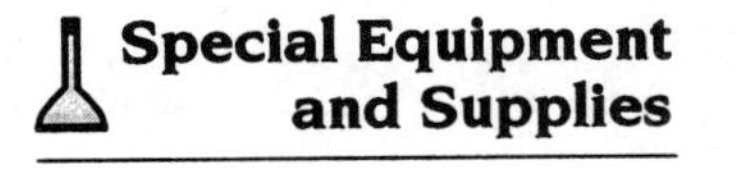

Special Equipment and Supplies

Test wires	6 M ammonia
Microspatulas	6 M nitric acid
Centrifuges	6 M hydrochloric acid
Litmus paper (red and blue)	6 M sodium hydroxide
Iron nail	0.05 M silver nitrate
Sandpaper	1 M barium chloride
Cobalt blue glass	3 M acetic acid
	Dimethylglyoxime (DMG)
	Sodium chloride (s)
	Unknown salts

Safety

Avoid getting acids or bases in your eyes or on your skin.

First Aid

Following skin contact with any of the reagents, wash the area with copious amounts of water. If acid or base enters your eyes, use the eyewash fountain to flush the chemical away, then consult a doctor.

In the past, qualitative analysis formed a major part of the laboratory program in introductory chemistry courses. A number of factors have led to a decrease in the amount of the time spent on qualitative analysis—concern over the toxicity of reagents, development of alternative methods of analysis (such as atomic absorption spectroscopy), a perception that students did not understand the complicated chemistry needed to separate the various ions, and so on. This experiment is a simple one, from which the need to perform separations has been eliminated. Nevertheless, it illustrates many interesting tests and introduces students to the analysis of anions, an area often ignored in traditional qualitative analysis. Depending on the needs of individual chemistry programs, it may be performed in place of Experiments 40 through 42 or in addition to them.

PRINCIPLES

Suppose someone handed you a vial containing a crystalline solid and said, "You're taking chemistry. Find out what's in this vial, please." Although you might like to oblige the petitioner, you'd have to admit that the request is fairly unreasonable, even impossible.

Yet, in this experiment, you will indeed be asked to identify the contents of several vials. However, the problem will be made more tractable by the formulation of some ground rules. All solids you will be asked to identify will be pure salts that are soluble in water. Each will contain only one type of cation and one type of anion. In fact, only six positive ions and five negative ions will be represented.

PROCEDURE

Obtain an unknown and record its identification code on the Summary Report Sheet. Perform the cation analyses described below to determine whether your unknown is a salt of NH_4^+, Ba^{2+}, Cu^{2+}, Ni^{2+}, K^+, or Na^+. Then perform the anion analyses to determine whether your unknown contains CO_3^{2-}, Cl^-, CrO_4^{2-}, I^-, or SO_4^{2-} ions. Once you have identified both the anion and cation in one salt, repeat the process with another unknown. Try to do as many analyses as time permits. Prepare a new Summary Report Sheet for each unknown. Use the Summary Report Sheet provided as a guide for your extra report forms.

Use your data and the supplementary information found in Table 43.1 to identify the cation and anion present in each salt you analyze. Use the equations given in Table 43.2 as a guide in writing balanced net ionic equations for the tests involved in making each identification. Give the correct chemical formula for each salt you analyze.

Cation Analyses

Dissolve a spatula-tipful of unknown in 5 mL of distilled water. Stir the mixture. If the solid does not dissolve completely, consult your instructor.

Pour one-half of the solution containing your unknown into a small beaker. Cautiously add 6 M NaOH, until the solution is just basic. Then add an additional 3 drops of sodium hydroxide. Observe the contents of the beaker. If a precipitate has formed, your unknown is a barium, copper(II), or nickel(II) salt. Proceed to the tests for Ba^{2+}, Cu^{2+}, and Ni^{2+} described below.

TABLE 43.1 Solubility Rules

1. All common sodium, potassium, and ammonium salts are soluble in water.
2. The chloride and iodide salts of all common metals, except silver, lead, and mercury(I), are soluble in water. Lead chloride is soluble in hot water.
3. The sulfates of all metals, except lead, mercury(I), barium, strontium, and calcium, are soluble in water.
4. The carbonates of all metals, except those of Group IA and ammonium carbonate, are insoluble in water.
5. Most metal hydroxides are insoluble in water. However, the hydroxides of Group IA are soluble and those of Group IIA are moderately soluble.

TABLE 43.2 Reactions

Ammonium

$NH_4^+ + OH^- \longrightarrow NH_3\uparrow + H_2O$

$\underset{\text{red}}{NH_3 + HIn} \longrightarrow \underset{\text{blue}}{NH_4^+ + In^-}$

Barium

$Ba^{2+} + SO_4^{2-} \longrightarrow BaSO_4\downarrow$

Carbonate

$CO_3^{2-} + 2H^+ \longrightarrow H_2O + CO_2\uparrow$

$Ba^{2+} + CO_3^{2-} \longrightarrow BaCO_3\downarrow$

$BaCO_3 + 2H^+ \longrightarrow H_2O + CO_2\uparrow + Ba^{2+}$

Chromate

$CrO_4^{2-} + 2H^+ \longrightarrow Cr_2O_7^{2-} + H_2O$

Chloride

$Ag^+ + Cl^- \longrightarrow AgCl\downarrow$

$AgCl + 2NH_3 \longrightarrow Ag(NH_3)_2^+ + Cl^-$

Copper

$Cu^{2+} + 2OH^- \longrightarrow Cu(OH)_2\downarrow$

$Cu(OH)_2 + 4NH_3 \longrightarrow Cu(NH_3)_4^{2+} + 2OH^-$

$Cu(NH_3)_4^{2+} + 2\ HC_2H_3O_2 \longrightarrow Cu^{2+} + 2NH_4C_2H_3O_2$

$Cu^{2+} + Fe \longrightarrow Cu + Fe^{2+}$

Iodide

$Ag^+ + I^- \longrightarrow AgI\downarrow$

Nickel

$Ni^{2+} + 2OH^- \longrightarrow Ni(OH)_2$

$Ni(OH)_2 + 6NH_3 \longrightarrow Ni(NH_3)_6{}^{2+} + 2OH^-$

$Ni(NH_3)_4{}^{2+} + 2\ HC_2H_3O_2 \longrightarrow Ni^{2+} + 2\ NH_4C_2H_3O_2$

$Ni^{2+} + 2\ DMG \longrightarrow Ni(DMG)_2\downarrow$

Sulfate

$Ba^{2+} + SO_4^{2-} \longrightarrow BaSO_4\downarrow$

$BaSO_4 + H^+ \longrightarrow NR$

If no precipitate has formed, your unknown is a salt of NH_4^+, K^+, or Na^+. To determine if your unknown is an ammonium salt, proceed as follows. Moisten a strip of red litmus paper and place it securely on the convex side of a watch glass (see Figure 7.2). Place the watch glass on top of the beaker that contains the unknown solution and NaOH. If the color of the litmus paper has not changed after 2–3 minutes, heat the beaker gently. If the litmus paper turns completely blue, your unknown contains the ammonium ion. Proceed to the anion tests.

If the test for NH_4^+ is negative, do a flame test on the portion of your unknown solution to which NaOH has not been added. Clean a test wire by alternately dipping it in 6 M HCl and heating it in the burner flame. When the wire does not impart any color to the flame, dip it into the unknown solution. Be sure a thin layer of solution fills the test wire loop and bring the loop to the edge of the outer cone of the burner flame, at the height of the inner cone. If a yellow-orange flame is observed, your unknown is a sodium salt. If a violet flame is seen, your unknown is a potassium salt. (Viewing the flame through a piece of cobalt glass may aid in the detection of the violet flame.) Proceed to the anion analyses.

To test for the presence of Ba^{2+}, Cu^{2+}, or Ni^{2+}, decant the liquid from the beaker containing the precipitate that formed when NaOH was added to the unknown solution. While mixing, add 5 mL of 3 M NH_3. If the precipitate dissolves, producing a royal-blue solution, your unknown is probably a salt of copper(II). Confirm that the cation is the copper(II) ion by adding 3 M $HC_2H_3O_2$, dropwise, to the royal-blue solution until it is nearly colorless. Then add a few grains of NaCl. Gently sand an iron nail and place it in the solution. Leave the nail in the solution, undisturbed, for 15–30 minutes. If a uniform brown coating forms on the nail, the cation is definitely copper(II). Proceed to the anion analyses.

If the hydroxide precipitate dissolves to give a green solution, your unknown is probably a salt of nickel(II). Confirm that the cation is nickel(II) by adding 3 M $HC_2H_3O_2$, by drops, until the solution is nearly colorless. Then, add a few grains of DMG. If a red precipitate is produced, the cation is nickel(II). Proceed to the anion analyses.

Anion Analyses

Dissolve a spatula-tipful of unknown in 5 mL of distilled water. Stir the mixture. Pour roughly equal portions of solution into each of three test tubes.

Add 6 M HNO_3 to the first of the three test tubes, by drops, while stirring, until the solution is just acid to litmus. If your solution was originally yellow and it turns orange upon addition of acid, your unknown is a chromate salt. If bubbles form or fizzing is heard, your salt is a carbonate. Proceed to another unknown.

If your solution did not turn orange or fizz when acid was added, add 10 drops of 1 M $BaCl_2$ to the second test tube. If a white precipitate forms, centrifuge the tube, using a tube filled with distilled water as a counterbalance. Decant the centrifugate (the liquid portion) and discard it. Treat the precipitate with several drops of 6 M HNO_3. If the precipitate fizzes and dissolves, your unknown is a carbonate. If the precipitate does not dissolve, your unknown is a sulfate. Proceed to another unknown.

If your salt is not a chromate, carbonate, or sulfate, add 10 drops of 0.05 M

$AgNO_3$ to the third test tube. If a white precipitate forms, your unknown is a chloride. If a yellow precipitate forms, your unknown is an iodide. If you cannot determine the color of the precipitate, centrifuge the tube and discard the centrifugate. Add 1 mL of 6 M NH_3 to the precipitate, while stirring. Silver chloride will dissolve, silver iodide will not. Proceed to another unknown.

Disposal of Reagents

Any solutions containing silver or nickel(II) ions should be placed in labeled collection bottles. All other reagents can be neutralized, diluted, and flushed down the drain.

Questions

1. Glenn Student dissolved a small amount of unknown salt in water and poured one-half of the resulting solution into a beaker. Next, Glenn made the solution basic with NaOH. No precipitate formed, so Glenn dipped a piece of red litmus paper into the solution. When the litmus paper turned blue, Glenn concluded the unknown was an ammonium salt. Why is Glenn's method for finding ammonium inconclusive? What, if anything, can you say about the identity of Glenn's salt?

2. Chris Student found that an unknown was a barium salt. Chris concluded that the unknown could not be a carbonate or sulfate. Do you concur with Chris? Explain your reasoning.

Date ______ Name ______________________________ Section ______ Desk Number ______

PRE-LAB EXERCISES FOR EXPERIMENT 43

These exercises are to be completed after you have read the experiment but before you come to the laboratory to perform it.

1. Name a single reagent and/or physical property that could be used to distinguish between:

   a. K_2CrO_4 (s) and KCl (s)

   b. AgCl (s) and Ag_2CO_3 (s)

   c. NaI (aq) and NaCl (aq)

   d. $Ni(OH)_2$ (s) and KOH (s)

2. A white, crystalline solid dissolved in water forming a pale blue solution. A precipitate resulted when the solution was treated with 6 M sodium hydroxide. The precipitate dissolved in ammonia, giving a royal-blue solution.

   A second portion of the solid was dissolved in water. Treatment of a portion of that solution with silver nitrate resulted in a white precipitate. The precipitate dissolved in ammonia.

   Identify the original salt, giving its chemical formula. Briefly explain your reasoning.

Date ______ Name ______________________________ Section ______ Desk Number ______

SUMMARY REPORT ON EXPERIMENT 43

Unknown identification code ______________________

Description of Unknown ______________________

Cation Analysis

Material Tested	*Reagent*	*Observation*	*Equation*

Identity of cation ______________________

Anion Analysis

Material Tested	*Reagent*	*Observation*	*Equation*

Identity of cation ______________________

Formula of salt ______________________

Date ______ Name ______________________________ Section ______ Desk Number ______

Unknown identification code ______________________

Description of Unknown ______________________

Cation Analysis

Material Tested	*Reagent*	*Observation*	*Equation*

Identity of cation ______________________

Anion Analysis

Material Tested	*Reagent*	*Observation*	*Equation*

Identity of cation ______________________

Formula of salt ______________________

Date ______ Name ______________________ Section ______ Desk Number ______

Unknown identification code ______________________

Description of Unknown ______________________

Cation Analysis

Material Tested	*Reagent*	*Observation*	*Equation*

Identity of cation ______________________

Anion Analysis

Material Tested	*Reagent*	*Observation*	*Equation*

Identity of cation ______________________

Formula of salt ______________________

Date ______ Name ________________________________ Section ______ Desk Number ______

Unknown identification code ____________________

Description of Unknown ____________________

Cation Analysis

Material Tested	*Reagent*	*Observation*	*Equation*

Identity of cation ____________________

Anion Analysis

Material Tested	*Reagent*	*Observation*	*Equation*

Identity of cation ____________________

Formula of salt ____________________

Date ______ Name ______________________________ Section ______ Desk Number ______

Unknown identification code ______________________

Description of Unknown ______________________

Cation Analysis

Material Tested	*Reagent*	*Observation*	*Equation*

Identity of cation ______________________

Anion Analysis

Material Tested	*Reagent*	*Observation*	*Equation*

Identity of cation ______________________

Formula of salt ______________________

APPENDIX A Mathematical Operations

LOGARITHMS

Common Logarithms

The logarithm X of a number N is defined by the relations

$$\log_a N = X \tag{A.1}$$

where

$$N = a^x \tag{A.2}$$

For a = 10, the common logarithm is obtained, while for a = 2.71828 (represented by the symbol e), the natural logarithm is obtained. Common logarithms are generally used in the mathematical operations of multiplication, division, and finding powers and roots. Natural logarithms are used in certain thermodynamic relations, distribution functions, and so on. In this laboratory manual the common logarithm of a number is represented by log N; the natural logarithm is represented by ln N.

Most calculators have keys for both log$N$ and ln$N$. If your calculator has only one of these functions, you may use Equation A.3 to convert values from common to natural logarithms, or vice versa.

$$\ln N = 2.303 \log N \tag{A.3}$$

In the days before inexpensive calculators were available, logarithms were frequently used to simplify difficult calculations. The relations shown below reveal the simplifying power of logarithms.

$$\log XY = \log X + \log Y \tag{A.4}$$

$$\log (X/Y) = \log X - \log Y \tag{A.5}$$

$$\log Y^X = X \log Y \tag{A.6}$$

Although you will most often use your calculator to obtain logarithms, it is still useful for you to know how to obtain them by using a log table (in case you lose your calculator or you want to check for malfunctions). To find the common logarithm of a number N, it is convenient to write N as an exponential number:

$$N = M \times 10^Z \tag{A.7}$$

where Z is an integer and M is a number between 1 and 10. For example, 456 would be writeen 4.56×10^2, while 0.0456 would be written as 4.56×10^{-2}. The logarithm of N is then given by Equation A.8.

$$\log N = \log M + Z \tag{A.8}$$

for example,

$$\log 4.56 \times 10^2 = \log 4.56 + 2$$

and

$$\log 4.56 \times 10^{-2} = \log 4.56 - 2$$

If this convention is adopted, it is obviously necessary to tabulate only the logarithms of numbers between 1 and 10 inclusive. Such a tabulation is given in Table A.1. The logarithm of a two-digit number may be read directly from the table; the logarithm of a three-digit number may be extimated by interpolation.

TABLE A.1 Three-Place Common Logarithms

N	**0.1**	**.1**	**.2**	**.3**	**.4**	**.5**	**.6**	**.7**	**.8**	**.9**
1	0.000	0.041	0.079	0.114	0.146	0.176	0.204	0.230	0.255	0.279
2	0.301	0.322	0.342	0.362	0.380	0.398	0.415	0.431	0.447	0.462
3	0.477	0.491	0.505	0.518	0.531	0.544	0.556	0.568	0.580	0.591
4	0.602	0.613	0.623	0.633	0.643	0.653	0.663	0.672	0.682	0.690
5	0.699	0.708	0.716	0.724	0.732	0.740	0.748	0.756	0.763	0.771
6	0.778	0.785	0.792	0.799	0.806	0.813	0.820	0.826	0.833	0.839
7	0.845	0.851	0.857	0.863	0.869	0.875	0.881	0.886	0.892	0.898
8	0.903	0.908	0.914	0.919	0.924	0.929	0.934	0.940	0.944	0.949
9	0.954	0.959	0.964	0.968	0.973	0.978	0.982	0.987	0.991	0.996
10	1.000									

Thus, the logarithm of 4.56 lies between the logarithm of 4.5 (0.653) and that of 4.6 (0.663). Because six-tenths of the difference between 0.663 and 0.653 is easily determined to be 0.006, the logarithm of 4.56 is 0.653 plus 0.006, or 0.659. By using Equation A.8, we find the logarithm of 4.56×10^2 is 0.659 + 2, or 2.659. Similarly, the logarithm of 4.56×10^{-2} is 0.659 – 2 or –1.341.

By reversing the process, any number whose logarithm is given may be determined. Such a number is called the antilog, or $\log^{-1}$, of the number X. The antilog of 2.659 is $10^{2.659}$, which is equal to $10^{0.659} \times 10^2$. Referring to Table A.1, you will find the antilog of 0.659 to be 4.56. Therefore the antilog of 2.659 is 4.56×10^2. Similarly, the antilog of –1.341 is $10^{-1.341}$, which is equal to $10^{0.659} \times 10^{-2}$. By using the table, you will find 4.56×10^{-2} to be the antilog of –1.341.

To multiply or divide using logarithms, you use Equations A.4 and A.5. For example, to multiple 2.1 by 3.3, add the logarithms of 2.1 and 3.3.

$$\log 2.1 = 0.322$$

$$\log 3.3 = 0.518$$

$$(\log 2.1 + \log 3.3) = 0.840$$

$$\text{antilog } 0.840 = 6.9$$

$$2.1 \times 3.3 = 6.9$$

To divide 2.1 by 3.3, subtract the logarithm of 3.3 from the logarith of 2.1.

$$\log 2.1 = 0.332$$
$$-\log 3.3 = -0.518$$
$$(\log 2.1 - \log 3.3) = -0.186 = 0.804 - 1$$
$$\text{antilog } (0.804 - 1) = 6.4 \times 10^{-1}$$
$$(2.1 \div 3.3) = 0.64$$

Note the somewhat unusual third step, in which we added and subtracted 1 from –0.186. This was done to obtain a positive number that could be found in the log table. This step is necessary only when a number is divided by a larger number.

To raise a number to a power using logarithms, use Equation A.6. Note that X may be an integer or a fraction, so that Equation A.6 may be used for "raising to a power" or for "taking roots." We illustrate this below by finding the value of 2^3 and $\sqrt[3]{8}$ $(= 8^{1/3})$.

$$\log 2^3 = 3 \log 2 = 3(0.301) = 0.903$$
$$\text{antilog } 0.903 = 8$$
$$2^3 = 8$$
$$\log 8^{1/3} = 1.3 \log 8 = 1.3(0.903) = 0.301$$
$$\text{antilog } 0.301 = 2$$
$$8^{1/3} = 2$$

Natural Logarithms

Natural logarithms follow the same combination laws as do common logarithms but are somewhat more difficult to work with because of the change of base from 10 to 2.71828. Although Equations A.4, A.5, and A.6 apply to natural logarithms, it is generally easier to perform simple mathematical operations using common logarithms and to use natural logarithms only when the natural log appears in the equation of interest. The natural logarithm may then be obtained by (1) consulting a table of natural logarithms, (2) using the ln key on your calculator, or (3) calculating the natural log from the common logarithm of the number using Equation A.3.

GRAPHS

General Principles

Graphs are frequently used to portray experimental data because such features as maxima, minima, and inflection points are more easily recognized in a graphical representation than in a tabular display. Graphs also facilitate such mathematical operations as interpolation, averaging, and integration, which would be more difficult using numerical methods. To be useful for portraying experimental data and for evaluating results, a graph should be constructed with care, according to the following rules.

1. Use a good grade of graph paper. The precisely ruled grid is well worth the nominal cost.
2. Clearly specify the variable to be plotted along each axis. Note that your choice of axes is not arbitrary. For instance, if you expect your data to fit the linear equation, $y = mx + b$, you should plot y on the vertical axis and x on the horizontal axis.
3. Mark the scales clearly along each axis. Choose the scale so that the coordinates of any point on the graph may be easily and accurately determined without computation.
4. Choose the scale so that the coordinates of any point may be estimated to the same number of significant figures as the original data. When observance of this rule requires a graph of unreasonably large size, it may be possible to portray all the data on a graph of reasonable size by using a relatively coarse scale. The portion of greatest interest can then be shown using a finer scale, either on a separate graph or as an insert on the main graph.
5. Surround each experimental point by a circle, diamond, or square, so that it is easily recognized.
6. Draw straight lines through the experimental points using a ruler. Use a French curve in constructing curved lines. Do not zigzag between points. Note that the best experimental line may not pass through every point.
7. Label the finished graph with a suitable title and mount it permanently in your notebook with glue or transparent tape.

Calculating the Slope

The slope of line *AB*, shown in Figure A.1, may be easily calculated using the coordinates of any two points that lie on the line. For example, the points P_1 and P_2 having the coordinates X_1, Y_1, and X_2, Y_2 may be used. The slope, m, of the line *AB* is then given by Equation A.9:

$$m = \frac{Y_2 - Y_1}{X_2 - X_1} \tag{A.9}$$

Note that the coordinates of P_1 and P_2 do not depend on the scale used. Therefore, the slope is independent of the scale used.

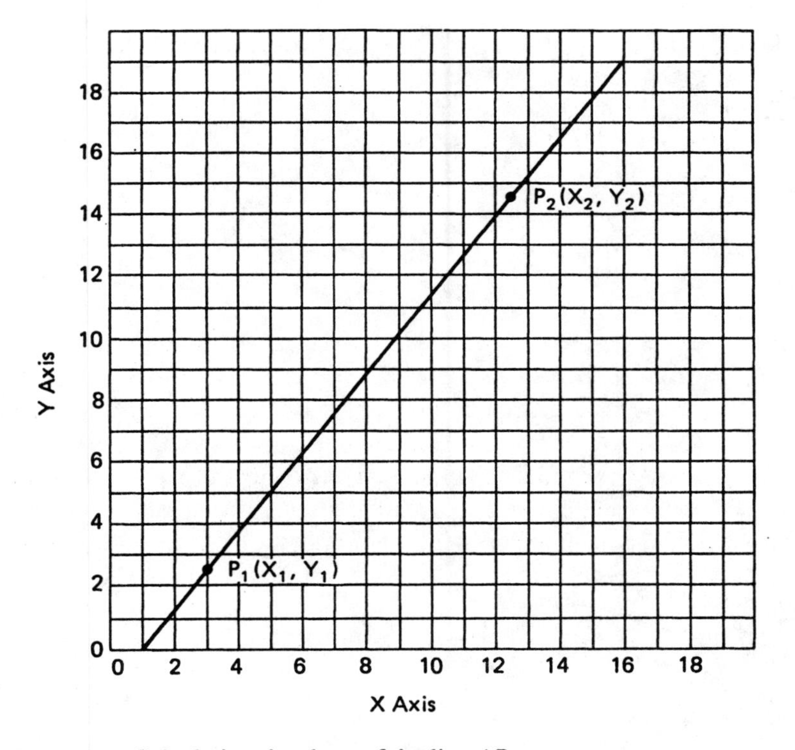

FIGURE A.1 Calculating the slope of the line *AB*.

Properties of Water

TABLE B.1 Density of Water at Various Temperatures

T(°C)	d(g/mL)	T(°C)	d(g/mL)
0	0.99984	26	0.99679
5	0.99997	27	0.99652
10	0.99970	28	0.99624
15	0.99910	29	0.99595
16	0.99895	30	0.99565
17	0.99878	31	0.99534
18	0.99860	32	0.99503
19	0.99840	33	0.99471
20	0.99821	34	0.99437
21	0.99800	35	0.99403
22	0.99777	40	0.99222
23	0.99754	45	0.99022
24	0.99730	50	0.98804
25	0.99705	55	0.98570

TABLE B.1 Density of Water at Various Temperatures

T(°C)	Vapor Pressure (torr)	T(°C)	Vapor Pressure (torr)
0	4.6	31	33.7
5	6.5	32	35.7
10	9.2	33	37.7
15	12.8	34	39.9
20	17.5	35	42.2
21	18.6	40	55.3
22	19.8	45	71.9
23	21.1	50	92.5
24	22.4	55	118.0
25	23.8	60	149.4
26	25.2	65	187.5
27	26.7	70	233.7
28	28.3	80	355.1
29	30.0	90	525.8
30	31.8	100	760.0